Record Label Marketing

Record Label Marketing
Second Edition

Tom Hutchison

Amy Macy

Paul Allen

AMSTERDAM • BOSTON • HEIDELBERG • LONDON • NEW YORK • OXFORD
PARIS • SAN DIEGO • SAN FRANCISCO • SINGAPORE • SYDNEY • TOKYO

Focal Press is an imprint of Elsevier

Focal Press is an imprint of Elsevier
30 Corporate Drive, Suite 400, Burlington, MA 01803, USA
Linacre House, Jordan Hill, Oxford OX2 8DP, UK

This book is printed on acid-free paper.

Library of Congress Cataloging-in-Publication Data
Hutchison, Thomas W. (Thomas William)
    Record label marketing / Tom Hutchison, Amy Macy, Paul Allen. – 2nd ed.
        p. cm.
    Includes bibliographical references and index.
    ISBN 978-0-240-81238-0 (pbk. : alk. paper) 1.Music trade. 2. Sound
recordings – Marketing. 3. Sound recording industry. I. Macy, Amy. II. Allen,
Paul. III. Title.
    ML3790.H985 2009
    780.68'8 – dc22

                                                2009025114

British Library Cataloguing-in-Publication Data
A catalogue record for this book is available from the British Library.

ISBN: 978-0-240-81238-0

For information on all Focal Press publications
visit our Web site at www.elsevierdirect.com

Printed in the United States
09 10 11   9 8 7 6 5 4 3 2 1

Working together to grow
libraries in developing countries

www.elsevier.com | www.bookaid.org | www.sabre.org

ELSEVIER   BOOK AID International   Sabre Foundation

Contents

Acknowledgements

Tom Hutchison:

I would like to thank my friends who contributed to this book, including those who helped secure permission for cover photos: Thad Keim, Vice President, Sales & Marketing at Compass Records, Bill DeMain of Swan Dive, Bill "Sauce Boss" Wharton, and The Karg Brothers. Much praise and thanks go out to our colleagues who served as technical reviewers and contributors: Clyde Rolston, Ph.D., Associate Professor at Belmont University's Mike Curb College of Entertainment and Music Business; Storm Gloor, Assistant Professor in the Department of Music and Entertainment Industry Studies at the University of Colorado Denver; and Trudy Lartz, VP, Sales & Service at Nielsen Entertainment. We appreciate the patience of our dean, Roy Moore and our department chair, Chris Haseleu, for allowing us the flexibility to work on this book. Thanks to our colleague Charlie Dahan who provided the marketing plan; to Geoffrey Hull for his advice—legal and personal; to fellow musician Bill Wharton for allowing me to use his website as an example, and to jazz vocalist Inga Swearingen for allowing me a glimpse inside her career. Thanks to my co-writers Amy Macy and Paul Allen-Weese, who made this all possible. And most of all, to my wife Lynne who not only had to put up with a book project for a third time in four years, but who has also taught me the process of writing and editing books.

Amy Macy:

I'm so grateful for my industry friends, old and new, who are (still) navigating these very murky waters that we call the record business. I'm thankful to my sage mentors that include Joe Galante, Randy Goodman, Ron Howie, Mike Dungan, and Mike Shallett who taught me SO much. Thanks to my new pals Kyla Fairchild of "No Depression" and Paul Roper and RIM Alum Joey Luscinski of Dualtone Records for enlightening me about the Independent world. And to the ongoing relationships that pick-up after long lapses of time, I thank Debbie Lynn of the RCA Label Group, Kelly Rich of Big Machine Records, David Macias of Thirty Tigers, and Jim Weatherson of Disney Records — all of whom have taken a moment to share with me a "snippet" or "morsel" of their world…which is moving at a swift pace these days. And a special "thanks" to Trudy Lartz, VP of Sales and Service for Nielsen SoundScan, who made Chapter 7 possible.

I stay inspired by my colleagues at Middle Tennessee State University and Department of Recording Industry who maintain an educational integrity that I so admire, especially my co-writers Tom Hutchison and Paul Allen – you guys are great! To other educational colleagues who keep me (and us) on our toes: Storm Gloor of the University of Colorado and Clyde Rolston of Belmont University – whose insight and critique are mightily embraced. And lastly, thanks to my Red Dirt Family: Doug, Millie Mae, Emma Jo, Grammy, and Choo Choo – who have charged on without me (thank goodness) while I have taken "time out" to write this book – but I'm catching up now!

Paul Allen:

Special thanks go out to Tom Hutchison and Amy Macy whose experience, research, and writing have made this book standard fare for people who want to learn about the music business – especially marketing by record labels. This second edition captures a moving target in ways no other book on the subject does. Some of the people who made special contributions to its continuing success include Jeff Walker, Lon Helton, Jared Forrester, Jennifer Harbin, Christian Haseleu, Jeff Leeds, and Chris Palmer. Continued thanks to Marc Singer and Troy Festervand for their support and encouragement, and thanks to J. William Denny for opening the door to the music business for me. And thanks to Cindy Campbell Weese whose career in music gave me a close-up and personal window into the music business.

CHAPTER 1

Marketing Concepts and Definitions

Tom Hutchison

SELLING RECORDED MUSIC

The concept of selling recorded music has been around for more than a century. While the actual storage medium for music has evolved, from cylinders to vinyl discs, magnetic tape, digital discs, and now downloads, the basic notion has remained the same: a musical performance is captured to be played back at a later time, at the convenience of the consumer. Music fans continue to enjoy the ability to develop music collections, whether in physical compact disc format or in collections of digital files on their computer hard drives. Consumers also enjoy the portability afforded by contemporary music listening devices, allowing the convenience of determining the time and place for listening to music. The ways consumers select to access music have been undergoing changes in the past few years, as physical sales have diminished and the industry scrambles to find new business models for underwriting the cost of developing new creative products.

Music consumption should not be confused with music purchases. The consumption of music by consumers has increased (Hefflinger, 2008), despite the fact that sales of recorded music albums have decreased over the past decade from 785 billion units in the United States in 2000 to 535 billion units in 2008 (SoundScan). In response to the decline in sales of recorded music, record labels are experimenting with new ways to monetize the consumption of their music. For example, new services like LastFM, Pandora, and MySpace Music offer music fans the opportunity to listen to music without actually purchasing it. Record labels are compensated through advertisement revenue sharing.

CONTENTS

Thus, recorded music is finding ways to make money much the same as television programming has done for over 50 years. For much of this time, the television programming industry relied solely upon advertising revenue to fund some of the most popular television shows in history. Other, more recent forms of revenue have come from premium (fee-based) programs and physical sales of shows (DVDs). Fee-based programming did not occur until the premium channels such as HBO and Showtime started developing their own proprietary shows. The physical sale of this commodity did not occur until consumers started collections of videotapes and DVDs. Even with these other forms of income, the bulk of revenue for producing television content still comes from advertising. Perhaps the recording industry can look to the television industry for ideas in developing new models to create revenue from creative products.

As the paradigm shifts from the physical sale of recordings to a more complex model of generating revenue, marketing efforts must also evolve to respond to the plethora of income possibilities.

WHAT IS MARKETING?

In today's marketplace, the consumer is showered with an array of entertainment products from which to choose, making the process of marketing more important than ever. Competition is fierce for the consumers' entertainment budget. But before explaining how records (recorded music) are marketed to consumers, it is first necessary to gain a basic understanding of marketing. Even the concept of a record or album has undergone changes recently. *Record* is short for recorded music—that much has not changed. An album is still a collection of songs released as a unit, whether it's in CD format or something else.

Marketing is simply the performance of business activities that direct the flow of goods and services from the producer to the consumer (American Marketing Association, 1960). Marketing involves satisfying customer needs or desires. To study marketing, one must first understand the notions of *product* and *consumer* (or *market*). The first questions a marketer should answer are, "What markets are we trying to serve?" and "What are their needs?" Marketers must understand the consumer's needs and develop products to satisfy those needs. Then they must price the products effectively, make the products available in the marketplace, and inform, motivate, and remind the customer. In the music business, this involves supplying consumers with the recorded music they desire.

The market is defined as consumers who want or need your product and who are willing and able to buy or pay. This definition emphasizes that the consumer wants or needs something. A product is defined as something that will satisfy the customer's want or need. You may want a candy bar, but not necessarily need one. You may need surgery, but not necessarily want it.

THE MARKETING MIX

The **marketing mix** refers to a blend of product, distribution, promotion, and pricing strategies designed to produce mutually satisfying exchanges with the target market. This is often referred to as the *four P's*, which are:

Product – goods or services designed to satisfy a customer's need
Price – what customers will exchange for the product
Promotion – informing and motivating the customer
Place – how to deliver and distribute the product

Product

The marketing mix begins with the product. It would be difficult to create a strategy for the other components without a clear understanding of the product to be marketed. New products are developed by identifying a market that is underserved, meaning there is a demand for products that is not being adequately met.

The product aspect of marketing refers to all activities relating to the product development, ensuring that

- There is a market for the product.
- It has appeal.
- It sufficiently differs from other products already in the marketplace.
- It can be produced at an affordable and competitive price.

An array of products may be considered to supply a particular market. Then the field of potential products is narrowed to those most likely to perform well in the marketplace. In the music business, the artist and repertoire (A&R) department performs this task by searching for new talent and helping decide which songs will have the most consumer appeal. In other industries, this function is performed by the research and development (R&D) arm of the company.

New products introduced into the marketplace must somehow identify themselves as different from those that currently exist. Marketers go to great lengths to position their products to ensure that their customers understand why their product is more suitable for them than the competitor's product.

Product positioning is defined as the customer's perception of a product in comparison with the competition.

Consumer tastes change over time. As a result, new products must constantly be introduced into the marketplace. New technologies render products obsolete and encourage growth in the marketplace. For example, the introduction of the compact disc (CD) in 1983 created opportunities for the record industry to sell older catalog product to customers who were converting their music collections from long-playing records (LPs) to CDs. Similarly, when a recording artist releases a new recording, marketing efforts are geared toward selling the new release, rather than older recordings (although the new release may create some consumer interest in earlier works and they may be featured alongside the newer release at retail).

The Product Life Cycle

The **product life cycle** (**PLC**) is a concept used to describe the course that a product's sales and profits take over what is referred to as the *lifetime of the product*—the sales window and market for a particular product, from its inception to its demise.

It is characterized by four distinct stages: introduction, growth, maturity, and decline. Preceding this is the *product development* stage, before the product is introduced into the marketplace, and following the four stages we have a withdrawal of the product from the marketplace. The *introduction* stage is a period of slow growth as the product is introduced into the marketplace. Profits are nonexistent because of heavy marketing expenses. The *growth* stage is a period of rapid acceptance into the marketplace and profits increase. *Maturity* is a period of leveling in sales mainly because the market is saturated—most consumers have already purchased the product. Marketing is more expensive (to the point of diminishing returns) as efforts are made to reach resistant customers and to stave off competition. *Decline* is the period when sales fall off and profits are reduced. At this point, prices are cut to maintain market share (Kotler and Armstrong, 1996).

The PLC can apply to a variety of situations such as products (a particular album), **product forms** (artists and music genres), and even **product classes** (cassettes, CDs, and vinyl). Product classes generally have the longest life cycle—the compact disc has been around since the early 1980s (and is currently in the decline phase). However, the life cycle of an average album release is 12 to 18 months. It is at this point that the label will generally terminate most marketing efforts and rely on catalog sales to deplete remaining inventory.

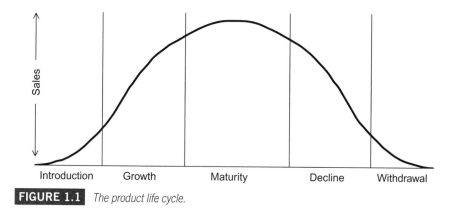

Introduction	Growth	Maturity	Decline	Withdrawal

FIGURE 1.1   *The product life cycle.*

Diffusion of Innovations

When a product is introduced into the marketplace, its consumption is expected to follow a pattern of diffusion. **Diffusion of innovations** is the process by which the use of an innovation is spread within a market group, over time and over various categories of adopters (American Marketing Association, 2004). The concept of diffusion of innovations describes how a product typically is adopted by the marketplace and what factors can influence the rate (how fast) or level (how widespread) of adoption. The rate of adoption is dependent on *consumer traits*, the *product*, and the company's marketing efforts.

Consumers are considered *adopters* if they have purchased and used the product. Potential adopters go through distinct stages when deciding whether to adopt (purchase) or reject a new product. These stages are referred to as *AIDA*, which under one model (affective or driven by emotion) is represented as attention, interest, desire, and action. Another model (cognitive or driven by thought) uses awareness, information, decision, and action. There can be elements of both models present in the decision-making process for most products, which tend to have an emotional and rational basis for their purchase. These stages describe the psychological progress a buyer must go through in order to get to the actual purchase. First, a consumer becomes *aware* that she needs to make a purchase in this product category. Perhaps the music consumer has grown tired of her collection and directs her *attention* toward buying more music. The consumer then seeks out *information* on new releases and begins to gain an *interest* in something in particular, perhaps after hearing a song on the radio or attending a concert. The consumer then makes the *decision* (and *desires*) to purchase a particular recording. The **action** is the actual purchase.

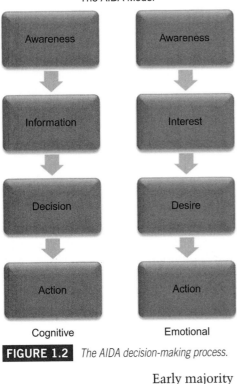

The AIDA Model

FIGURE 1.2 *The AIDA decision-making process.*

This process may take only a few minutes or several weeks, depending on the importance of the decision and the risk involved in making a wrong decision. **Involvement** refers to the amount of time and effort a buyer invests in the search, evaluation, and decision processes of consumer behavior. If the consumer is not discriminating and dissatisfaction with the decision is not a setback, then the process may take only moments and is thus a low-involvement situation. On the other hand, if the item is expensive, such as a new car, and if the consequences of a wrong decision are severe, then the process may take much longer. Several factors influence the consumer's level of involvement in the purchase process, including previous experience, ease of purchase, interest, and perceived risk of negative consequences. As price increases, so does the level of involvement.

Consumers who adopt a new product are segmented into five categories:

Innovators
Early adopters
Early majority
Late majority
Laggards

Innovators are the first 2.5% of the market (Rogers, 1995) and are eager to try new products. Innovators are above average in income and thus the cost of the product is not of much concern. *Early adopters* are the next 13.5% of the market and adopt once the innovators have demonstrated that the new product is viable. Early adopters are more socially involved and are considered opinion leaders. Their enthusiasm for the new product will do much to assist its diffusion to the majority. The *early majority* is the next 34% and will weigh the merits before deciding to adopt. They rely on the opinions of the early adopters. The *late majority* represents the next 34%, and these consumers adopt when most of their friends have. The *laggards* are the last 16% of the market and generally adopt only when they feel they have no choice. Laggards adopt a product when it has reached the maturity stage and is being "deep discounted" or is widely available at discount stores. When introducing a new product, marketers target the innovators and early adopters. They will help promote the product through word of mouth.

Products (or innovations) also possess characteristics that influence the rate and level of adoption. Those include:

> Relative Advantage: the degree to which an innovation is perceived as better than what it supersedes. Compatibility: the degree to which an innovation is perceived as consistent with existing values, past experiences, and needs. Complexity: the degree to which an innovation is perceived as difficult to understand and use. Trialability: the degree to which an innovation may be experimented with on a limited basis. Observability: the degree to which the results of an innovation are visible to the receiver and others. (Rogers, 1995)

One way marketers can increase the potential for success is by allowing customers to "try before they buy." Listening stations in retail stores and online music samples have increased the level of trialability of new music. Marketers can improve sales numbers by ensuring the product has a relative advantage, that it is compatible, that it is not complex, and that consumers can observe and try it before they purchase.

Hedonic Responses to Music

Researchers in the fields of psychology and marketing strive to understand the hedonic responses to music. **Hedonic** is defined as "of, relating to, or marked by pleasure" (www.thefreedictionary.com). Hedonic products are those whose consumption is primarily characterized by an affective or emotional experience. Marketers use this principle to study why people enjoy listening to music and what motivates them to seek out music for this emotion-altering experience. Recorded music is considered a tangible hedonic product, compared with viewing a movie or attending a concert, which is an intangible hedonic product. "The purchase of a tangible hedonic portfolio product, such as a CD, gives the consumer something to take home and experience at her convenience, possibly repeatedly" (Moe and Fader, 1999). This convenience factor has fueled the increase in *consumption* of music, despite the fact that sales of recorded music have been falling this century.

Consumers like music for a variety of reasons, mostly connected to emotions or emotional responses to social situations involving music. A brief glance through marketing articles and blogs offers the following reasons.

- To evoke an emotional feeling or regulate a mood
- To evoke a memory or reminisce
- As a distraction or escape from reality

- To create a mood in an environmental setting or a cultural/sporting event
- To combat loneliness and provide companionship
- To foster social interaction with peers
- To calm and relax
- To stimulate (such as to stay awake while driving)
- For dancing or other aerobic performances
- To enhance/reinforce religious or cultural experiences
- For therapeutic purposes (such as pain reduction)
- To pass the time while working or waiting

By understanding the situations that drive consumers to purchase or consume music, marketers can be more effective in providing the right music to the right customers at the right time. Even retail placement of music can benefit from understanding the context in which the music will be consumed. For example, music designed to inspire sports fans may be made available in locations that fans are likely to visit on their way to or while attending a sporting event.

FIGURE 1.3

Altissimo! Recordings is dedicated to the continuing production and distribution of the vast array of music from the wonderful musicians of U.S. military bands, orchestras, ensembles, and choruses. The album *A Patriotic Salute to Military Families* is a collection of marches, bugle calls, and ceremonial and Americana music featuring performances from the bands of ALL branches of the U.S. military. The songs are used for many military and patriotic events, and the album charted at number seven in the *Billboard* Top Classical Albums Chart in July 2002.

The Boston Consulting Group Growth-Share Matrix

Most companies produce a glut of products, some of which perform better than others. The Boston Consulting Group has come up with a method for classifying a company's products based on market growth and market share. The resulting growth-share matrix can quickly indicate the potential for future sales of a product. Market growth is indicated by the vertical axis, with high growth at the top. Relative market share (relative to the competition) is indicated by the horizontal axis, with high market share located on the left side. Products are placed in quadrants based on their relative market share and growth potential in that market.

Stars: Stars are high-growth, high-share entities or products. Because the market for such products is still growing, they require heavy investment to finance their rapid growth. In the recording industry, a genre of music showing growth (such as rock during the late 2000s) would be considered in the high market-growth category. An artist with a high relative market share (such as Coldplay) would be considered a star.

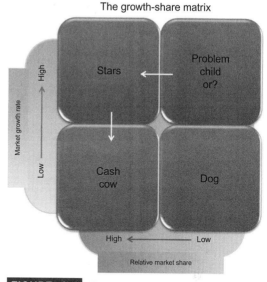

The growth-share matrix

FIGURE 1.4 *The growth-share matrix. (Source: The Boston Consulting Group)*

Cash Cow: These business entities generate a high rate of return for little ongoing investment and therefore are very profitable for the company. These are established products with brand recognition and require fewer marketing dollars per unit sold. In the recording industry, established artists such as Metallica and U2 fall into this category.

Problem Child: The problem child or question mark is a business unit that requires lots of cash and generates little return. These are low-share products in a high-growth market. The goal is to build these into stars by increasing market share while the market is still growing.

Dogs: These are low-growth, low-market share units with no promise of success. For record labels, this category may represent artists who are no longer successful (or never had much success) in producing music in a genre that has stopped growing or an artist who never "breaks." The dogs should be relegated to niche markets at best.

The overall goal for a company after reviewing the portfolio of products is to attempt to move the problem children into the star category and to groom the star products to become tomorrow's cash cows, while discontinuing the dogs.

Price

Pricing is more complex than "How much should we charge?" Pricing structure must be based on maximizing marketing success. Once the wholesale and retail pricing is determined, price-based incentives must then be considered. For example, should the product be put on sale? Should coupons be issued? Should the retailers receive wholesale price breaks for quantity orders or other considerations?

There are generally three methods for deciding the retail price of a product: cost-based pricing, competition-based pricing, and consumer-based (value-based) pricing.

Cost-Based Pricing

Cost-based pricing is achieved by determining the cost of product development and manufacturing, marketing and distribution, and company overhead. Then an amount is added to cover the company's profit goal. The weakness with this method is that it does not take into account competition and consumer demand.

$$\text{Breakeven Point (units)} = \frac{\text{Total Fixed Costs}}{\text{Price} - \text{Variable Costs per Unit}}$$

When determining cost-based pricing, consideration must be given to the fixed costs of running a business, the variable costs of manufacturing products, and the semivariable costs related to marketing and product development. Fixed costs include items such as overhead, salaries, utilities, mortgages, and other costs that are not related to the quantity of goods produced and sold. In most business situations, variable costs are usually associated with manufacturing and vary depending on the number of units produced. For recorded music, that would include discs, CD booklets, jewel cases, and other packaging. *However, within the marketing budget of a record label, the costs of manufacturing and mechanical royalties are considered fixed once the number of units to be manufactured is determined and the recording costs have been computed*. Then there are other (semi) variable costs, which can vary widely, but which are not directly related to the quantity of product manufactured. They would include recording costs (R&D) and marketing costs. For purposes of the following formula, the costs of marketing are considered variable costs, and the recording and manufacturing costs, having already been determined for the project, are considered fixed at this point.

Table 1.1	Standard business model vs. record label business model		
	Variable costs	**Fixed costs**	**Semi-variable**
Standard business model	Manufacturing cost per unit, royalties, licensing, or patent fees per unit, packaging per unit, shipping and handling per unit	Overhead, salaries, utilities, mortgages, insurance	Marketing, advertising, research and development (R&D)
Record label business model	All marketing aspects, including advertising, discounts, promotion, publicity expenses, sales promotions, etc.	Overhead, salaries, utilities, mortgages, insurance, manufacturing costs, production costs (A&R)	

$$\text{Breakeven Point (units)} = \frac{\text{Total Fixed Costs}}{\text{Price} - \text{Variable Costs per Unit}}$$

Competition-Based Pricing

Competition-based pricing attempts to set prices based on those charged by the company's competitors—rather than demand or cost considerations. The company may charge more, the same, or less than its competitors, depending on its customers, product image, consumer loyalty, and other factors.

Companies may use an attractive price on one product to lure consumers away from the competition to stimulate demand for their other products. This is known as *leader pricing*. The item is priced below the optimum price in hopes that sales of other products with higher margins will make up the difference or that purchase will create brand loyalty. The most extreme case of this is **loss leader pricing**. Under loss leader pricing, the company actually loses money on one product in an attempt to bring in customers who will purchase other products that are more profitable. Brand new high-profile compact discs are commonly used by retailers as *loss leaders* because they will generate traffic to the store and increase sales of other items.

Companies are prohibited from coordinating pricing policies to maximize profits and reduce competitive pricing. Such **price fixing** is illegal under the Sherman Act (1890), which prevents businesses from conspiring to set prices (Finch, 1996). The Robinson-Patman Act (1936) prohibits any form of price discrimination, making it illegal to sell products to competing buyers at different prices unless those price differentials can be justified.

Value-Based (Consumer-Based) Pricing

Compact discs are among those items that can be priced based upon the consumer's perceived value rather than on any cost-based or competition-based factors. **Consumer-based pricing** uses the buyer's perceptions of value, a reversal from the cost-plus approach. The music business has at times sought to maintain a high perceived value for compact discs by implementing controversial pricing programs such as the *minimum advertised price (MAP)*. In the 1990s, labels sought to discourage certain retailers from using CDs as loss leaders over concerns that (1) consumers would balk at paying full retail price for CDs that were not offered at the reduced price, and (2) traditional retail stores would be driven out of business because loss leader pricing is not a part of their business model.

In an effort to maintain the perceived value of recorded music, record labels are now focusing on how to add value to music downloads, which provide a limited amount of value-added materials compared to a physical CD. Electronic "digital booklets" are available to consumers who download albums. These digital booklets, usually in PDF file form, contain the artwork, liner notes, and lyrics normally found in a CD booklet.

Promotion

Promotion includes the activities of advertising, personal selling, sales promotion, and public relations. It involves informing, motivating, and reminding the consumer to purchase the product. In the recording industry, the four traditional methods of promotion include radio promotion (getting airplay), advertising, sales promotion (working with retailers), and publicity. More recently, record labels have become more aggressive in marketing through street teams, Internet marketing, digital distribution, tour support, and tie-ins with other products.

Push vs. Pull Strategy

There are two basic promotion strategies: *push promotion* and *pull promotion*. A **push** strategy involves pushing the product through the distribution channel to its final destination in the hands of consumers. Marketing activities are directed at motivating channel members (wholesalers, distributors, and retailers) to carry the product and promote it to the next level in the channel. In other words, wholesalers would be motivated to inspire retailers

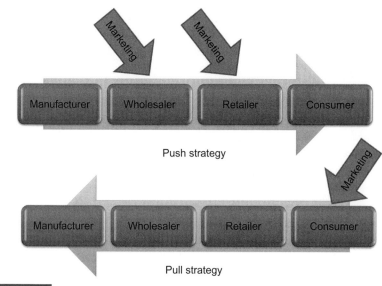

Push strategy

Pull strategy

FIGURE 1.5   *Push vs. pull marketing strategy.*

to order and sell more product. This can be achieved through offering monetary incentives, discounts, free goods, advertising allowances, contests, and display allowances. All marketing activities are directed toward these channel members and are regarded as "trade" promotion and "trade" advertising. With the push strategy, channel members are motivated to "push" the product through the channel and ultimately on to the consumer.

Under the **pull** strategy, the company directs its marketing activities toward the final consumer, creating a demand for the product that will ultimately be fulfilled as requests for product are made from the consumer to the retailer, and then from the retailer to the wholesaler. This is achieved by targeting consumers through advertising in consumer publications and creating "consumer promotions." With a pull strategy, consumer demand pulls the product through the channels.

Most large companies employ both strategies in a combination of consumer advertising and consumer promotions (coupons and sale items) and trade promotion (incentives).

Place

This aspect of marketing involves the process of distributing and delivering the products to the consumer. Distribution strategies entail making products available to consumers when they want them, at their convenience. The various methods of delivery are referred to as **channels of distribution**.

The process of distribution in the record business is currently in a state of evolution, as digital distribution continues to encroach upon the market share for physical recorded music products. Digital delivery has had a dramatic effect on the physical marketplace, causing erosion in market share and closing of some retail music chains, and causing big box stores to reexamine their commitment to offering recorded music.

In an attempt by physical retailers to regain some of the market share from digital downloads, companies have begun offering recorded music on other tangible storage devices, starting with the SanDisk slotMusic campaign. SanDisk partnered with the major record labels to offer recorded music on a memory or slot card, sold through Best Buy and Wal-Mart. These MP3 files are without copy protection and are thus compatible with a wide range of hardware products and systems, making them comparable to compact discs rather than music downloads. In early studies from JupiterResearch, 40% of music consumers said they still preferred CDs to the new media cards. But among heavy music users and young fans, almost none prefer CDs to other digital versions.

Distribution Systems

Most distribution systems are made up of **channel intermediaries** such as wholesalers and retailers. These channel members are responsible for processing large quantities of merchandise and dispersing smaller quantities to the next level in the channel (such as from manufacturer to wholesaler to retailer). Manufacturers engage the services of distributors or wholesalers because of their superior efficiency in making goods widely accessible to target markets (Kotler, 1980).

The effectiveness of intermediaries can be demonstrated in Figure 1.6. In the first example, no intermediary exists and each manufacturer must engage with each retailer on an ongoing basis. Thus, the number of contacts equals the number of manufacturers multiplied by the number of customers or retailers (M × C). In the second example, a distributor is included with each manufacturer and each customer contacting only the distributor. The total number of contacts is the number of manufacturers plus the number of customers (M + C).

Types of Distribution Systems

The three basic types of physical distribution systems are corporate, contractual, and administered. All three are employed in the record business.

In **corporate systems**, one company owns all distribution members at the next level in the channel. A record label would use a distribution system owned by the parent company, as is the case in the relationship between

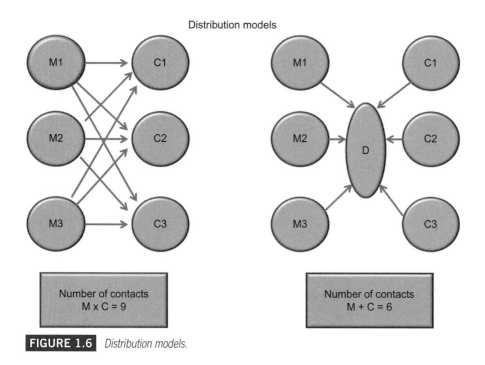

Distribution models

FIGURE 1.6 *Distribution models.*

Mercury Records and Universal Music and Video Distribution. Both companies are owned by Universal. This type of ownership is often referred to as **vertical integration**. When a manufacturer owns its distributors, it is called **forward vertical integration**. When a retailer such as Wal-Mart develops its own distribution or manufacturing firms, it is called **backward vertical integration**.

Contractual distribution systems are formed by independent members who contract with each other, setting up an agreement for one company to distribute goods made by the other company. Independent record labels commonly set up such agreements with independent record distributors. Before the major record labels developed in-house distribution systems in the 1960s and '70s, nearly all recorded music was handled through this type of arrangement.

In **administered distribution systems**, arrangements are made for a dominant channel member to distribute products developed by an independent manufacturer. This type of arrangement is common for independent record labels whose agreements are handled by the distribution branch of one of the major labels.

Digital distribution systems will be addressed in the chapter on distribution and retail.

Distribution chain

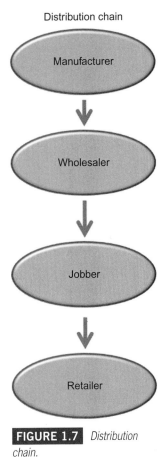

FIGURE 1.7 *Distribution chain.*

Retail

Retailing consists of all the activities related to the sale of the product to the final consumer for personal use. Retailers are the final link in the chain of distribution. There are numerous types of retailers, each serving a special niche in the retail environment. Independent stores focus on specialty products and customer service, while mass merchandisers such as Wal-Mart concentrate on low pricing. Online retailing of physical and digital download product through Amazon.com, CD Baby, and others has been growing in the past few years, with iTunes now claiming the number one spot in music retailing.

Several decades ago, the retail industry developed marketing tools to measure the effectiveness of product location in their stores. **Universal Product Codes (UPC)**, or bar codes, and computerized scanners have helped retailers to determine the optimal arrangement of products within the store. They can move products around to various locations and note its effects on sales. As a result, retailers are charging manufacturers a rental fee for prime space in the store, including counter space, endcaps, and special displays. Retail stores sometimes charge **slotting fees**—a flat fee charged to the manufacturer for placement of (mostly introductory) products on the shelves for a limited period of time. If the product fails to sell, the retail store has reduced its risk. A similar concept has developed in the record business, but it is also tied to promotion and favorable location for the product. It is referred to as *price and position* and will be discussed in the chapter on retailing.

MARKETING STRATEGY

The marketing strategy is the fundamental marketing plan by which the company intends to achieve its marketing objectives. It consists of a coordinated set of decisions based on target markets, the marketing mix, and the marketing budget. The marketing strategy is the "road map," or guide, to growth and success.

The marketing plan for a company should start with an assessment of the company's internal strengths and weaknesses, followed by an examination of the market environment. This **situation analysis** is sometimes referred to as a **SWOT analysis** (strengths, weaknesses, opportunities, and threats). It is a review of the company's present state and an evaluation of the external and internal factors that can affect future success.

As a part of the SWOT analysis, the **internal marketing audit** is designed to examine the company's areas of strength (the company's competitive

Table 1.2 The SWOT analysis	
Strengths	**Weaknesses**
▪ Strong roster ▪ Strong catalog ▪ Skilled label personnel ▪ Effective distribution ▪ Abundant finances ▪ Reputation	▪ Underperforming artists on roster ▪ Expenses are too high ▪ Unhappy artists leaving label (or suing) ▪ Artists' legal problems ▪ Loss of label personnel
Opportunities	**Threats**
▪ New technologies ▪ New markets (international or domestic) ▪ Piracy controls ▪ Trends swing your direction ▪ International trade agreements ▪ New ways to monetize music consumption	▪ Piracy and peer-to-peer (P2P) file sharing ▪ Censorship ▪ Competition for entertainment dollars ▪ Economic downturn ▪ Loss of retail opportunities ▪ New technology

advantages) and weaknesses (the company's vulnerable areas and underperforming units). Then, an assessment is conducted on the external opportunities (areas for growth) and threats (roadblocks and challenges) that exist for the company. The resulting investigation yields the SWOT analysis—strengths, weaknesses, opportunities, and threats. A SWOT assessment allows the marketing department to determine:

- **Where the company is right now:** What is our position in the marketplace? Our market share? Which products are currently successful? What are our areas of strengths and what are our weaknesses?

- **Where the company needs to go:** Where are the opportunities? Where are we likely to hold a competitive advantage in the marketplace in the future? What is our game plan for the future? What threats should we be aware of?

- **How to best get there:** What action needs to be taken to achieve our goals?

A SWOT analysis for a record label may yield some of the following findings.

The SWOT helps the marketing department determine marketing strategy, including the identification of suitable markets and appropriate products to meet the demand. The company can then engage in sales forecasting, budgeting, and projection of future profits.

FIGURE 1.8   *SWOT vs. marketing plan.*

In the book *Marketing: Relationships, Quality, Value*, Nickels and Wood (1997) use Motown Records in their example of a SWOT analysis. Motown's main strength is its strong catalog of 1960s and '70s music, including The Supremes, The Four Tops, The Jackson Five, Michael Jackson, and Boyz II Men. A weakness is their lack of start-up urban and rap artists compared to other labels such as Interscope. Opportunities exist because of the nostalgia fads that periodically emerge, such as a resurgence of interest in old recordings based on the release of a movie or movie soundtrack. In January 2007, after the movie *Dreamgirls*, based on The Supremes, was released, the soundtrack slipped into the number 1 position on the *Billboard* chart. A threat Nickels and Wood mention is "a shift in popular interest away from the label's music" as old-school rhythm and blues (R&B) gives way to newer artists.

The SWOT analysis helps the company discover its core competencies, which enables the company to surpass the competition. However, it is the product *marketing plan* that helps to coordinate the strategic efforts to promote a particular product. It is a carefully planned strategy "with specialists in the areas of artist development, sales, distribution, advertising, promotion, and publicity joining forces in a coordinated effort to break an artist and generate sales" (Lathrop and Pettigrew, 1999). The term *marketing plan* is used by record labels to refer exclusively to the plan designed for each album release and is generally not associated with an overall company strategy.

GLOSSARY

administered distribution—The distribution of products from a smaller independent label is handled by one of the major labels.

backward vertical integration—A retail company develops its own in-house manufacturing facilities for products sold in its stores.

call to action—A statement usually found near the conclusion of a commercial message that summons the consumer to act, such as "Call today" or "Watch tonight at 11."

channel intermediaries—Commonly known as middlemen, these players are involved in the intermediate steps of distribution, between the manufacturer and the retailer.

channels of distribution—The various methods of distributing and ultimately delivering products to the consumer.

competition-based pricing—Attempts to set prices based on those charged by the company's competitors.

contractual distribution systems—Arrangements for distribution are made between an independent manufacturer and an independent distributor and usually involves a contract.

consumer-based pricing—Using the buyer's perceptions of value to determine the retail price, a reversal from the cost-plus approach.

corporate systems of distribution—A manufacturer handles its distribution internally.

cost-based pricing—Determining retail price based on the cost of product development and manufacturing, marketing and distribution, and company overhead, and then adding the desired profit.

diffusion of innovations—The process by which the use of an innovation (or product) is spread within a market group, over time and over various categories of adopters.

forward vertical integration—A manufacturer buys or develops means of distribution or retailing.

hedonic—Of, relating to, or marked by pleasure.

internal marketing audit—The first two steps of the SWOT analysis whereby the internal strengths and weaknesses of the organization are analyzed.

involvement—The amount of time and effort a buyer invests in the search, evaluation, and decision processes of consumer behavior.

loss leader pricing—The featuring of items priced below cost or at relatively low prices to attract customers to the retail store.

marketing—The performance of business activities that direct the flow of goods and services from the producer to the consumer (American Marketing Association, 1960). Marketing involves satisfying customer needs or desires.

marketing mix—A blend of product, distribution, promotion, and pricing strategies designed to produce mutually satisfying exchanges with the target market. Often referred to as "the four P's."

price fixing—The practice of two or more sellers agreeing on the price to charge for similar products.

product class—A group of products that are homogeneous or generally considered as substitutes for each other.

product form—Products of the same form make up a group within a product class.

product life cycle (PLC)—The course that a product's sales and profits take over what is referred to as the "lifetime" of the product.

product positioning—The customer's perception of a product in comparison with the competition's product.

pull strategy—The company directs its marketing activities toward the final consumer, creating a demand for the product that will ultimately be fulfilled as requests for product are made from the consumer.

push strategy—Pushing the product through the distribution channel to its final destination through incentives aimed at retail and distribution.

Situation analysis—The systematic collection and study of past and present data to identify trends, forces, and conditions with the potential to influence the performance of the business and the choice of appropriate strategies.

SWOT analysis—Stands for strengths, weaknesses, opportunities, and threats. An examination of both the internal factors (to identify strengths and weaknesses) and external factors (to identify opportunities and threats).

slotting fees—Fees paid by manufacturers to have their products made available in retail stores. The fees guarantee that the product will be stocked on the shelves for the specified period of time.

vertical integration—Merger of firms at different stages of production and/or distribution.

Universal Product Code (UPC)—An American and Canadian coordinated system of product identification by which a 10-digit number is assigned to products. The UPC is designed so that at the checkout counter an electronic scanner will read the symbol on the product and automatically transmit the information to a computer that controls the sales register.

BIBLIOGRAPHY

American Marketing Association. (1960). *Marketing definitions: A glossary of marketing terms, committee on definitions*. Chicago: American Marketing Association.

Boston Consulting Group. www.bcg.com.

American Marketing Association. (2004). *Dictionary of marketing terms*. www.marketingpower.com.

Committee on Definitions. (1960). *Marketing definitions: A glossary of marketing terms*. Chicago: American Marketing Association.

Finch, J. E. (1996). *The Essentials of Marketing Principles*. Piscataway, NJ: REA.

Hefflinger, M. (2008). Report: U.S. music consumption up in 2007, spending down. DMW Daily. http://www.dmwmedia.com/news/2008/02/26/report:-u.s.-music-consumption-2007,-spending-down.

Kotler, P. (1980). *Principles of marketing*. Englewood Cliffs: Prentice-Hall.

Kotler, P., & Armstrong, G. (1996). *Marketing: An introduction*. Upper Saddle River, NJ: Prentice Hall.

Lathrop, T., & Pettigrew, J. (1999). *This business of music marketing and promotion*. New York: Billboard Books.

Moe, W., & Fader, P. (1999). *Tangible hedonic portfolio products: A joint segmentation model of music CD sales*. http://knowledge.wharton.upenn.edu/papers/840.pdf.

Nickels, W. G., & Wood, M. B. (1997). *Marketing: Relationships, quality, value*. New York: Worth Publishers.

Nielsen SoundScan data. (2008–2009).

Rogers, E. M. (1995). *Diffusion of innovations* (4th ed.). New York: The Free Press.

The Free Dictionary. www.thefreedictionary.com/hedonic.

van Goethem, A. (n.d.). Sad—in a Nice Way! The Royal Philharmonic Society web site. http://www.classicfm.co.uk/music/hear-here/Article.asp?id=856556.

CHAPTER 2

Markets, Market Segmentation, and Consumer Behavior

Tom Hutchison

Before a marketing plan can be designed, it is necessary to fully understand the **market**—who your customers are. It is not possible to make decisions about where to advertise, where to distribute the product, how to position the product, and so forth, until a thorough examination of the market is conducted. This chapter will explain how markets are identified and segmented, and how marketers learn to understand groups of consumers and their shopping behavior. A market is defined as a set of all actual and potential buyers of a product (Kotler, 1980). The market includes anyone who has an interest in the product and the ability and willingness to buy. Markets are identified by characteristics that are measurable. The basic goal of market segmentation (subdividing a market) is to determine the target market.

On the most basic level, markets can be segmented into three sections: (1) fans and current users, (2) potential fans and users, and (3) those people who are not considered part of the target market. Perhaps this third group includes people who cannot or will not consume your products. For music, that may mean people who do not particularly care for the genre that your artist represents. It may include people who do not consume music, people who are unwilling to pay, and those without access to become consumers. Target marketing focuses on the first two groups.

CONTENTS

TARGET MARKETS

Simply stated, a **target market** consists of a set of buyers who share common needs or characteristics that the company decides to serve (www.marketing-teacher.com, 2004). Once market segments have been identified, the next step is choosing the target markets within those segments. Target marketing is choosing the segment or segments that will allow the organization to most effectively achieve its marketing goals. To get a product or service to the right person or company, a marketer would initially segment the market, then target a single segment or series of segments, and finally position the product within the segment (Finch, 1996).

MARKET SEGMENTATION

Because some markets are so complex and composed of people with different needs and preferences, markets are typically subdivided so that promotional efforts can be customized—tailored to fit the particular submarket. For most products, the total potential market is too diverse or heterogeneous to be treated as a single market. To solve this problem, markets are divided into submarkets called *market segments*. **Market segmentation** is defined as "the process of dividing a large market into smaller segments of consumers that are similar in characteristics, behavior, wants, or needs." The resulting segments are homogeneous with respect to characteristics that are most vital to the marketing efforts. This segmentation may be made based on gender, age group, purchase occasion, or benefits sought. Or they may be segmented strictly according to their needs or preferences for particular products.

For example, the fast food chain McDonald's caters to a wide variety of customers—from kids to grandparents. But McDonald's does not use the same marketing messages and techniques to reach children as they do to reach adults. For children, McDonald's has developed special meal packages, recreational areas, and their signature clown. To reach young adults with healthy appetites and meager finances, they developed the "supersize" menu and the 99-cent menu. They reach working adults who need fast food fast by providing packaged breakfast items and hot coffee through their drive-through service windows.

Marketers segment markets for several reasons:

1. It enables marketers to identify groups of consumers with similar needs and interests and get to know the characteristics and buying behavior of the group members.

2. It provides marketers with information to help design custom marketing mixes to speak to the particular market segment.

3. It is consistent with the marketing concept of satisfying customer wants and needs.

In order to be successful, segmentation must meet these criteria:

1. **Substantiality** – The segments must be large enough to justify the costs of marketing to that particular segment. Costs are measured by how much it costs to reach each member of the market and the *conversion rate*—the percent of those you reach who follow through on the purchase.

2. **Measurable** – Marketers must be able to conduct an analysis of the segment and develop an understanding of their characteristics. The result is that marketing decisions are made based on knowledge gained from analyzing the segment. The Internet has allowed for data mining and behavioral targeting, creating market segments based on what Internet users purchase online and what types of sites they visit. For example, based on weather reports and restaurant listings online, a search engine company can determine where someone lives. And based on searches they have conducted on the web and what keywords they have used, the company can determine what products that person might be interested in receiving information about (Jesdanun, 2007).

3. **Accessible** – The segment must be reachable through existing channels of communication and distribution. The Internet has opened up accessibility to all marketers to reach members of their target market—as long as they are Internet users. It has also lowered the costs to reach target members, lowering the barriers to entry and allowing small, undercapitalized companies to compete with major players.

4. **Responsiveness** – The segment must have the potential to respond to the marketing efforts in a positive way, by purchasing the product. The use of credit cards, PayPal, and Google Checkout has opened up new payment methods, allowing businesses to expand their customer base to Internet shoppers who may otherwise be reluctant to give out their credit card numbers.

5. **Unique** – The segments must be unique enough to justify separate offerings, whether the uniqueness calls for variety in product features or simply variety in marketing efforts. Media fragmentation has allowed for more tightly defined market segments. The explosion of specialty magazines and specialty TV channels, along with the innate characteristics of the Internet, has allowed marketers to more tightly target their messages to different audiences.

The process of segmenting markets is done in stages. In the first step, segmentation variables are selected and the market is separated along those partitions. The most appropriate variables for segmentation will vary from product to product. The appropriateness of each segmentation factor is determined by its relevance to the situation. After this is determined and the market is segmented, each segment is then profiled to determine its distinctive demographic and behavioral characteristics. Then the segment is analyzed to determine its potential for sales. The company's target markets are chosen from among the segments determined at this stage.

MARKET SEGMENTS

There is no single correct way to segment markets. Segmentation must be done in a way that maximizes marketing potential. This is done by successfully targeting each market segment with a uniquely tailored plan—one that addresses the particular needs of the segment. Markets are most commonly segmented based upon a combination of geographics, demographics, personality or psychographics, and actual purchase behavior. The bases for determining geographic and demographic characteristics are quite standardized in the field of marketing. Psychographics, lifestyle, personality, behavioristic, and purchase characteristics are not as standardized, and the categorization of these variables differs from textbook to textbook. Evans and Berman (1992) combine these into one category of consumer lifestyles. Kotler (1980) breaks these down into psychographics and behavioristic. Psychographics includes personality, beliefs, lifestyle, and social class. Behavioristic includes both attitudes toward the product and actual purchase behavior. Hall and Taylor segment according to demographics, geographics, psychographics, buyer thoughts and feelings, and purchase behavior (Hall and Taylor, 2000).

Traditional marketing has relied on demographic, geographic, and psychographic segmentation or combinations of them (using some demographics combined with psychographics), but it has been evolving to include more purchase behavior as technology provides a means for more measurement of consumer behavior. Behavior segmentation is a more effective way to segment markets because it is more closely aligned with propensity to consume the product of interest.

Geographics

Geographic segmentation involves dividing the market into different geographical units such as towns, cities, states, regions, and countries. Markets also may be segmented depending upon population density, such as urban, suburban, or rural. Location may reflect a difference in income, culture, social values, and types of media outlets or other consumer factors (Evans and Berman, 1992). Media research companies such as Arbitron and Nielsen use geographic units called the **area of dominant influence** (**ADI**) or **designated market area** (**DMA**). DMA is defined by Nielsen as an exclusive geographic area of counties in which the home market television stations hold a dominance of total hours viewed. The American Marketing Association describes both ADI and DMA as the geographic area surrounding a city in which the broadcasting stations based in that city account for a greater share of the listening or viewing households than do broadcasting stations based in other nearby cities (The American Marketing Association, 2004). Following is an index chart for Contemporary Hit Radio (CHR) listening by geographic region.

This chart shows the American national listenership to CHR radio by geographic region. With the national index, or average, being 100, it is easy to see which regions of the country tend to prefer CHR and which ones do not listen to the format as much as the average.

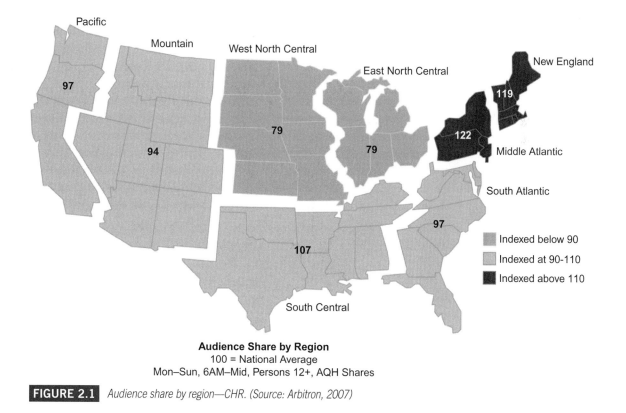

Audience Share by Region
100 = National Average
Mon–Sun, 6AM–Mid, Persons 12+, AQH Shares

FIGURE 2.1 *Audience share by region—CHR. (Source: Arbitron, 2007)*

When the marketer of recorded music and live performances uses geographic segmentation in this manner, it can help develop the logic for the use of resources to support the marketing plan. Marketing strategies are then easily tailored to particular geographic segments. For example, tour support relies on geographics to determine which cities to include in the promotion. As the label's act tours from city to city, the marketing department geographically shifts the focus of their efforts ahead of the actual performance date in each market.

Personal Demographics

Personal demographics are basic measurable characteristics of individual consumers and groups such as age, gender, ethnic background, education, income, occupation, and marital status. Demographics are the most popular method for segmenting markets because groups of people with similar demographics have similar needs and interests that are distinct from other demographic segments. Also, demographics are easier and cheaper to measure than more complex segmentation variables such as personality and consumer behavior.

Age is probably the demographic most associated with changing needs and interests. Consumers can be divided into several age categories such as children, teens, young adults, adults, and older adults. Recently, the segment of "tweens" has been added to the mix to account for the enormous spending power of this group of 23 million preadolescents between the ages of 8 and 13. This concept of segmenting based on age has even been extended to dog food, with companies positioning their various products to appeal to different stages of canine life.

Gender is also a popular variable for segmentation, as the preferences and needs of males are perceived as differing from those of females. Differential needs based upon gender are obvious for product categories such as clothing, cosmetics, hairdressing, and even magazines. But even in the area of music preferences, differences in taste exist for males and females. Arbitron reports that males are more likely to listen to alternative, rock or news/talk radio, while women prefer top 40, country, adult contemporary, and urban radio (Arbitron, 2004).

Income segmentation is popular among certain product categories such as automobiles, clothing, cosmetics, and travel but is not as useful in the recording industry. *Educational* level is sometimes used to segment markets. Well-educated consumers are likely to spend more time shopping, read more, and are more willing to experiment with new brands and products (Kotler, 1980). Segmentation by *race* is also important as certain

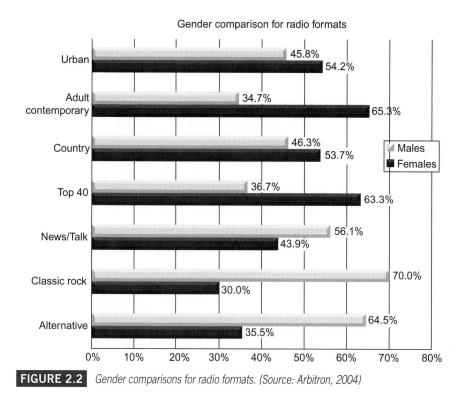

FIGURE 2.2   *Gender comparisons for radio formats. (Source: Arbitron, 2004)*

product classes are targeted toward one race over another, including specific genres of music.

Multivariable Segmentation

The process of combining two or more demographic or other variables to further segment the market has proven effective in accurately targeting consumers. By combining age, gender, and income, marketers can better tailor the marketing messages to reach each group. For example, older males may prefer news/talk radio, whereas younger males prefer modern rock radio.

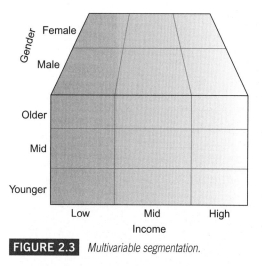

FIGURE 2.3   *Multivariable segmentation.*

Psychographic Segmentation

Psychographic segmentation involves dividing consumers into groups based upon lifestyles, personality, opinions, motives, or interests. Psychographic segmentation is designed to provide additional information about what goes on in consumers' minds. While demographics may paint a picture of what consumers are like, psychographics adds vivid detail, enabling marketers to shape very specific marketing messages to appeal to the target market.

Genres	Personality
Classical, jazz, blues, folk	Reflective and complex
Alternative, rock, heavy metal	Intense and rebellious
Country, pop, religious, soundtracks	Upbeat and conventional
Rap/hip-hop, soul/funk, electronica/dance	Energetic and rhythmic

By understanding the motive for making purchases, marketers can emphasize the product attributes that attract buyers. For example, if consumers are driven by price, pricing factors such as coupons can be emphasized. If another segment is driven by convenience, this issue can be addressed through widespread product distribution. Lifestyle segmentation divides consumers into groups according to the way they spend their time and the relative importance of things in their life.

Psychographics are more difficult to measure and are a moving target. Marketers are constantly updating information to stay abreast of changes in the marketplace. One such system is the VALS2 (values and lifestyles) segmentation system, developed by the Stanford Research Institute. VALS2 is a marketing and consulting tool that identifies current and future marketing opportunities by segmenting the consumer markets on the basis of the personality traits that drive consumer behavior. The system places consumers into three self-orientation categories of people primarily inspired either by ideals, achievement, or self-expression. The market is further subdivided into eight segments by adding the variables of resources and innovation. Those consumers who possess abundant resources are placed in the upper categories. For example, *thinkers* are conservative and motivated by ideals; *experiencers* are young and enthusiastic and motivated by self-expression, excitement, and innovation. More information is available at www.sric-bi.com/VALS/.

Researchers at the University of Texas have found that personal music preferences can be linked to personality traits. Rentfrow and Gosling (Rentfrow and Gosling, 2003) found that people's music preferences typically classify them into one of four basic dimensions: (1) reflective and complex, (2) intense and rebellious, (3) upbeat and conventional, or (4) energetic and rhythmic. Preference for each of the following music dimensions is differentially related to one of these basic personality traits.

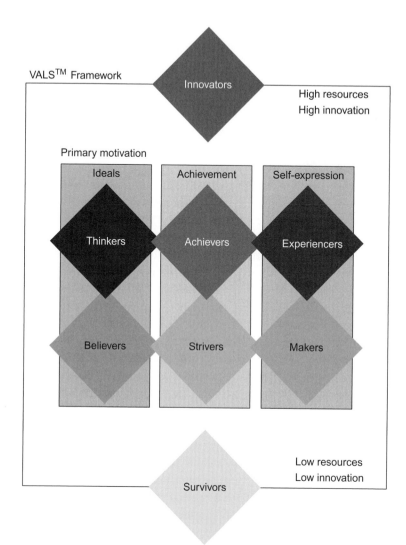

FIGURE 2.4   *Values and lifestyle segments. (Source: SRI Consulting Business Intelligence)*

The Quest for Cool

Marketers use a variety of research techniques to spot lifestyle trends as they develop so that their products will be in the marketplace ahead of demand and competition. Known in vernacular terms as "cool hunting" or the "quest for cool," these marketers are hired by well-known brands to track and understand youth culture. Then the principles of diffusion theory are applied to target opinion leaders in the hopes of penetrating the market (Grossman, 2003). Cool hunting involves observing alpha consumers (cool people) talking, eating, dressing, or shopping, and then predicting what the rest of the market will be doing a year from now. One such company, Look-Look, uses a variety of research techniques including surveys, field observations, ethnographies, mall intercepts, and focus groups. They document their findings with photographs and video recordings and report their conclusions to a variety of major companies who are targeting the youth market (Look-Look, 2004).

The Millennial Generation

Neil Howe and William Strauss discuss the emergence of a millennial generation, which they define as the generation of young adults, the first of whom "came of age" around the time of the millennium (Howe and Strauss, 2000). This segment includes people born between 1982 and 1995. The estimates range between 60 and 74 million, and the **millennials** are becoming a huge economic force. This segment is characterized by a sharp break from Generation X, and hold values that are at odds with the baby boomers. They are described as optimists—team players who follow the rules and accept authority. For this market segment, technology is a part of life, and staying connected to others is important (Bock, 2002).

R. Craig Lefebvre, a Professor at George Washington University, characterizes the millennial generation as more involved in peer-to-peer communication than previous generations. He states ". . . reliance and trust in nontraditional sources—meaning everyday people, their friends, their networks, the network they've created around them—has a much greater influence on their behaviors than traditional advertising." He goes on to state ". . . the challenge for marketers is how to create peer2group exchanges that feature their brands, products, services, and behaviors." The question is no longer "What motivates someone to change" but rather "What motivates someone to share something they find intrinsically useful and valuable with their most trusted friends and colleagues?"

Behavioristic Segmentation

Behavioral segmentation is based on actual customer behavior toward products. Behavioral segmentation has the advantage of using variables that are closely related to the product and its attributes. Some of the more common behavioral segmentation variables are benefits sought, user rate, brand loyalty, user status, readiness to buy, and purchase occasion.

Product Usage

Product usage involves dividing the market based on those who use a product, potential users of the product, and nonusers—those who have no usage or need for the product. How a product is used (and for what purposes) is also of importance. For example, Kodak found that disposable cameras were being placed on tables for wedding guests to help themselves to. As a result, Kodak modified the product by offering five-pack sets of cameras in festive packaging (Nickels and Wood, 1997). The recording industry has adapted to usage situations by repackaging music that is customized for usage occasions, such as wedding music compilations, party mixes, romantic mood music, and so forth. (See the section on hedonic responses to music in Chapter 1.)

Benefits Sought

It is said that people do not buy products; they buy benefits. They buy aspirin to alleviate pain; toothpaste to whiten or protect teeth; and music to elevate mood, combat boredom, provide companionship, or create atmosphere. Benefit segmentation divides the market according to benefits sought by consumers. In the toothpaste market, some consumers are seeking a product that whitens teeth, while others may consider fresh breath most important. Still others may seek fluoride protection against cavities or protection against gum disease. To address these segments, toothpaste companies have all created product variations to accommodate each segment.

User Status

User status defines the consumer's relationship with the product and brand. It involves level of loyalty and propensity to become a repeat buyer. Stephan and Tannenholz (2002) identify six categories of consumers based on user status: sole users, semi-sole users, discount users, aware non-triers, trial/rejecters, and repertoire users.

Seeking Behavior

The Internet has made it easy to track what consumers do, where they go, and what interests they have. Consumers can now be measured and segmented based upon their online behavior, which sites they visit, which pages they look at, and so forth. Then assumptions can be made based upon the particular content that user tends to seek out. One way of keeping track of that data is through the use

---

User status

Sole users are the most brand loyal and require the least amount of advertising and promotion.

Semi-sole users typically use brand A, but have an alternate selection if it is not available or if the alternate is promoted with a discount.

Discount users are the semisole users of competing brand B. They don't buy brand A at full price, but perceive it well enough to buy it at a discount.

Aware non-triers are category users, but haven't bought into brand A's message.

Trial/rejecters bought brand A's advertising message, but didn't like the product.

Repertoire users perceive two or more brands to have superior attributes and will buy at full price. These are the primary brand switchers; therefore, the primary target for brand advertising.

Stephan and Tannenholz

FIGURE 2.5 *User status. (Source: Stephan and Tannenholz)*

of cookies. Webopedia (2009) defines **cookies** as a "message given to a web browser by a web server."

The browser stores the message in a text file. The message is then sent back to the server each time the browser requests a page from the server

Webopedia goes on to explain: "The main purpose of cookies is to identify users and possibly prepare customized web pages for them. When you enter a web site using cookies, you may be asked to fill out a form providing such information as your name and interests. This information is packaged into a cookie and sent to your web browser, which stores it for later use.

The measurement and tracking of Internet usage has opened up a new world of research into consumer behavior, all with the use of **web analytics**—programs that measure a web user's journey online and throughout a company's web site. The Web Analytics Association defines analytics as "the measurement, collection, analysis and reporting of Internet data for the purposes of understanding and optimizing Web usage" (www.webanalyticsassociation.org).

Web data mining is the process of compiling information from analytics reports and noting patterns and developing profiles. Wikipedia states, "This technology has enabled ecommerce to do personalized marketing. . . . Companies can understand the needs of the customer better and they can react to customer needs faster. The companies can find, attract, and retain customers" (www.wikipedia.org). Data mining employs computer algorithms to find buried patterns in large databases of information. These patterns are then analyzed to identify visitor characteristics. In his article *Data Mining on the Web*, Dan Greening states that data mining can be applied in a number of ways to further marketing goals. It can provide characteristics about web visitors that include demographics, psychographics, and technographics. (Technographic information includes what types of technology web visitors are using.) Other information can help determine what web visitors are seeking based upon which links they click on and how much time they spend on each page; then that information is compared with the content to determine interest.

Greening explains that this information can be used for several marketing goals.

1. **Targeting** – By determining a profile for a group of people who may have a particular interest or need, external promotion and advertising can help steer these potential customers to a company's products and web site.

> *When visitors interact with your site, they provide information about themselves and how they respond to your content. . . . Some visitors may even fill out a lifestyle survey or provide names and addresses.*
>
> *Dan Greening @ www.webtechniques.com.*

FIGURE 2.6   *Quote from Dan Greening.*

2. **Personalization** – Greening points out that personalization is the converse of targeting. While targeting determines the types of people that will be exposed to a marketing campaign, personalization offers them specialized content and provides streamlined pathways for them to achieve their purchase goals.

3. **Association** – This identifies products that are likely to be viewed and/or purchased in the same shopping session, thus allowing for segmentation based upon association of purchases.

4. **Clustering** – Clustering identifies people who share common characteristics and provides common and shared access to pertinent information and products.

5. **Estimation** – Estimation guesses an unknown value based upon other information known about that consumer. For example, income may be estimated based upon other factors such as education, area of residence, and previous purchase behavior.

6. **Prediction** – This guesses or predicts a future behavior such as the probability of making a purchase based upon other forms of information. For example, it may be determined that visitors to a music-based site are 90% likely to purchase an album if they spend over 8 minutes on that page and listen to at least four music samples.

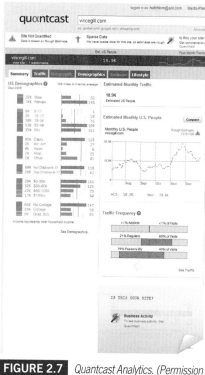

FIGURE 2.7 *Quantcast Analytics. (Permission from Quantcast)*

Thus, segmentation can be based upon particular online behaviors.

CONSUMER BEHAVIOR AND PURCHASING DECISIONS

Consumer behavior is defined as the buying habits and patterns of consumers in the acquisition and usage of goods and services. According to the American Marketing Association (1960), consumer behavior is defined as "the dynamic interaction of affect and cognition, behavior, and environmental events by which human beings conduct the exchange aspects of their lives." One consumer behavior construct is the hierarchy of effects model that combines involvement and left–right brain processing to determine the process a consumer goes through in making a purchase decision.

High- and Low-Involvement Decision Making

Involvement refers to the amount of time and effort a consumer invests in the search, evaluation, and decision process of consumer behavior (see AIDA models in Chapter 1). The level of involvement in the purchase depends on the risk of making the wrong decision and the product's economic and social importance to the consumer. Consumers who search for information about products and brands in order to evaluate them thoroughly are engaging in *high-involvement decision making*. They want to know as much as possible about all choices before making their decision.

In *low-involvement decision making*, the consumer perceives very little risk, low identification with the product, and little or no personal relevance. Low-involvement decision making is often habitual, associated with consumer package goods (prone to repeat purchases) and characterized by brand familiarity and loyalty.

Several factors can influence a consumer's level of involvement in a purchase process:

- **Previous experience** – If consumers have previous experience with the product or brand, they are more likely to have a low level of involvement because they have already formed a perception of the advantages and disadvantages of selecting a particular product.

- **Interest** – The more personal interest consumers have in the product, product class, or benefits associated with the product, the more likely they are to engage in high-involvement behavior.

- **Perceived risk of negative consequences** – If the stakes are high and the consequences of making the wrong decision are dire, consumers engage in high-involvement behavior. Price is a variable associated with risk. High-priced purchases bear more risk. As price increases, so does the level of involvement.

- **Situation** – Circumstance may play a role in temporarily raising the level of involvement for a particular product. A food or beverage purchase casually made under most circumstances may call for more scrutiny if guests are coming to dinner. A CD bought for personal use is a rather low-involvement purchase; when given as a gift, it has higher involvement because of the risk of a bad decision.

- **Social visibility** – Involvement increases as the social visibility of a product increases. Designer clothing falls into this category.

Decision-Making Process

Chapter 1 mentioned the stages involved in the consumer's decision-making process. They were identified as awareness, information, decision, and action (AIDA). This is part of the hierarchy of effects theory, which involves a series of steps by which consumers receive and use information in reaching decisions about what actions they will take (e.g., whether or not to buy a product).

A more elaborate model involves the following six steps:

1. **Problem recognition** – The consumer realizes that a purchase must be made. Perhaps a new need has emerged, or perhaps the consumer's supply of a commonly used product has run low. For example, you have just been told your car will cost more to repair than to replace.

2. **Information search** – The consumer will seek out information about the various options available for satisfying the want, need, or desire. In the car example, you are now entering the phase of doing research: finding out how much you can afford for a new car and then what is available in your price range.

3. **Alternative evaluation** – The consumer will look for alternate opinions of the product under consideration (comparative shopping).

4. **Purchase or acquisition** – The consumer makes the purchase.

5. **Use** – The consumer uses the product, thereby gaining personal experience.

6. **Postpurchase evaluation** – Experience with the product leads to evaluation to determine whether the product meets expectations and should be purchased again next time the need arises.

Cognitive vs. Emotional Decisions

Much research has been done in the field of advertising on the relative roles of affective factors vs. cognitive factors in motivating consumers. Cognition-based attitudes are thoughts and beliefs about an attitude object. Affect-based attitudes are emotional reactions to the attitude object. Attitudes can have both of these components and can be based more or less on either cognitions or affect. While there is probably an element of both in each purchase decision, the relative contribution of each, as well as the process involved, may differ depending upon the type of product and the level of involvement.

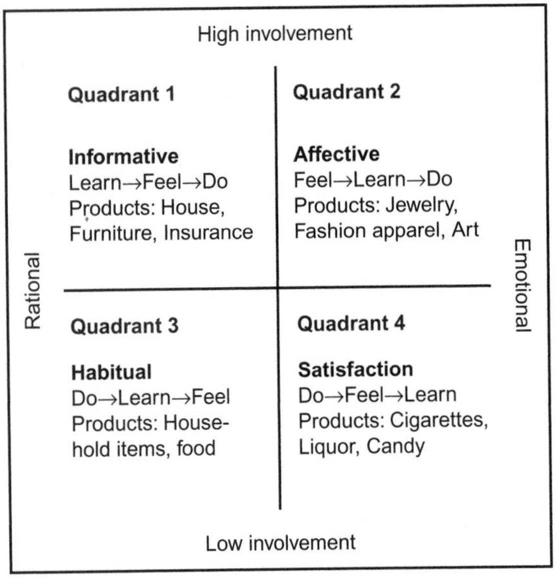

FIGURE 2.8   *Hierarchy of effects. (Source: Foote, Cone, and Belding)*

The model of the hierarchy of effects has been subsequently modified to include three constructs: **cognition** (awareness or learning), **affect** (feeling, interest, or desire), and **conation** (action or behavior)—CAB. The order in which these steps are taken is subject to the type of product and the level of involvement. According to Richard Vaughn, product category and level of involvement may determine the order of effects as well as the strength of each effect (Vaughn, 1980). His model uses involvement (high/low) and think/feel (cognitive or affective components) as the two dimensions for classifying product categories and ordering these three steps.

To begin with, high-involvement situations suggest that the *cognitive* stage and the *affective* stage usually appear first, and these two stages are followed by the *conation* (behavioral) stage. In low-involvement situations, advertising may create awareness first, but attitudes or brand preferences are formed after product trial or experience. Thus, conation or action occurs before opinions are formed about the product. In other words, the thinking and/or feeling occur before the action in high-involvement situations, whereas opinions are formed only after trial of the product in low-involvement situations. Low-involvement products are more subject to impulse purchases.

The resulting model is a four-quadrant grid with quadrant 1 (informative) containing high-involvement cognitive products, quadrant 2 (affective) for high-involvement emotional purchases, quadrant 3 (habitual) for low-involvement rational purchases, and quadrant 4 (satisfaction) for low-involvement emotional purchases. Music purchases generally fall into the satisfaction quadrant, although as prices increase, so does the level of involvement. Some music purchases can be considered impulse buys, if correctly positioned at retail. Other purchases may require creating the emotional (affective) response in the customer. Because music by its very nature evokes emotional responses, exposure to the product is known to increase purchase rate.

NEEDS AND MOTIVES

When a product is purchased, it is usually to fill some sort of need or desire. There is some discrepancy between the consumers' actual state and their

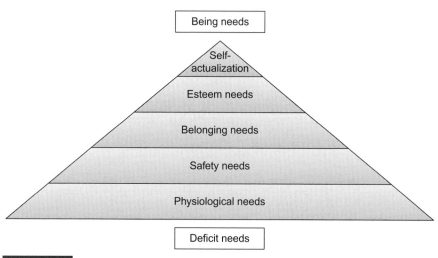

FIGURE 2.9 *Maslow's hierarchy of needs. (Source: Abraham Maslow)*

desired state. In response to a need, consumers are motivated to make purchases. Motives are internal factors that activate and direct behavior toward some goal. In the case of shopping, the consumer is motivated to satisfy the want or need, as outlined in Chapter 1. Understanding motives is a critical step in creating effective marketing programs. Psychologist Abraham Maslow developed a systematic approach of looking at needs and motives.

Abraham Maslow's Hierarchy of Needs

Maslow states that there is a hierarchy of human needs. More advanced needs are not evident until basic needs are met first. Maslow arranges these needs into five categories of physiological, safety, love, esteem, and self-actu-alization. Marketers believe that it is important to understand where in the hierarchy the consumer is before designing an effective marketing program.

Physiological needs are the basic survival needs that include air, food, water, sleep, and sex. Until these needs are met, there is little or no interest in fulfilling higher needs. Survivors in the VALS2 system fit into this category. Safety needs deal with establishing a sense of security. Safety needs motivate consumers to seek political and religious solutions. Until these needs are met, the need for love and belongingness is absent. Love, approval, and belongingness are the next level along with feeling needed and a sense of camaraderie. Products oriented toward social events sometimes appeal to this need and may include music. Esteem needs drive people to seek validation and status. Self-actualization involves seeking knowledge, self-fulfillment, and spiritual attainment.

CONVERTING BROWSERS TO BUYERS

One of the greatest challenges facing marketers is converting potential customers who are browsing the merchandise into actual purchasers. Retail stores refer to a *conversion rate*—the percentage of shoppers who actually make a purchase. The actual rate varies and depends upon the level of involvement, with small-ticket items having a higher conversion rate than expensive items such as appliances or jewelry. A mall store will have more walk-in traffic and thus a lower conversion rate than a freestanding "destination" store. Conversion rates can be improved by having the right combination of merchandise, presenting the merchandise correctly, and offering a level of service that meets or exceeds expectations. When shoppers don't buy, it's often because they were unable to find what they wanted.

Online, conversion rates are improved by offering purchase opportunities with a minimum amount of consumer effort. This is why many online stores such as Amazon.com offer accounts to customers, with their address and credit card information stored in the system. The consumer can purchase with one mouse click.

GLOSSARY

ADI—Area of dominant influence (see *DMA*).

affect—A fairly general term for feelings, emotions, or moods.

behavioral segmentation—Based on actual customer behavior toward products.

cognition—All the mental activities associated with thinking, knowing, and remembering.

conation—Represents intention, behavior, and action.

consumer behavior—The dynamic interaction of affect and cognition, behavior, and environmental events by which human beings conduct the exchange aspects of their lives.

cookies—Parcels of text sent by a server to a web browser and then sent back unchanged by the browser each time it accesses that server. The main purpose of cookies is to identify users and possibly prepare customized web pages for them.

DMA—Designated marketing area. The geographic area surrounding a city in which the broadcasting stations based in that city account for a greater share of the listening or viewing households than do broadcasting stations based in other nearby cities.

geographic segmentation—Dividing the market into different geographical units such as town, cities, states, regions, and countries. Markets also may be segmented depending upon population density, such as urban, suburban, or rural.

involvement—The amount of time and effort a consumer invests in the search, evaluation, and decision process of consumer behavior.

market—A set of all actual and potential buyers of a product. The market includes anyone who has an interest in the product and has the ability and willingness to buy.

market segmentation—The process of dividing a large market into smaller segments of consumers that are similar in characteristics, behavior, wants, or needs. The resulting segments are homogeneous with respect to characteristics that are most vital to the marketing efforts.

millennials—People born between the years 1982 and 1995. Also called the "Net Generation" because they grew up with the Internet.

personal demographics—Basic measurable characteristics of individual consumers and groups such as age, gender, ethnic background, education, income, occupation, and marital status.

psychographic segmentation—Dividing consumers into groups based upon lifestyles, personality, opinions, motives, or interests.

target market—A set of buyers who share common needs or characteristics that the company decides to serve.

user status—The consumer's relationship with the product and brand. It involves level of loyalty.

web analytics—The use of data collected from a web site to determine which aspects of the web site work toward the business objectives.

web data mining—Web usage mining is the application that uses data mining to analyze and discover interesting patterns of user's usage data on the web. The usage data records the user's behavior when the user browses or makes transactions on the web site.

BIBLIOGRAPHY

American Marketing Association. (1960). *Marketing definitions: A glossary of marketing terms, committee on definitions*. Chicago: American Marketing Association.

Arbitron Radio Today. (2004). How America listens to radio. www.arbitron.com/downloads/radiotoday04.pdf.

Bock, W. (2002). www.mondaymemo.net/010702feature.htm.

Evans, J. R., & Berman, B. (1992). *Marketing* (5th ed.). New York: Macmillan.

Finch, J. E. (1996). *The essentials of marketing principles*. Piscataway: Research and Education Association.

Foote, Cone, and Belding. www.draftfcb.com/flash/index.html.

Greening, D. R. (2000). Data mining on the web: There's gold in that mountain of data. New Architect—Internet Strategies for Technology Leaders. www.webtechniques.com/archives/2000/01/greening/ (no longer available).

Grossman, L. (2003). The quest for cool (no longer available). *Time*, August 31.

Hall, C., & Taylor, F. (2000). *Marketing in the music industry* (4th ed.). Boston: Pearson Custom Publishing.

Howe, N., & Strauss, W. (2000). *Millennials rising: The next great generation*. New York: Vintage Books.

Jesdanun, A. (2007). Ad targeting grows as sites amass data on web surfing habits. *The Tennessean*, December 1.

Kotler, P. (1980). *Principles of Marketing*. Englewood Cliffs: Prentice-Hall.

Lefebvre, R. C. (2006). Communication patterns of the millennium generation. http://socialmarketing.blogs.com/r_craiig_lefebvres_social/2006/01/communication_p.html.

Look-Look. (2004). www.look-look.com/.

Marketing Teacher. (2004). www.marketingteacher.com/Lessons/lesson_positioning.htm.

Maslow, A. www.businessballs.com/maslow.htm.

Nickels, W. G., & Wood, M. (1997). *Marketing*. New York: Worth Publishers.

Rentfrow, P. J., & Gosling, S. D. (2003). The do-re-mi's of everyday life: Examining the structure and personality correlates of music preferences. *Journal of Personality and Social Psychology, 84*, 1236–1256.

SRI Consulting Business Intelligence. www.sric-bi.com/VALS/.

Stephan and Tannenholz. (2002). *Market segmentation and the marketing mix: Determinants of advertising strategy*. McGraw-Hill Online Learning Center http://highered.mcgraw-hill.com/0072415444/student_view0/chapter5/els.html.

The American Marketing Association. (2004). www.marketingpower.com.

Vaughn, R. (1980). How advertising works: A planning model. *Journal of Advertising Research, 20*(5), 27–33.

Webopedia. (2009). www.webopedia.com/TERM/C/cookie.html.

CHAPTER 3

The U.S. Industry Numbers

Tom Hutchison

SALES TRENDS

In the early 1990s, the recording industry was in the midst of a **replacement cycle**, where consumers were in the process of replacing their old vinyl and cassette collections with the first digital format—the compact disc (CD). In 1994, there was a 20% increase in sales, to over $12 billion. This boom was fueled by discount retailers such as Best Buy and Circuit City aggressively entering the music market and discounting CD prices. A calculation of the Recording Industry of America Association (RIAA) 1995 figures indicates that the average selling price of a CD dropped from $13.14 in 1993 to $12.78 in 1994. As the retail sector contracted and discount stores backed off from using CDs as loss leaders, the average price crept up to around $15 by 2002 before dropping again in 2007 to $14.58 and to $14.22 in 2008. In an effort to prop up CD sales, in 2009 the industry began to offer more product extensions—CDs with extra features for a premium price sold through particular retailers (see Chapter 18 on special products).

By 1995, the CD replacement cycle was nearing completion, and the impact of the changing retail landscape was beginning to take its toll on the bottom line. Traditional retail stores were struggling to compete with discount stores, and sales gains in the industry overall were modest (from $12.068 billion in 1994 to $12.320 billion in 1995). Blockbuster Music closed hundreds of stores, causing a massive rush of returned product. There was slight improvement in 1996, with sales at $12.5 billion as the

CONTENTS

Table 3.1	RIAA data on annual shipments (Source: RIAA 2008)		
	RIAA data on annual shipments		
Year	**CD dollar value***	**CD units shipped***	**Price per unit**
1993	$6511.4	495.4	$13.14
1994	$8464.5	662.1	$12.78
1995	$9377.4	722.9	$12.97
1996	$9934.7	778.9	$12.75
1997	$9915.1	753.1	$13.17
1998	$11,416.0	847.0	$13.48
1999	$12,816.3	938.9	$13.65
2000	$13,214.5	942.5	$14.02
2001	$12,909.4	881.9	$14.64
2002	$12,044.1	803.3	$14.99
2003	$11,232.9	745.9	$15.06
2004	$11,446.5	767.0	$14.92
2005	$10,520.2	705.4	$14.91
2006	$9372.6	619.7	$15.12
2007	$7452.3	511.1	$14.58
2008	$5471.3	384.7	$14.22

*In millions.

industry sought to reexamine its strategy for selling prerecorded music (RIAA, 1997), while traditional retailers continued to struggle. The industry experienced a decline in 1997 as the RIAA commented that "the industry was responding to a smaller but healthier retail base" (RIAA, 1998). But it was an industry faced with bankruptcy filings and consolidation. One anomaly in 1997 occurred with a 54.4% increase in the sale of CD singles, attributed to Elton John's remake of *Candle in the Wind*, a tribute to Princess Diana. Growth returned briefly in 1998 as shipments grew by 5.7%. While shipments of singles dropped, CD units and music video units showed a healthy increase. The RIAA attributed the increase to a steady flow of releases by top artists throughout the year and an increase in the diversity of offerings. The moderate growth continued through 1999, with a 3.2% increase in units, fueled by strong growth in the full-length CD format (12.3% in value), despite a drop-off in music video sales. Credit is given to retailers and suppliers for improved efficiency in inventory management. The RIAA stated that the music industry was successfully competing against the "ever-increasing competition for the consumer's entertainment dollar" (RIAA, 2000).

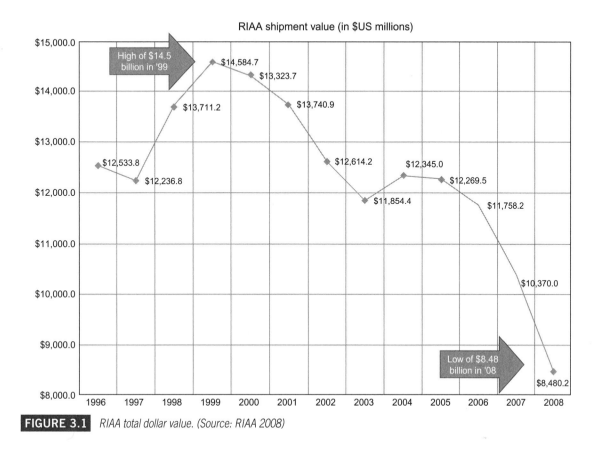

RIAA shipment value (in $US millions)

High of $14.5 billion in '99

$14,584.7

$13,323.7

$13,711.2

$13,740.9

$12,533.8

$12,614.2

$12,345.0

$12,236.8

$12,269.5

$11,854.4

$11,758.2

$10,370.0

Low of $8.48 billion in '08

$8,480.2

FIGURE 3.1 *RIAA total dollar value. (Source: RIAA 2008)*

A Period of Decline

The new millennium ushered in a period of decline in recorded music sales as the industry struggled against threats on numerous fronts. The market for CD singles plummeted as the Internet took over, and peer-to-peer file-sharing services such as Napster were blamed for much of the downturn. From an industry high of $14.58 billion in 1999, the industry dropped nearly 30% in 8 years to end at $10.37 billion in 2007. Total *album* sales in the United States fell another 18% in 2008 to end the year at $8.48 billion.

This period of decline began in the first half of 2000 and was attributed to the rapid drop in sales of CD singles, brought on by the Internet. (The U.S. economy also experienced a downturn at that time.) Cassettes continued to decline, while full-length CDs grew slightly in **dollar value**. The decline was more drastic in 2001 as total units dropped 10%, with a 6.4% drop in full-length CDs for the first time. This downturn coincided with the rapid adoption of music downloading from illegal **peer-to-peer** (**P2P**) services and ownership of CD burners, up drastically in two years.

In 2002, the market dropped another 11.2% in units and 8% in value. Every configuration saw a decrease in sales except DVD. At this point, the RIAA began an aggressive campaign to discourage consumers from using illegal file-sharing services, but other factors also contributed to the downturn. Young consumers were spending their entertainment budget on cell phones, computers, video games, DVD collections, and other forms of entertainment. Consumers began to report (through research studies) that they perceived the cost of CDs to be too high. Upon further analysis, young consumers, who had grown accustomed to cherry-picking songs from albums through P2P services, were opposed to paying retail price for an album just to own the one or two songs they wanted. These consumers were beginning to burn their own personal mix CDs either from their own collection or from P2P services. As a result, they preferred music à la carte (by the track) and were a receptive market for licensed music downloading services. But these services were slow to develop and the offerings were not sufficient to entice droves of consumers into subscription services.

Things did not improve for 2003 as the industry fell another 7.2% in units and 6% in value. Again, full-length CDs, the industry's moneymaker, fell 6.7% in value and 7% in units. The industry began to see some evidence of bottoming out and even a slight turnaround in 2004, with total sales of $12.154 billion. This turnaround was fueled by a slight increase in CD album sales (1.9% in dollar value), digital singles (139.4 million units), and DVD videos (66% unit increase).

In 2005, the RIAA added new format categories of mobile, digital subscription, digital music video, and kiosks. "Counting all formats and all distribution channels (retail and special markets distribution), overall unit shipments of physical product decreased by 8.0 percent in 2005" (RIAA, 2006). Mobile formats (such as ringtones) shipped 170 million units, representing $421.6 million in retail value. Full-track downloads were not a significant sector at that time.

The drop continued through 2006 and 2007, falling to $10.3 billion in 2007. For 2008, there was another 14% drop to 428 million total albums sold in the United States, while sales of digital downloads increased 27% to just over a billion songs. CD sales were down almost 20% from the previous year (Sisario, 2009).

SoundScan Sales Trends

Whereas the RIAA measures units shipped for their annual industry numbers, SoundScan measures actual music sales. In an effort to monitor recorded music sales and determine trends and patterns, the industry in general and SoundScan in particular have come up with a way to measure digital album sales and compare them with music sales in previous years. When SoundScan first started tracking digital download sales, the unit of measurement for downloads was the single track, or in cases where the customer purchased the entire album, the unit of measurement was an album. But this did not give an accurate reflection of how music sales volume had changed, because most customers who download buy songs à la carte instead of in album form. In an attempt to more accurately compare previous years with the current sales trend, SoundScan came up with a unit of measurement called **track equivalent albums** (**TEA**), which means that 10 track downloads are counted as a single album. Thus, the total of all the downloaded singles is divided by 10 and the resulting figure is added to album downloads and physical album units to give a total picture of "album" sales. Here is an example of how this works from Billboard.biz.

> When albums are tallied using the formula of 10 digital track downloads equaling one album, the 582 million digital track downloads last year translates into 58.2 million albums, giving overall albums a total of 646.4 million units. The overall 2006 total of 646.4 million is a drop of 1.2% from 2005's overall album sales of 654.1 million. (Christman, 2007)

Having established the TEA as a new unit of measurement, industry trends show the following: U.S. album sales have continued to slide every year from a high in 2000 of 785 million units to 646.4 million units in 2006 (including TEA). (Without the addition of the TEA, album units in 2006 were 588 million.) This represents a drop of 1.2% from the 2005 overall album sales of 654.1 million. In 2006, the digital single format grew 60% and digital album sales grew 84% to $770 million, while the overall U.S. industry declined 6.2%. The year 2007 saw a dramatic decrease in overall sales numbers, with the total U.S. industry down nearly 12% from the previous year. Sales of physical product dropped nearly 20% in dollar value from the previous year. The value of digital downloads grew nearly 45%.

By the close of 2007, album sales were down 15% from the previous year and down 20% for the holiday season. With track-equivalent albums factored in, the slide was 9.5% from 2006. The one bright spot for physical units has been online sales. Meanwhile, retail stores continued to reduce the amount of shelf space they devoted to recorded music as online sales and digital downloads continued to erode at the physical unit **market share**. Sales for 2008 for all album products (CDs, LPs, and digital albums) reached only 428.4 million units. When digital tracks (singles) are added using

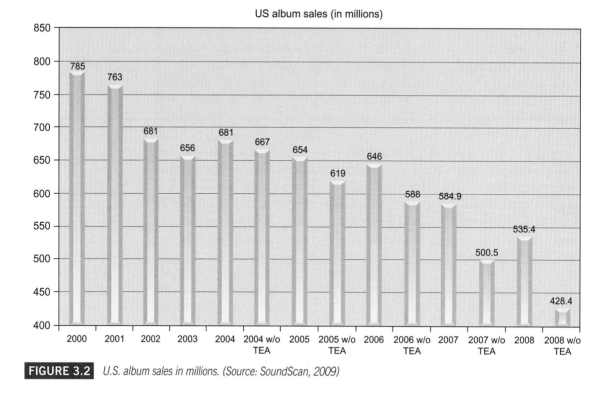

US album sales (in millions)

FIGURE 3.2 *U.S. album sales in millions. (Source: SoundScan, 2009)*

the TEA formula, sales were at 535.4 million. In 2008, digital track sales were up 25% and digital album sales were up 32%. Also in 2008, SoundScan reported that digital music accounted for 32% of music unit purchases.

Annual Sales Trends

Music sales are seasonal, with the greatest majority of sales occurring during the fourth quarter holiday season. For this reason, many of the superstars wait until the fourth quarter to release a new album so they can take advantage of the holiday shopping season. And conversely, many newer artists will avoid releasing an album in this same time period because competition for retail space is more intense. Valentine's Day is the second largest record purchasing holiday (if Thanksgiving weekend is considered a part of the Christmas holiday season). Note that in 2008, digital track sales caught up to album sales right after the Christmas holiday as music fans cashed in their iTunes gift cards and purchased music online.

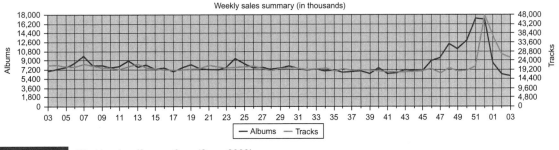

FIGURE 3.3 *Weekly sales. (Source: SoundScan, 2009)*

MUSIC GENRE TRENDS

While the category of rock music has dominated the marketplace for many years, the double-digit lead held by rock music for decades has been reduced from 35% of the market share to almost 25% for nearly a decade. The major shift occurred in 1998 as the steady rise of rap and the return of pop began to siphon off market share, but this trend reversed in the second half of the decade between 1996 and 2004. Rock music began a comeback in 2004, rising through 2006. Even the popular show *American Idol* began to feature more rock artists among the top contestants in 2007–2008, with rocker David Cook winning the 2008 contest and rock singers edging out country and pop in 2009.

Country music reached a market share peak in 1993 when Garth Brooks was its top seller. Pop music suffered in the mid-1990s but came back strong with the emergence of the "boy bands" and teen pop idols Britney Spears and Jessica Simpson. That trend peaked in 2001, and after that, pop music suffered a drop in market share through 2006. Rap music was on a steady rise throughout the 1990s until emerging as the number 2 genre in 2000. After a setback in 2001, rap began to rise again in 2002 to hold on to the number 2 spot for the first half of the decade. R&B music enjoyed a growth in market share until reaching a peak in 1998; after a precipitous drop to less than 10%, R&B has been making a slow, moderate comeback. Religious music enjoyed a banner year in 1998 with the *Prince of Egypt* soundtrack and took another upturn after the attacks of September 11, 2001. Religious music has since dropped to below 5% of the market. By 2007, there was a convergence of market share as the genres of country, R&B urban, rap, and pop each held about a 12% market share.

The following chart shows the relative market share of these genres. To examine the other genres with smaller market share, a **logarithmic scale** was created to expand the lower end of the spectrum (see Figure 3.5).

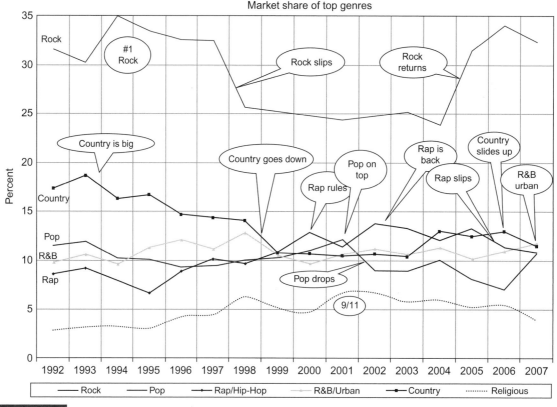

FIGURE 3.4   *Top genres: market share. (Source: RIAA 2009)*

In the smaller genres, a breakout release (one that sells outside the niche market) can elevate the market share for the genre for the duration of the release. For example, a blockbuster album from Enya in 2001 finished the year at number 4 on the *Billboard 2001* year-end chart. Movie soundtracks had strong showings in 1998 and 2001, assisted by the top-selling soundtrack to the *Titanic* in 1998 and *O Brother, Where Art Thou?* in 2001. Soundtracks had a strong showing again in 2003 despite the fact that there were no soundtrack titles in the top 10 albums for the year. However, there were several titles in the top 50, including the soundtracks to the movies *Chicago, Lizzie McGuire,* and *Bad Boys II*. The category of soundtracks then took a slide through 2007.

The growth of Latin music has contributed to the rise in the "other" category, posting a 12% gain in 1998 and another 8% in 1999 before falling 7% in 2000; but rising again in mid-decade. The "other" category also includes big band, swing, ethnic, spoken word, holiday, comedy, exercise,

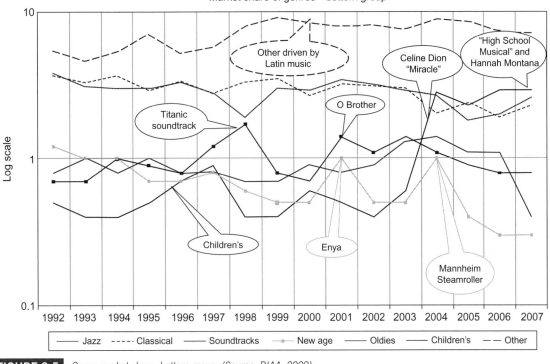

FIGURE 3.5   *Genre market share: bottom group. (Source: RIAA, 2009)*

electronic, standards, and folk. The "oldies" category has fallen off dramatically, starting in 2004. Children's music was boosted by the introduction of Radio Disney in 1996 and has been supported in the late 2000s by the efforts of Disney to reach younger consumers with recorded music and music-related products such as the Jonas Brothers, *High School Musical*, and *Hannah Montana*.

DEMOGRAPHIC TRENDS

The RIAA only measures age and gender for their consumer profile studies.

Age

The RIAA reports that record buyers are getting older. The 45+ group expanded from 15.1% in 1994 to 26.6% in 2003, and it continued to grow through the mid-decade. Some slide has been evident toward the end of the 2000s. Perhaps it is time to further divide this group to gain more insight into differences between what people buy in their 40s to what they buy in their 60s. College-age consumers (ages 20–24) began to decline in 1997, a trend that was assisted by peer-to-peer file-sharing services starting in 1999. The decline continued until 2005 and has not fully recovered. The P2P services have also been blamed for a decline in the teenage group of 15- to 19-year-olds, starting in 1999, despite the popularity of acts geared toward teens during this time. A gradual decline in the 25- to 29-year-old group was halted in 2003, while the 10- to 14-year-old group remained relatively steady. Since 2006, market share for the oldest two categories has been dropping, but the 45+ group still maintains the largest market share. Meanwhile, many of the other categories have started to converge, to about 12% to 14% by 2007. The 10- to 14-year-old group rose dramatically in 2007—*Hannah Montana* and *High School Musical 2* soundtrack finished in the top 10 sellers for the year. For 2008, the oldest age group once again

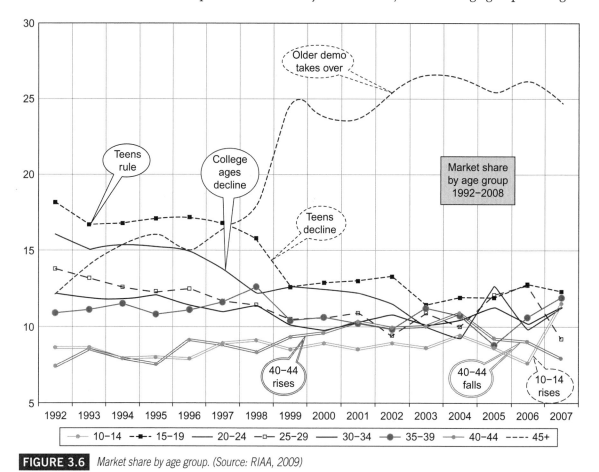

FIGURE 3.6 *Market share by age group. (Source: RIAA, 2009)*

showed the most growth in market share expanding from nearly one-fourth to over one-third.

Gender

Although males did not exactly dominate the market in the early '90s, they comprised more than half of all record buyers until 1997. The metal, rock, and grunge acts of the early '90s sent males to the record stores. This started to change in 1997 as many female artists who appeal to females began to dominate the charts. The Lillith Fair brought female fans to concerts and inspired sales for many of the artists appearing on the tour. The Spice Girls were also popular with younger females, as were teen pop acts such as N'Sync, Britney Spears, and Backstreet Boys. This changed again in 1999 and 2000 when rap music and rage rock brought male fans back to the record stores despite the fact that Britney Spears finished in the top five both years. Females were once again in the majority in 2001–2003 with blockbusters from Enya, Destiny's Child, Alicia Keys, Dixie Chicks, Shania Twain, Norah Jones, and Beyonce. Since 2004, there has been a slight correlation between

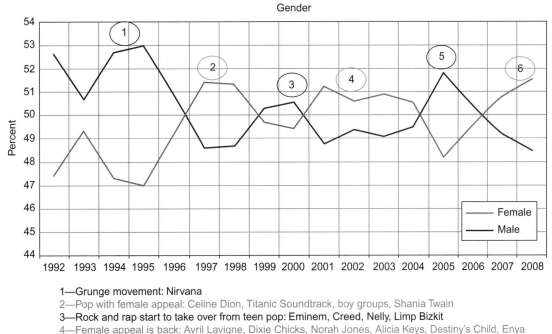

1—Grunge movement: Nirvana
2—Pop with female appeal: Celine Dion, Titanic Soundtrack, boy groups, Shania Twain
3—Rock and rap start to take over from teen pop: Eminem, Creed, Nelly, Limp Bizkit
4—Female appeal is back: Avril Lavigne, Dixie Chicks, Norah Jones, Alicia Keys, Destiny's Child, Enya
5—Rock (favored by males) returns to popularity as a genre
6—Pop returns and rock drops

FIGURE 3.7   *Market share by gender. (Source: RIAA, 2009)*

gender market share and the relative market shares of rock and pop. Pop music is stronger in years that sales are stronger among women, with males showing stronger market share in years that rock music is stronger.

FORMAT – CONFIGURATION

The category of configurations, as reported by the RIAA, deals with the different types of recorded music product and has undergone a drastic change recently. Gone are categories such as cassettes, CD singles, cassette singles, the extended-play EP, super audio CD (SACD), and DVD-audio. The new categories include CD, music video, other albums, and other singles in the physical section. The digital section includes downloaded singles, downloaded albums, music from kiosks (such as those found in movie theaters), downloaded music videos, mobile, and subscription.

Shipments are reported as market share in the RIAA consumer profile and as units and dollar value in the year-end sales table. The product life cycle (from Chapter 1) can be used to examine the rise and fall of

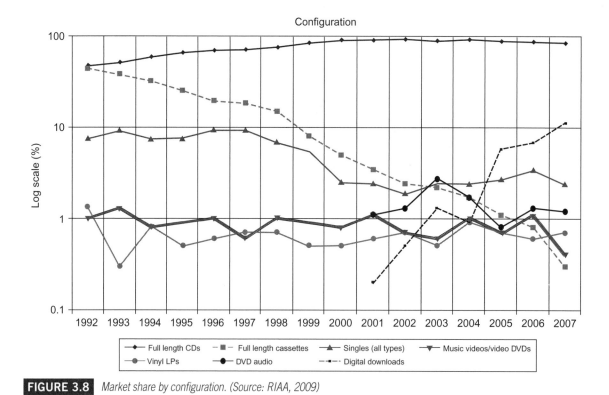

FIGURE 3.8   *Market share by configuration. (Source: RIAA, 2009)*

configurations. As of the end of the first decade of the millennium, the staple of the industry—the CD album—is in the decline stage of the product life cycle. The format of DVD-audio, introduced in 2001, seems all but forgotten. All physical formats for singles are at the end of the decline stage, with the market for downloaded singles in mid-growth stage. The growth in digital downloads has created a recent turnaround in the category of singles—a format that was near oblivion just a few years earlier.

Digital Sales

The RIAA and SoundScan began tracking digital sales in 2004. From 2005 to 2006, there was a 65% increase in digital download sales, with a 45% increase by 2007. In the United States, more than 844 million digital tracks were sold (SoundScan, 2009) in 2007 compared to 582 million in 2006. By the end of 2008, over 1 billion tracks had been sold that year. Digital album sales reached the 50 million mark in 2007, a 53.5% increase over 2006. Digital album sales accounted for 10% of total album sales (without TEA included). Sales of music downloads have continued to grow, although that growth is slowing from the rapid pace earlier in the decade. In 2007, digital sales in the United States accounted for 30% of all music (units) sold according to the International Federation of the Phonographic Industries (IFPI, 2008), but only 23% according to SoundScan. Digital *album* sales

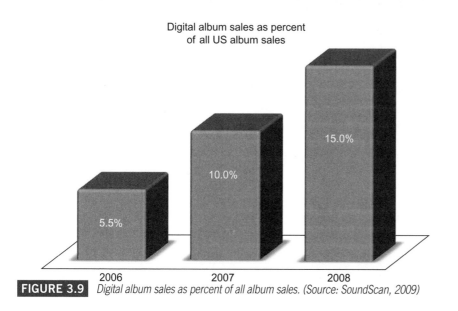

Digital album sales as percent
of all US album sales

FIGURE 3.9 *Digital album sales as percent of all album sales. (Source: SoundScan, 2009)*

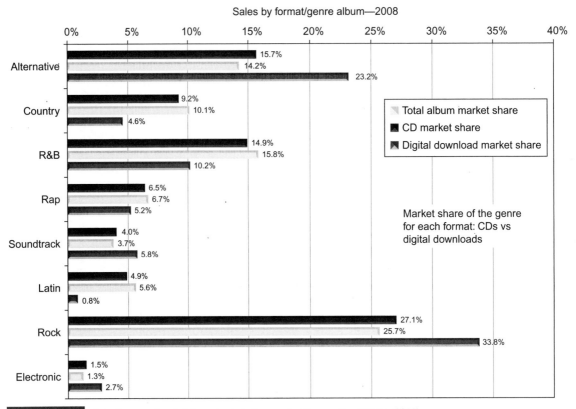

Sales by format/genre album—2008

	Total album market share	CD market share	Digital download market share
Alternative	14.2%	15.7%	23.2%
Country	10.1%	9.2%	4.6%
R&B	15.8%	14.9%	10.2%
Rap	6.7%	6.5%	5.2%
Soundtrack	3.7%	4.0%	5.8%
Latin	5.6%	4.9%	0.8%
Rock	25.7%	27.1%	33.8%
Electronic	1.3%	1.5%	2.7%

Market share of the genre for each format: CDs vs digital downloads

FIGURE 3.10 *Sales by genre for digital vs. physical album units. (Source: SoundScan, 2009)*

accounted for 5.5% of album sales in the United States in 2006, jumping to 10% in 2007 and up to 15% by 2008.

The switch from CDs to digital downloading varies when separated out by genres. According to Nielsen SoundScan (2009), digital album sales for both 2007 and 2008 were as follows: the genres of alternative, rock, soundtracks, and electronic showed an active downloading market, with fans of these genres more likely to select to download an album rather than buy the CD. Fans of country, R&B, rap, and Latin are more likely to buy CDs than download albums.

OUTLETS

The RIAA reports on where records are sold. Trends indicate that traditional retail record stores are no longer the source of the majority of record

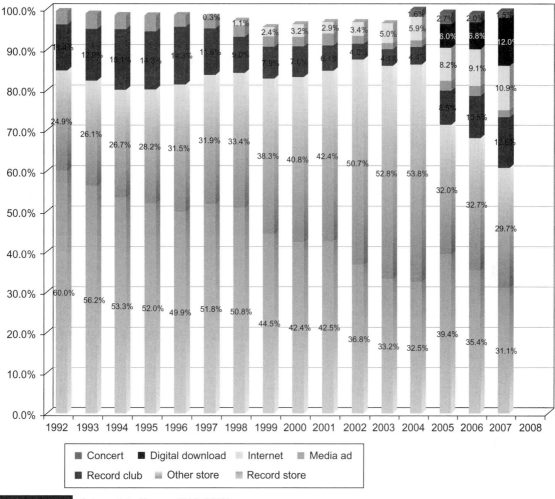

FIGURE 3.11   *Sales outlets. (Source: RIAA, 2009)*

sales. This loss of market share spells trouble for new artists and indie labels that have traditionally relied on record retail stores to introduce new artists to consumers. Traditional record stores have historically carried a larger variety of titles and are more likely to carry titles that are obscure or geared toward a niche market. With the loss of these stores, many indies and self-promoted artists have turned to the Internet to generate sales and provide product to consumers. Traditional record stores held 60% of the market in 1992, and this dwindled to just 30% by 2008.

The category of "other stores" includes mass merchants such as Wal-Mart, Target, and K-Mart, as well as "big box" electronic superstores like Best Buy. This segment grew from less than one-quarter of the market in 1992 to over half in 2003. These retailers are less likely to carry the depth of product of the record stores, and mass merchants concentrate on the top sellers. In recent years, mass merchants and big box retailers have cut back on the floor space they designate for music sales and are turning to other products to generate store traffic. In 2004, these "other stores" held almost 54% of the market share; since then, their market share has been reduced to less than 30%.

Record clubs remained popular through the mid-1990s, but fell off during the first part of this decade as consumers finished up their catalog CD collections. There has now been a resurgence in the popularity of record clubs, which are now also selling DVD movies and using the Internet to recruit and manage members.

Meanwhile, the categories of Internet sales of physical units, digital downloads, and the newly monitored category concert sales have all shown growth in the late 2000s.

MARKET SHARE OF THE MAJORS

Label market share is measured for both new releases and for a combination of new releases and recent sales of catalog titles. Universal Music Group (**UMG**) has dominated market share since acquiring PolyGram Records in 1998 for $10.4 billion. It is impressive that Universal has managed to grow additional market share since the purchase of PolyGram. They have grown from about one-quarter to nearly one-third of the industry market share. The only other segment of the industry showing consistent growth over 10 years is the indie sector.

The 2004 merger of Sony and Bertelsmann Music Group (**BMG**) has created a second behemoth in the industry and put the joint venture just behind Universal at number 2. BMG held a strong market share in the late '90s and early 2000s in partnership with Jive Records, who, at that time, was responsible for the teen hit sensations N'Sync, Britney Spears, and Backstreet Boys. Sony enjoyed success in the late '90s with Celine Dion, Mariah Carey, and other pop artists but saw their market share slip when the pop movement subsided. The merger of these two companies, with Sony ultimately purchasing the music division from Bertelsmann, has failed to yield any advantages in market share; yet the new company, Sony Music Entertainment (**SME**), has managed to shed expenses and eliminate redundant departments.

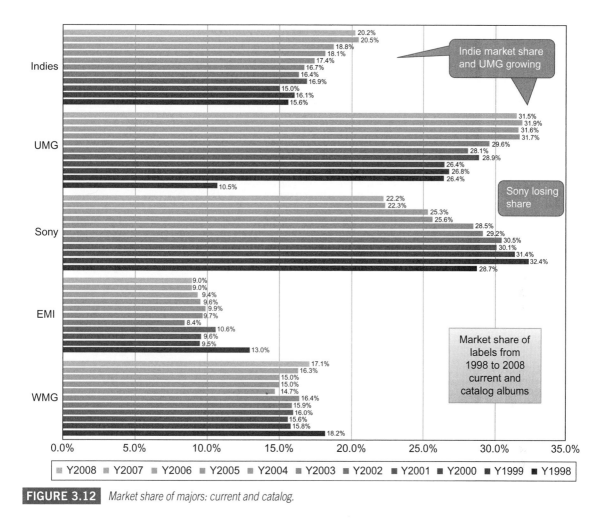

FIGURE 3.12 *Market share of majors: current and catalog.*

One of the strengths of a record label can be measured by the relative market share for new releases and indicates the vitality of its A&R skills. Labels that are increasing market share from sales of new releases are those with success in finding and developing new acts and releasing successful new recordings from their established acts. The following table shows that only Universal and the indies have been successfully increasing market share of new releases over the past several years, again with Sony surging for a while and then dropping back. Sony holds a smaller market share for new releases than it does for the combination of catalog and new releases (see Figure 3.13). After several years of turmoil, Warner Music Group (**WMG**) is beginning to gain market share for both new releases and total product.

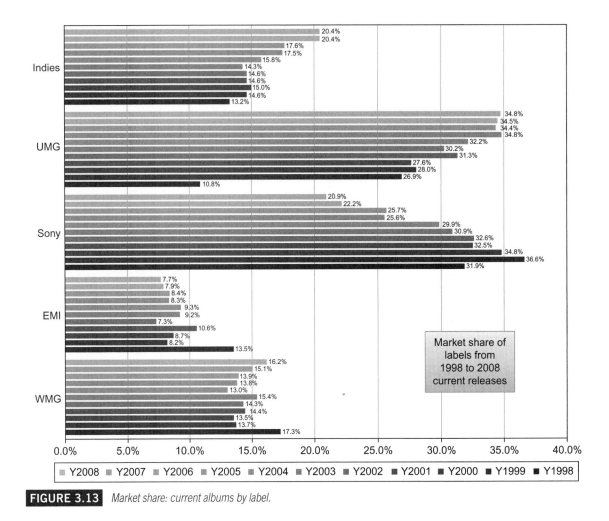

FIGURE 3.13 *Market share: current albums by label.*

Comparison of All and Current Albums

In the following charts, a comparison is made for market share of all albums and market share of new releases for each label. So not only is an increase or decrease in current album market share an indicator, but a comparison of all albums and current albums indicates how much a label relies on catalog sales compared to sales from its new releases.

Of the major labels, only UMG has a larger market share for current releases than for all album releases. The indie sector shows strong market share in new releases, but SME, Electrical and Mechanical Industries (**EMI**), and WMG are all more dependent on catalog sales than Universal and the indies.

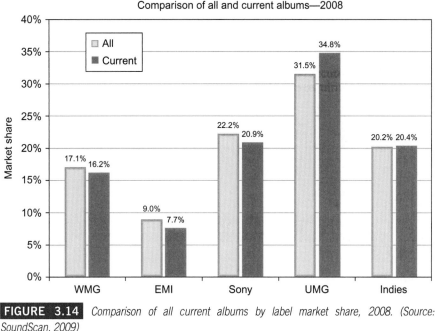

Comparison of all and current albums—2008

FIGURE 3.14 *Comparison of all current albums by label market share, 2008. (Source: SoundScan, 2009)*

CATALOG SALES

Catalog sales are defined as sales of records that have been in the marketplace for over 18 months. Current catalog titles are those over 18 months old but less than 36 months old. Deep catalog albums are those over 36 months since the release date.

When the compact disc was first introduced in the early 1980s, it fueled the sales of older catalog albums as consumers replaced their old vinyl and cassette collections with CDs. This windfall allowed labels to enjoy huge profits and led to the industry expansion of the 1980s and early '90s. However, catalog sales started to diminish in the mid-1990s as consumers finished replacing their collections. The closure of traditional retail stores also contributed to the decline in catalog sales, with customers having fewer opportunities to be exposed to the older titles.

Pricing strategies to sustain catalog sales were introduced as early as 1996. In the late '90s labels sought new ways to promote catalog sales through reissues, compilations, and

The business was saved by the introduction of the compact disc, or the CD, in the early '80s. That reversed the decline, and created another 15 years of growth and profits for the business as, over the course of that period of time, people converted from buying...what was then a $6.98 cassette to $13.98 CDs, you know.

And for the first couple of years that most people got a CD player they would actually replace a lot of their collection. So classical sales went through the roof and old recordings went through the roof, and the business therefore bounced back, and had another 10 to 15 years of double-digit growth. . . .

–Danny Goldberg
Chairman and CEO, Artemis Records

Published with the permission of FRONTLINE, a documentary series on PBS. This material originally was published on FRONTLINE's web site "The Way the Music Died"
www.pbs.org/frontline/shows/music.

FIGURE 3.15 *Impact of CDs on sales. (Source: PBS* Frontline*)*

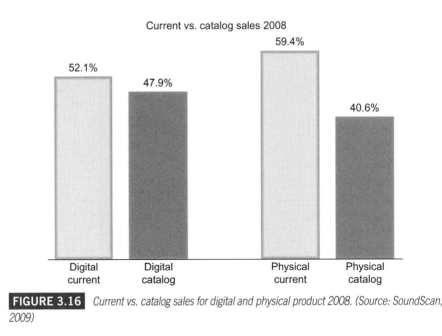

Current vs. catalog sales 2008

Current vs. catalog sales for digital and physical product 2008. (Source: SoundScan, 2009)

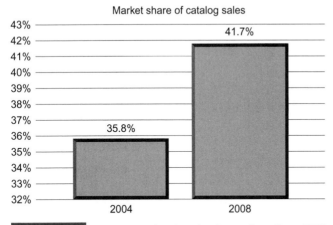

Market share of catalog sales

Market share of catalog sales. (Source: SoundScan, 2009)

new formats. Michael Omansky, senior VP of strategic marketing at RCA, stated in 1999, "We [now] put out what I think is substantially better product [on Elvis], with substantially more unreleased material" (Morris, 1999). An article in the *LA Times* in 2001 stated, "Companies have watched catalog sales slip from about 50% to 38% over the last decade" (Phillips, 2001).

During the mid-2000s, catalog sales declined overall, but the introduction of digital downloads saw a spurt in catalog sales as consumers sought to fill in their collections with catalog singles and albums that had not been available for some time. Catalog sales accounted for 46% of digital downloads in 2004, while they accounted for only 35% of sales of physical product. By 2008, catalog sales accounted for almost 48% of digital downloads and only 40% of physical sales.

In 2008, the sales share of catalog albums had reversed course, climbing to account for 41.7% of album sales versus the 35.8% the category comprised in 2004.

INDUSTRY CONCENTRATION

During the 1990s and early 2000s, the industry relied on just a few massive-selling hits to drive the industry rather than spreading the wealth around with a plethora of profitable releases. Ed Christman of *Billboard* magazine reported in 2001 that less than one-half of 1% of releases accounted for more than 50% of all units sold and that "about 3% of the total universe of available albums in the U.S. last year accounted for more than four-fifths of all album sales" (Christman, 2001). Of the

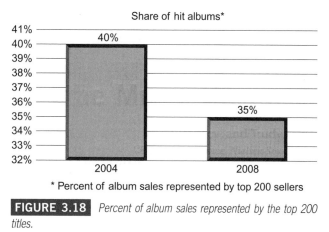

FIGURE 3.18 *Percent of album sales represented by the top 200 titles.*

265 million unit sales of albums released in 2004, only 100 titles made up nearly 50% of unit sales. Over 70% of the 265 million album sales in 2004 came from less than 1% of the releases.

According to Christman (2003), new releases for 2002 averaged 7871 sales per title, down from 9291 sales per title in 2001 for a 15.3% drop. (In 2000, the average was 8350 units scanned per title.) More than half of albums released in 2004 sold less than 100 units each, and 81% sold less than 1000 units each (SoundScan, 2005).

However, starting in the middle of the decade, the top 200 albums began to account for a shrinking share of all albums sold. In 2009, Christman wrote, "Hit album releases still account for a large but shrinking share of total sales. The 200 best-selling titles of the year have seen their share of annual sales fall from 40.1% in 2004 to 35% in 2008" (Christman, 2009).

The number 1 album in 2000 (N'Sync's *No Strings Attached*) sold 9.9 million copies that year; by 2004, the top-selling album (Usher's *Confessions*) sold 7.9 million units. By 2007, the number 1 album (Josh Grobin's *Noel*) only sold 3.7 million copies that year. In 2008, the top-selling album (Lil Wayne's *Tha Carter III*) sold less than 2.9 million copies. Meanwhile, the number of titles released each year has jumped from 33,433 in 2002 to nearly 80,000 in 2007.

CONCLUSIONS

This chapter is meant to illustrate trends within the U.S. recording industry. While the numbers and percentages will change from year to year, the goal here is to present the tools from which to draw conclusions in the future as new data become available. By looking at trends for music genres, demographics of buyers, configurations, outlets where music is sold, label market share, the proportion of catalog to current releases, and the proportion of blockbuster hits to the total number of releases, one can make inferences about the state of the industry. Updated information can be found annually at www.RecordLabelMarketing.com or www.riaa.com.

GLOSSARY

BMG—Bertelsmann Music Group. One of the major labels and is now in partnership with Sony. (Parent company is no longer involved in the music industry)

catalog—Older album releases that still have some sales potential. Recent catalog titles are those released for over 18 months but less than 36 months. Deep catalog: those titles over 36 months.

catalog sales—As sales of records that have been in the marketplace for over 18 months.

dollar value—The monetary worth of a stated quantity of shipped product multiplied by the manufacturers suggested retail price of a single unit. The value of shipments is given in U.S. dollars.

EMI—Electrical and Mechanical Industries, one of the major labels.

logarithmic scale—A scale of measurement that varies logarithmically with the quantity being measured. A constant ratio scale in which equal distances on the scale represent equal ratios of increase. A method of displaying data (in powers of 10) to yield maximum range, while keeping resolution at the low end of the scale.

market share—A brand's share of the total sales of all products within the product category in which the brand competes. Market share is determined by dividing a brand's sales volume by the total category sales volume.

peer-to-peer (P2P)—Electronic file swapping systems that allow users to share files, computing capabilities, networks, bandwidth, and storage.

product configuration—Any variety of "delivery system" on which prerecorded music is stored. Various music storage/delivery mediums include the full-length CD album, CD single, cassette album or single, vinyl album or single, mini-disc, dual-disc, DVD-audio, DVD, or VHS music video.

replacement cycle—Consumers replacing obsolete collections of vinyl records and cassettes with a newer compact disc format.

SME—Sony Music Entertainment, one of the major labels. SME purchased the music division BMG (Bertelsmann Music Group) from Bertelsmann AG (now known as Sony Music, with no affiliation with BMG).

track equivalent albums (TEA)—Ten track downloads are counted as a single album. The total sales of all downloaded singles are divided by 10 and the resulting figure is added to album downloads and physical album units to give a total picture of "album" sales. This unit of measurement was implemented to be able to compare current unit sales with sales prior to digital track downloads, when the industry was almost exclusively album-oriented.

UMG—Universal Music Group, one of the major labels.

units shipped—The quantity of product delivered by a recording manufacturer to retailers, record clubs, and direct and special markets, minus any returns for credit on unsold product.

WEA—The distribution arm of Warner Music Group (stands for Warner, Elektra, Atlantic).

WMG—Warner Music Group, one of the major labels.

BIBLIOGRAPHY

Christman, E. (2001). SoundScan numbers show 35% of albums account for more than half of all units sold. *Billboard, April 28.*

Christman, E. (2003). Average sale of albums dropped in '02 as labels released more, sold less. *Billboard, April 26.*

Christman, E. (2007). Nielsen SoundScan releases year-end data www.billboard.biz/bbbiz/content_display/industry/e3iXZLO0IdrWuAOeIRwz3vtYA%3D%3D. *Billboard.biz, Jan. 4.*

Christman, E. (2009). Running the numbers www.billboard.biz/bbbiz/search/article_display.jsp?vnu_content_id=1003928717. *Billboard.biz, Jan. 17.*

Morris, C. (1999). Labels search for new ways to jump-start catalog sales. *Billboard, Dec. 11.*

PBS *Frontline.* From Fig. 3.15.

Phillips, C. (2001). Record label chorus: High risk, low margin. *Los Angeles Times, May 31.*

RIAA. (1997). 1996 yearend marketing report on U.S. recording shipments press release. *RIAA.*

RIAA. (1998). 1997 yearend marketing report on U.S. recording shipments press release. *RIAA.*

RIAA. (2000). Recording industry releases 1999 yearend marketing report press release. *RIAA.*

RIAA. (2006). Recording industry releases 2005 yearend marketing report press release. *RIAA.*

RIAA. (2008). Recording industry releases 2007 yearend marketing report press release. *RIAA.*

RIAA. (2009). Recording industry releases 2008 yearend marketing report press release. *RIAA.*

SoundScan. (2009). www.soundscan.com.

Sisario, Ben. (2009). Music sales fell in 2008, but climbed on the web. http://www.nytimes.com/2009/01/01/arts/music/01indu.html?_r=3&scp=2&sq=music%20online&st=cse. *New York Times, Jan. 1.*

CHAPTER 4

Marketing Research

Tom Hutchison

Marketing research is important in every aspect of marketing recorded music. Research is used to find out about the marketplace, who is buying what products, and what they like and dislike about existing products in the marketplace. Basically, marketing research is conducted to learn more about the market and to learn how to improve the product. Chapter 1 outlined how diffusion theory can be applied to improve upon products and make them more appealing in the marketplace. Marketers learn from research what the most popular product attributes are for each market segment, and then they retool the product and the promotional messages to focus on those attributes. For example, toothpaste users can be segmented into those whose choices are driven by ability to whiten teeth, freshen breath, fight cavities, fight gum disease, and so forth. It is research that gives marketers the insight into who wants what.

CONTENTS

WHAT IS MARKETING RESEARCH?

Marketing research is defined as the systematic design, collection, analysis, and reporting of data and findings relevant to a specific situation facing a company (Blankenship and Breen, 1993). The American Marketing Association describes it as follows:

> *Marketing research links the consumer, customer and public through information—information used to identify and define marketing opportunities and problems; generate, refine, and evaluate marketing actions; monitor marketing performance; and improve understanding of marketing as a process.*

> —American Marketing Association, 2009

Research is conducted for many reasons:

– To check out a new product or concept idea

– To ask consumers what they think of an artist's new singles, videos, image changes

– To touch bases with a particular consumer group, either genre- or demographic-based, to see what they are thinking about artists, radio, video play

– To ask questions about how music is fitting into people's lifestyles as compared to other entertainment items like videogames, movies, DVD, cell phones, etc.

– To understand better the relationship between an artist and other brands

- Linda Ury Greenberg
VP Consumer Research
Sony Music Entertainment

FIGURE 4.1 *Why marketing research is conducted.*

The information gathered with marketing research helps companies make marketing decisions: decisions about where to advertise, how to allocate the marketing budget, how to craft marketing messages, and how to present the product in the marketplace.

Much of what is done for research is designed to predict future behavior of consumers. Some research accomplishes this by analyzing past consumer behavior, looking for patterns or clues that will answer research questions (see the following paragraph). Research is commonly conducted using **samples**, a small representative subsection of the market segment. Then the findings are generalized to the entire market segment. Sampling is done because of the physical impossibility of interviewing, measuring, or monitoring every member of a market segment. Therefore, much of the information reported in research findings, being predictive in nature, has a preset level of **margin of error**: the level of uncertainty associated with the research. Margin of error is usually expressed as a plus or minus factor, for example 75% ± 2% (see the section on research validity).

An important aspect of marketing research involves collecting data on previous sales patterns and then projecting that to future sales, whether for the same products or similar products. SoundScan data are used to examine sales of one release to help guide and shape the marketing decisions of other releases. Researchers also use test marketing: introducing the product into a small portion of the marketplace and monitoring sales before financing a large-scale release.

Research techniques have changed in the past few years, with many of the traditional procedures giving way to use of the Internet to conduct surveys, track online activity of consumers, and receive more direct feedback from consumers. These new techniques are less costly; paying consumers to show up for a focus group can cost several thousand dollars, and mail-out surveys generally cost money and have a low response rate. We will address these new techniques, but first we need to outline the basic concepts of marketing research.

Syndicated vs. Custom Research

The two basic categories of marketing research are *syndicated* and *custom*. **Syndicated research** is the gathering of continuous or periodic information that is sold in standardized form to all companies involved in the industry. Examples include the work of Arbitron, Nielsen, BDS, and SoundScan. **Custom research** is usually tailored, proprietary, initiated by a company, and often specific to a particular product developed by that company. Examples include product surveys (found inside the product or attached to the warranty card), focus groups, customer satisfaction surveys, and personal interviews.

The Research Process

There are six stages to the research process, starting with identifying the problem that must be researched. Through the subsequent stages, the research project is shaped and designed to discover ways to overcome the marketing challenge or problem.

1. **Problem identification** – Before a survey is developed, or a research project is designed, the questions that will be addressed by the project must be identified. Stage 1 begins with a definition of the marketing problem. This leads to a preliminary statement of research objectives. The goals of the research may include a description of the market, a prediction of how the product will perform in the marketplace, or an evaluation of the product or a portion of the marketing plan. For example, before marketing the release of a new album, the marketing department may wish to learn more about the target market: what they read, where they hang out, how they find out about new music, and so on.

2. **Exploratory/secondary research** – Stage 2 involves finding out what has already been learned on the topic and what information is already available. This is known as **secondary research**, defined as going over data that have been previously collected for a project other than the one at hand (Zikmund, 1991). Examining previous research is necessary to avoid repeating mistakes and to gain an underlying knowledge of the area of study. This helps guide the

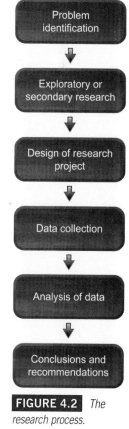

FIGURE 4.2 *The research process.*

researcher toward the type of project that will most likely address the research questions, so that in stage 3, the research project can be designed.

Exploratory research is defined as initial research conducted to clarify and define the nature of the problem. It is done to help diagnose the situation, allow for screening of alternatives, and discover new ideas. **Concept testing** is frequently done in exploratory research. Concept testing is a means of evaluating ideas by providing an early indicator of the merits of an idea prior to the commitment of resources to development and manufacturing—in other words, a prototype. Some record labels will concept test music earmarked for an upcoming album before deciding which songs should be included and which should not.

3. **Design of research project** – The three most popular research designs are experiments, descriptive, and qualitative (focus groups).

    Experiments are designed primarily to determine cause and effect. For example, experiments were conducted in the 1960s and '70s to determine whether smoking cigarettes caused lung cancer. Ongoing research experiments seek to determine whether observing violence on television or in music lyrics increases the likelihood of aggressive behavior. Experiments usually involve an experimental group and a control group. The experimental group gets the treatment (cause), or, in this case, is subjected to the marketing program, and the control group is not. In marketing, experiments may be conducted by placing product or promotional materials in particular locations in some stores and other locations in similar stores, and then monitoring the differences in sales; or, by finding similar geographic units or markets and implementing advertising plan A in one market and plan B in another.

    Descriptive research is designed to describe something and is often used to describe the market. Surveys are the most common **instrument** used in descriptive studies. Consumers are asked questions about who they are, what they like, and what they do. Record labels sometimes use consumer reply cards embedded in CD jewel cases to survey consumers. Questions are designed to provide information of value to marketers. For example, the survey questions

Table 4.1	Survey questions and research goals	
	The survey question(s)	**The research question or goal**
1	What is your age, gender, ethnicity, and hometown?	To know more about the demographic and geographic makeup of our market.
2	How did you acquire this CD? How did you hear about this CD? What influenced you to buy this CD?	To know which of our marketing efforts are most successful in reaching our customers and which were wasted money.
3	How familiar with this music were you before purchasing?	To learn how much of a risk customers will take before purchasing and how much exposure they need before deciding.
4	What are your favorite tracks? How many other albums do you own by this same artist?	Which songs should we be looking at for future release as singles? Which songs are most likely driving sales of the album?
5	How many other albums do you own by this artist?	Loyalty: How loyal are our customers?
6	What do you read? Watch? Listen to?	How can we best reach our customers with future advertising and media placement?
7	What genres and other artists do you like?	How should we position this artist? As what genre? As similar to what other artists?
8	Where and how do you buy music?	To what extent should we be using various distribution methods?
9	What is your favorite beverage? Sneaker company? Car or truck? Music store?	What companies should we team up with for promotional tie-ins?
10	Do you own/purchase DVDs, videogames? How much time do you spend on entertainment technology?	Technographics: How technology-driven are they? How much do they engage in entertainment activities at home?
11	How often you do go out for dinner, movies, sporting events, concerts?	How much do they engage in entertainment activities outside the home?

in Table 4.1 are designed to guide marketers or identify consumers in particular ways.

Focus groups, or other qualitative methods such as personal interviews, are used when marketers want more in-depth information from the market. **Focus groups** are unstructured, free-flowing group interviews designed to gain a deeper understanding of consumer feelings, beliefs, and attitudes. Responses to questions can be *probed* with further questioning to find out why consumers hold these attitudes. Focus groups are good for test marketing new music and finding out why consumers like or dislike certain songs. They are not good for generalizing and providing statistical data about a market.

Thank you for buying this album. We'd love to hear your feedback. After you've had time to listen to the album, please complete this card and return it for a chance to win valuable prizes. Thanks!

(1)

First Name _____

Last Name _____

Date of Birth ____/____/____   Gender: ❑ male ❑ female

Address _____

City: _____ State: _____ Zip _____

Email: _____

(2)

Where did you purchase this album?
❑ Store: _____
❑ Ordered from TV ad
❑ Internet: _____
❑ Received as gift
❑ Won in Contest

How did you find out about this album?
(check all that apply)
❑ Print
❑ Radio
❑ Internet
❑ Listening stations
❑ Friends
❑ TV/cable
❑ Fan club
❑ Store browsing
❑ At Artist's performance
❑ Ad on radio
❑ Ad on TV
❑ Ad in magazine/newspaper
❑ Other:

(3)

How many songs on this disc were you familiar with before purchasing?

1   2   3   4   5+

(4)

What are your three favorite songs on this disc?
1. _____
2. _____
3. _____

(5)

Not including this album, how many CDs, cassettes or LPs do you own by this artist?
0   1   2   3   4   5   6   7   8+

(6)

What TV shows that feature music do you watch regularly?
❑ Saturday Night Live
❑ David Letterman
❑ Jay Leno Show
❑ Conan O'Brien
❑ Jimmy Kimmel
❑ VH1 Top 20 Countdown
❑ America's Best Dance Crew
❑ Making the Band
❑ BET Live Countdown
❑ American Idol
❑ Other: _____

Which of the following music video channels/shows do you watch?

❑ MTV   ❑ Video Soul
❑ BET   ❑ Rap City
❑ VH-1   ❑ Video LP
❑ TNN   ❑ The Grind
❑ CMT   ❑ Jukebox Network
❑ GAC   ❑ Other

Which of the following daytime TV shows do you watch?

❑ CBS This Morning   ❑ Rachael Ray
❑ Montel Williams   ❑ Oprah Winfrey
❑ Maury Povich   ❑ Regis and Kelly
❑ Today Show   ❑ Dr.Phil
❑ Sally Jesse Rafael ❑ Other
❑ Ellen Degeneres Show
❑ Good Morning America

What are your three favorite TV shows?
1. _____
2. _____   **(6)**
3. _____

Which of these magazines do you read regularly?

❑ Spin   ❑ US
❑ Rolling Stone ❑ People
❑ Details   ❑ TV Guide
❑ N.M.E.   ❑ Melody Maker
❑ Time   ❑ Newsweek
❑ Metal Edge   ❑ Jet
❑ Ebony   ❑ Essence
❑ Down Beat   ❑ Entertainment Weekly
❑ Elle   ❑ Other

List three magazines that you read regularly:
1. _____
2. _____
3. _____

What are your favorite radio stations?
1. _____
2. _____

How many hours per week do you spend listening to the radio? _____

(7)

Aside from this artist, whose albums have you purchased recently?
1. _____
2. _____
3. _____

What are your favorite types of music?
(1-favorite, 2-next, 3-third)
❑ Pop   ❑ Jazz
❑ Rock   ❑ Club/Dance
❑ Rap   ❑ Classical
❑ Country   ❑ Gospel
❑ Blues   ❑ Alternative
❑ Metal   ❑ Soul/R&B
❑ Latin   ❑ Folk
❑ Reggae   ❑ Oldies
❑ Other

How much music downloads have you purchased in the past month?
Single tracks _____
Albums _____

What sites?
❑ iTunes   ❑ Amazon
❑ Walmart   ❑ _____

(8)

Do you belong to any music subscription services?
❑ Rhapsody   ❑ Napster
❑ Other _____

Where do you usually buy music?
	Usually	Sometimes	Rarely
Local/indie record store	❑	❑	❑
Internet/Mail Order	❑	❑	❑
Chain or Mall	❑	❑	❑
At Artist's Performance	❑	❑	❑
Download from the Internet	❑	❑	❑
Purchase from mobile device	❑	❑	❑

How many CDs have you purchased in the past 6 months? _____

(9)

What is your favorite...?
Soft drink: _____
Fast food chain: _____
Sneakers: _____
Jeans: _____
Clothing store: _____
Book store: _____
Car/Truck: _____
Software: _____
Videogame: _____
Sports event: _____

Do you own a DVD player?
❑ No   ❑ Yes
If yes, how long? _____

How many DVDs have you purchased in the past 12 months? _____

Do you own a home computer?
❑ No   ❑ Yes
If yes, how long? _____

How many hours per week do you spend online? _____

Do you have high-speed access at home? ❑ No   ❑ Yes

(10)

How many hours per week do you spend on...
____ MySpace   ____ imeem
____ Facebook   ____ Bebo
____ others: ____

Do you own a videogame player?
❑ No   ❑ Yes
If yes, how long? _____

How many hours per week do you spend playing videogames? _____

Do you own a cell phone?
❑ No   ❑ Yes
If yes, how long? _____

How many hours per week do you spend on your cell phone? _____

How often to you go to the movies?
_____ per _____

How often do you go out for dinner/ _____ per _____

(11)

What are your three favorite movies?
1. _____
2. _____
3. _____

Do you go to clubs to hear live music?
❑ No   ❑ Yes
If yes, how often? _____ per _____

How many concerts have you attended in the past 6 months? _____

Would you like to receive information and specific music offers via email?
❑ No   ❑ Yes

FIGURE 4.3   *Sample consumer survey.*

For greater accuracy, sometimes **triangulation** is used where two methods are employed; if findings are consistent, researchers are more confident in the results. Triangulation is the application and combination of several research methodologies in the study of the same phenomenon (Zulkardi, 2004). By looking at something from two different angles, more-accurate conclusions can be drawn.

FIGURE 4.4
Triangulation.

4. **Data collection** – Data are collected by conducting the experiment, survey, or focus groups. Once the design has been determined, a sample must be selected. Rarely does a research project rely on a census (a measure of everyone in the population or market) because of the costs involved in collecting the data. Therefore, a sample is selected to represent the market, and data are collected from this sample.

Sampling must be scientific and follow certain guidelines, known as **sample design**. This includes the method of selection, the sample structure, and plans for analyzing and interpreting the results (Tutor2U, 2005). Common folklore has people believing that population size is very important in determining sample size. In other words, if you wanted to generalize to the entire U.S. population, you would need a larger sample than if you wanted to generalize to all college students. This is not true. Population size does not normally affect sample size. But the sample does have to be proportionally representative of the population, known as a **sampling frame**. For example, if you wanted to know what high school males and females thought about algebra, it would not be wise to conduct a survey of female shoppers in the mall during

Table 4.2	Research design	
Research design	**Uses**	**Weaknesses**
Experiments	To determine cause and effect. To find out which of our marketing efforts are most effective. Good predictive validity.	You must have full control over all aspects of the experiment to isolate the cause and be certain the desired effects are not being caused by something else.
Description	Surveys are used to describe the market. Easy and cost-effective. Can be generalized.	Less predictive value than experiments.
Qualitative	Good for in-depth information and understanding why. Good for test marketing new products.	Not generalizable to the general market. Relies on just a few people to provide information about a much larger population.

regular school hours. Instead, you would need to sample a certain number of high school males and females; the proportion of males to females would be dependent upon the proportion in the population. The sampling frame must be representative of the population and is defined as a list of elements from which a sample may be drawn. Often these elements can be identified by existing information. For presidential election polls, the sampling frame would be all registered voters, or likely voters.

There are two phases in the process of gathering data: the *pretesting phase* and the *main study*. Pretesting is done to ensure the validity of the instrument (i.e., a survey) used to measure the research issues. After pretesting, the revised survey is administered, or the experiment performed, or the focus groups are conducted. The data must then be converted into a form that allows for systematic analysis, whether that be numerical data that are input into a program, or recordings and transcripts of interviews and focus groups for analysis.

5. **Analysis of data** – Data analysis begins with editing and coding. **Editing** involves checking the data collection forms for omissions, legibility, and consistency in classification, especially for fill-in-the-blank questions. The data are then coded. **Coding** refers to the systematic process of interpreting, categorizing, recording, and transferring the data to the data processing program. **Data analysis** is the application of logic to the data, looking for systematic patterns and details that will answer the research questions.

The appropriate analytical technique is determined by the survey design. Surveys are generally analyzed using a statistical software program. Interviews and focus groups are often transcribed, and sometimes software is used to look for key terms in the text and identify patterns of responses. In experiments, the experimental variable (e.g., the presence or absence of cancer in smokers, the presence or absence of aggressive behavior after watching violence on television) is measured with the appropriate instruments.

6. **Conclusions and recommendations** – The research report is the main tool for communicating the findings and recommendations from the research company to the client. A **research report** is an oral presentation and/or written report, and the purpose is to communicate the results, findings, and recommendations to the marketing client. The research report states the conclusions from the data analysis, first as straightforward responses to the measurements. These are then compared to the original research questions and

goals to make sweeping conclusions and recommendations. A typical written report contains an executive summary, an introduction (including objectives), methodology, the literal results, and conclusions and recommendations. The **executive summary** is a one- or two-page bulleted-list summary of the highlights of the findings. It is designed to be read by those people who need to know the main points but do not have time to read through the report to find them.

RESEARCH VALIDITY

Caution must be taken to ensure any research project is valid. **Validity** is a term that describes the ability of a research project to measure what is intended to be measured. If the goal is to measure *purchase behavior*, asking questions about *attitude* toward various products may not provide a valid measurement of their likelihood to actually purchase the product. Consumers may have a more positive attitude toward a product that they have no intention of buying because of the price or some other factor. Instead, questions should be asked about actual purchase behavior and purchase intention.

CONFIDENCE LEVEL AND MARGIN OF ERROR

Because almost all research involves taking a sample of the market rather than measuring everyone, there is the possibility of some error—that is, the findings from the sample are not exactly what occur in nature. One of the main questions asked by people without a background in statistics is "How large does my sample size need to be?" The answer is not a simple number and is not based on the size of the population to whom you will be applying the findings. Called the margin of error—the plus or minus 3 percentage points seen after percentages are listed in reporting results—this number has to do with the probability that the findings represented by the sample will be true for the population at large. In other words, if a result is presented as 36% ± 3%, it means that while the sample found a result of 36%, there is a strong likelihood that in the population, the true result is between 33% and 39%. The margin of error decreases as the size of the sample increases, but only to a point. Researchers generally weigh the risk of error against the additional costs required to fund research for a larger, more accurate sample. The level of confidence is a measure of how confident we are in a given margin of error, and the general acceptable standard is a 95% level of confidence with a margin of error of ±2% or 3%. (For example, if we ran this experiment 100 times, it would produce findings within the margin of error 95 times.)

Table 4.3	Margin of error
Survey sample size	**Margin of error percent***
2000	2
1500	3
1000	3
900	3
800	3
700	4
600	4
500	4
400	5
300	6
200	7
100	10
50	14

*Assumes a 95% level of confidence.

THE APPLICATION OF RESEARCH PRINCIPLES

Marketing researchers must have an understanding of when to use various research designs and measurements, and also when they may not be appropriate.

When Focus Groups Are Appropriate

Focus groups are considered exploratory research but can be conducted either before a full-scale survey is launched, in conjunction with a survey, subsequent to a survey, or in rare cases as a stand-alone research project. Focus groups are usually conducted before a major survey in order to gain an understanding of the target market for the survey. But care must be taken to select the correct participants who are typical of the market for which the findings will apply. The group interviews help the researcher understand context, vernacular, and how various survey questions will be interpreted.

When triangulation is advised, focus groups may be run in conjunction with a survey, with the survey providing the information to be generalized to the population and the focus groups providing depth of understanding to the responses. Occasionally, focus groups are conducted after a survey is completed, especially if there are lingering questions about the findings that need clarification from members of the market. Stand-alone focus groups are conducted when researchers want some insight into the minds of the consumers but do not need quantitative data and do not intend to generalize the findings to the larger population.

Focus groups are good for concept testing. Record labels use them to judge consumer opinions about artists and their music, their appearance, the quality of the recordings, the lyrics, and the artist image. In these situations, focus group attendees are exposed to elements of the marketing plan (the music, video, photos, etc.) in a controlled setting, and reactions to these elements are noted. The groups can be conducted early in the project, before the music has been finalized for the release and before imaging materials have been selected. The attendees may be presented with photos of the artist with different looks and then asked to comment on each one. They may be also presented with recordings and asked to judge and comment on them. The results may not guarantee acceptance in the marketplace, but at least the labels will be aware of the various "groups of thought" regarding their products.

Focus groups are not appropriate when attempting to analyze the market for an artist. The market must be identified prior to selecting participants for focus groups. Focus groups are not appropriate for generating quantitative information such as ratio of males to females, a comparison of market segments on an issue, or to identify traits in the market.

When Surveys Are Appropriate

Surveys are best suited to situations where the research questions can be answered in a straightforward manner, when more information about aggregate consumer groups is needed, and when that information will need to be generalized to a larger population. Surveys are good for identifying characteristics of target markets, describing consumer purchasing patterns, and measuring consumer attitudes. Surveys provide an inexpensive, efficient, and accurate means of evaluating information about a market by using a small sample and extrapolating the results to the total population or market.

Entire textbooks are devoted to survey design. The questions must be worded and presented in a clear, concise, and unbiased manner to ensure validity. **Measurement scales** are developed to measure concepts of a more abstract nature such as personality and attitudes. These scales are banks of questions that have been carefully tested and found valid, when applied appropriately, to measure the concept they

---

Case Study: Geffen Records

In the early to mid-1990s, Geffen Records was interested in learning more about young male music fans who were buying the metal and grunge recordings released by the label. Bounceback cards were placed inside specific releases and the results were analyzed to gain an understanding of the market. Label personnel were surprised that many of the fans read *Rolling Stone* and *Spin* magazines, rather than the heavy metal genre-specific magazines targeted to those consumers. A comparison of magazine preferences with geographic location (urban, rural, and suburban) led to the conclusion that these more specific magazines were not available on news racks in many locations and therefore not as popular. It was also learned for the first time that many of the grunge fans enjoyed skateboarding, and that tattoo parlors were a good place to promote heavy metal music. (In today's business climate, these findings are intuitive, but at the time they were not as obvious.) How did the label learn of the connection between their music and skateboarding? The skateboard magazine, *Thrasher*, kept showing up in the "top ten favorite magazines" listed by those who returned the bounceback cards. The label was also able to find out about the interest in vinyl recordings and usage of the Internet at that time. These types of findings help a record label make better decisions about where and how to market their products, and which media outlets to use.

FIGURE 4.5   *Case study: Geffen Records (Source: Hutchison)*

were designed to measure. If the survey has been judged to be valid and the sampling procedures are followed correctly, the results of the survey should convey a degree of confidence that the entire market or population exhibit the opinions and characteristics similar to those found in the sample.

Surveys are used extensively inside products to gather information on customers. Often the survey is part of the warranty card, but in the recording industry the cards are inserted in jewel boxes. The cards have different names depending upon the label; at some, they are *tossback* cards; at others, bounceback cards; and still others may refer to them as customer response cards or business reply cards. The sample base for these cards should not be considered random or representative of the market overall. Sample bias may occur because consumers who send in the cards may be systematically different from customers who purchase the product but do not send in the

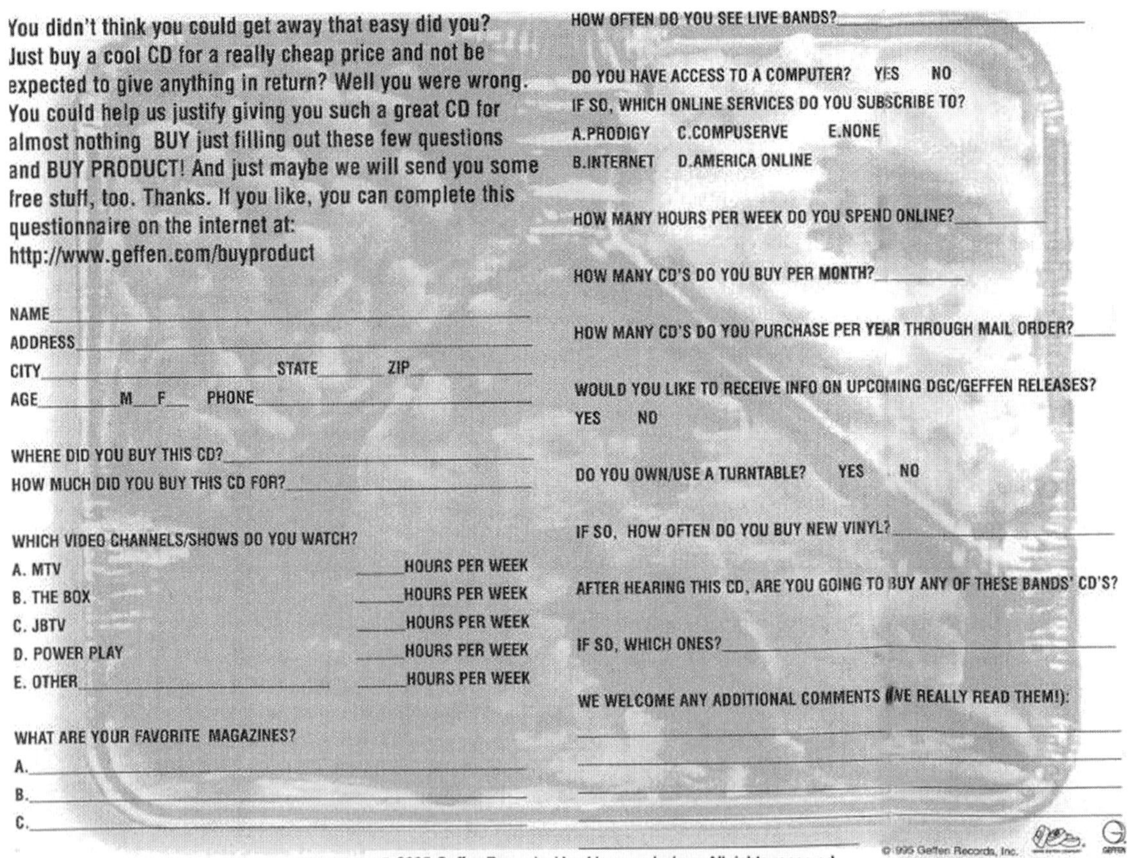

You didn't think you could get away that easy did you? Just buy a cool CD for a really cheap price and not be expected to give anything in return? Well you were wrong. You could help us justify giving you such a great CD for almost nothing BUY just filling out these few questions and BUY PRODUCT! And just maybe we will send you some free stuff, too. Thanks. If you like, you can complete this questionnaire on the internet at:
http://www.geffen.com/buyproduct

NAME_____
ADDRESS_____
CITY_____ STATE_____ ZIP_____
AGE_____ M__ F__ PHONE_____

WHERE DID YOU BUY THIS CD?_____
HOW MUCH DID YOU BUY THIS CD FOR?_____

WHICH VIDEO CHANNELS/SHOWS DO YOU WATCH?
A. MTV   _____ HOURS PER WEEK
B. THE BOX   _____ HOURS PER WEEK
C. JBTV   _____ HOURS PER WEEK
D. POWER PLAY   _____ HOURS PER WEEK
E. OTHER_____   _____ HOURS PER WEEK

WHAT ARE YOUR FAVORITE MAGAZINES?
A._____
B._____
C._____

HOW OFTEN DO YOU SEE LIVE BANDS?_____

DO YOU HAVE ACCESS TO A COMPUTER?   YES   NO
IF SO, WHICH ONLINE SERVICES DO YOU SUBSCRIBE TO?
A.PRODIGY   C.COMPUSERVE   E.NONE
B.INTERNET   D.AMERICA ONLINE

HOW MANY HOURS PER WEEK DO YOU SPEND ONLINE?_____

HOW MANY CD'S DO YOU BUY PER MONTH?_____

HOW MANY CD'S DO YOU PURCHASE PER YEAR THROUGH MAIL ORDER?_____

WOULD YOU LIKE TO RECEIVE INFO ON UPCOMING DGC/GEFFEN RELEASES?
YES   NO

DO YOU OWN/USE A TURNTABLE?   YES   NO

IF SO, HOW OFTEN DO YOU BUY NEW VINYL?_____

AFTER HEARING THIS CD, ARE YOU GOING TO BUY ANY OF THESE BANDS' CD'S?

IF SO, WHICH ONES?_____

WE WELCOME ANY ADDITIONAL COMMENTS (WE REALLY READ THEM!):
_____
_____
_____

FIGURE 4.6 *Bounceback card for Geffen Records. (Source: Geffen Records)*

cards. Nonetheless, the cards are beneficial in gaining an understanding of the market. As a cost-cutting measure, labels are switching over to using the more convenient online survey to measure customer attitudes and behaviors (see the section on conducting surveys online).

Conducting Surveys Online

The Internet has made it possible for any web designer to create and administer surveys online. While this greatly increases convenience and reduces costs, the same rigorous research design and sampling procedures must be followed to provide accurate findings that can be generalized to a particular market with a degree of accuracy. Survey design should be careful not to include "leading" questions that may influence response behavior. A poorly worded question may result in inaccurate responses; one would not phrase a question as "You wouldn't want to buy brand X, would you?" but would instead ask, "How likely are you to buy brand X?" The lack of proper sampling could result in findings that are biased in favor of the product because all members of the sample are already fans. Information gathered from such a survey would not be beneficial in trying to determine how to reach those who are not fans. However, if conducted correctly, online surveys can generate a wealth of research information, and the process is available and affordable to everyone.

Survey Monkey is currently the top service for providing online survey services. The process is simple and the software is designed to generate basic reports without further statistical analysis. If a more complex analysis is warranted, the raw data can be downloaded in comma-separated-values (CSV) form and ported into Excel or any statistical software program.

Other services in competition with Survey Monkey include

Zoomerang – www.zoomerang.com
Instant Survey – www.instantsurvey.com
E-customer Survey – http://ecustomersurvey.com
My Survey Lab – www.mysurveylab.com
Vovici – www.vovici.com/
Survey Gizmo – www.surveygizmo.com
Wufoo – http://wufoo.com/
Google Docs – http://docs.google.com/

Message from Trent Reznor of Nine Inch Nails

Hello everyone. I'd like to thank everyone for a very successfull year so far in the world of Nine Inch Nails. I'm enjoying my couple of weeks off between legs of our Lights In The Sky tour and got to thinking... "wouldn't it be fun to send out a survey to everyone that's shown interest in NIN?" Well, that's not exactly how it went, but regardless — here it is. As we've moved from the familiar world of record labels and BS into the unknown world of doing everything yourself, we've realized it would benefit us and our ability to interact with you if we knew more about what you want, what you like, what you look like naked, etc. I know it's a pain in the ass but we'd truly appreciate it if you'd take a minute and help us out. As an incentive, everyone who completes the survey will able to download a video of live performace from this most recent tour (and I know what's going through your little minds right now. "I'll just grab this off a torrent site and not have to fill out the survey!!!" and guess what? You will be able to do just that and BEAT THE SYSTEM!!!! NIN=pwn3d!!!)"

FIGURE 4.7 *Public email message from Trent Reznor about a survey.*

Consumers can be encouraged to log on to a web site or click on a link to complete the online survey. Surveys can easily be embedded into an artist web site. Often, an incentive is used to increase response rate: a coupon or a raffle.

When Experiments Are Appropriate

Scientific experiments require that all variables be controlled. **Variables** are any factors that might influence the outcome. In the case of marketing music, the outcome is sales of the recording. The various factors or variables that might influence this include any and all marketing efforts—radio airplay, TV performance, touring, in-store activities, local and national press coverage, word of mouth, and so forth. In a normal experimental setting, one or more of the variables would be *manipulated*, and the other variables would be held constant or controlled. For example, experimental research on the impact of television advertising on sales would involve at least two groups in the market that are identical in all aspects except exposure to the advertising variable: a flight of commercials. Other factors that might influence sales would have to be kept identical for both groups. This is much easier to accomplish in a laboratory setting, but not practical for marketing research because people do not shop in a vacuum.

Field experiments are conducted in a natural setting despite the fact that complete control of extraneous variables is not possible. Attempts are made to find two groups of consumers that are alike in as many aspects as possible. Then one group may be exposed to a particular marketing strategy, while the other is not. This is not popular in the recording industry because it involves withholding some marketing efforts to a portion of the market during the crucial window for generating sales. Controlled store tests (or test marketing) may be conducted by providing products to some stores to gauge sales before investing in a nationwide product rollout. Results from this type of testing do not automatically generalize to the entire market and guarantee national success, due to other factors that may not have been accounted for during the controlled store tests.

Tracking Consumer Behavior on the Web

Before the emergence of the Internet, marketing researchers used a variety of techniques to learn more about consumer behavior. Many of these studies were not comprehensive, meaning that shopping behavior may have been measured on one group of consumers while advertising exposure was measured on another. So the effects of advertising on a group of study subjects did not necessarily also measure their purchase behavior; that would be left to a different study. It was difficult to conduct a comprehensive measurement program without being intrusive. In other words, it was difficult to put one group of consumers under the microscope for study without influencing their normal purchase behavior. One company attempted to measure media consumption and consumer purchases in the same household. Generally, participants had to subject themselves to extensive monitoring and extraordinary procedures to collect the data. In some ways, that influenced the outcome and made them unlike the general marketplace to which the results would be generalized. So, the measuring had a tendency to get in the way of the natural consumer behavior. With the Internet, data collection is more transparent—web users are not really aware that their movements through the web are being recorded and analyzed.

An important aspect of gathering information on your consumer base involves monitoring traffic to your site. Web traffic refers to the number of visitors to your web site and the number of pages visited. Oftentimes, it is measured to analyze the importance of its individual pages and elements. By including a bit of programming code on each page of the web site, the webmaster can learn a lot about the visitors to the site. This helps the webmaster and other marketing professionals understand what product information and which products are considered valuable to its visitors and which are not.

Monitoring Web Traffic

Cookies

The Internet has made it easy to track what consumers do, where they go, and what interests they have. One way of keeping track of that data is through the use of *cookies*. Webopedia defines cookies as a "message given to a web browser by a web server. The browser stores the message in a text file. The message is then sent back to the server each time the browser requests a page from the server." David Whalen made this analogy on www.cookiecentral.com:

> You drop something off [at the dry cleaners], and get a ticket. When you return with the ticket, you get your clothes back. If you don't have the ticket, then the [dry cleaner] man doesn't know which clothes are yours. In fact, he won't be able to tell whether you are there to pick up clothes, or a brand new customer. As such, the ticket is critical to maintaining state between you and the laundry man.

Cookies identify users and prepare customized web pages for them. When a user enters a web site using cookies, he or she may be asked to fill out a form providing such information as name, demographics and interests.

The information is packaged into a cookie and sent to that user's web browser which stores it for later use. So when you return to that same web site, your browser will send the cookie to the web server letting it know who you are—it's your ID card or your frequent shopper card. Then, the server can use this information to load up personalized web pages that may include content that interests you, based on information the site collected the last time you visited. So, for example, instead of seeing just a generic welcome page, you might see a welcome page with your name and features on it. The use of cookies is frowned upon by privacy advocates but hailed by marketers and webmasters alike in its ability to offer customized information to visitors.

What to Measure

Some of the most important factors that are measured include the following:

1. *The number of visitors* – This is represented by the number of different people who access your web site over a period of time. From this information, you can determine which times are most popular for visitors. You can determine whether your traffic is influenced by any marketing campaigns that may be unfolding, the impact of promotional materials such as email blasts, and the impact of advertising. For example, you send out an email blast to members of a fan club announcing a new tour schedule and notice a jump in the number of visitors to the site and the tour schedule page for the next couple of days. The concept of unique visitors is also measured—where each visitor is counted only once despite the fact that he or she may revisit the site several times.

2. *Whether these visitors are new or returning* – The effects of advertising and other marketing efforts to expand the market can be measured by observing the number of new visitors to the site. The number of returning visitors indicates the success level of efforts designed to bring visitors back to the site and generate fan loyalty.

3. *The number of page views* – This is a measurement of how many pages each visitor looks at on the site. If the ratio is high, meaning that each visitor on average visits a fair number of pages, that is an indication of the "stickiness" of the site. Stickiness means that the site is so compelling that visitors are inclined to stick around and visit other sections. However, this could also indicate that they are not finding what they are looking for, so they keep going on to the next page hoping to find what they need. Determining which of these two factors is in play is covered by the next measurement.

4. *Time spent per page* – If visitors are spending a lot of time on particular pages, one could conclude that these pages contain something of interest to the visitor. If other pages are glossed over quickly, then perhaps they are not as meaningful to the visitor or the visitor has not yet found what he is looking for. If certain pages don't get much traffic or visitors tend to spend little time on them, they should be reviewed to determine whether the level of interest is appropriate (it may be a page designed for a subsection of visitors, such as the media) or whether the page should be revamped or combined with another page.

5. *Time spent on the site* – Visitors who spend a long time on the site are probably the most dedicated customers or fans, especially if they are returning visitors. The average amount of time spent on the site indicates the worthiness of the site in providing something of interest.

6. *Date and time* – It is helpful to know the most popular viewing times and days, to plan when updates will be made to the site and to know whether traffic is seasonal.

7. *Where visitors reside* – This information is not always accurate, as some visitors may use an Internet service provider (ISP) that reflects the location of the main servers instead of the visitor's hometown. But for most systems, country of origin and city are listed in the visitor statistics. You can determine whether there is more activity on the web site coming from areas where the artist is touring. Then, by combining that with information on page hits, you can determine how important or useful the tour information page is to visitors.

8. *Where visitors are coming from and which page they enter the site through* – This information can help you to determine which outside URLs are providing most of the traffic, whether it's other sites that link to yours, search engine traffic, or direct request (the user types in the domain name).

9. *Exit page* – Which page do visitors commonly view last before leaving your site? Sometimes the page content will help determine the reason people leave the site: they found what they were looking for, they didn't find what they wanted, you directed them elsewhere, or they made the purchase.

10. *The technology that visitors use* – This function indicates the resolution of the monitor, connection type, browser type, and

operating system of each visitor. It is helpful in determining whether users have the technology to handle the latest "bells and whistles" before deciding to add those features to the site.

How to Use That Information

How much time a visitor spends viewing particular information, as well as how often visitors make a purchase and what they purchase, gives marketing researchers feedback on their efforts. Other aspects of marketing research rely on input from customer feedback forms, surveys, and other devices. Often this requires effort on the part of the consumer to provide this valuable information to market research experts. One of the great advantages of the Internet is that it offers marketing analysts a rich body of marketing information based on where web visitors go, what they click on, and how long they engage with the marketing message. In the article "Five Reasons to Track Web Site Traffic," author Monte Enbysk (n.d.) pointed out that too many "small businesses build web sites, invest time in online marketing campaigns, and then devote little or no effort to analyzing the return on their investment." Here are some of the ways that *web analytics* tools can provide feedback on marketing strategy:

1. *Evaluate the effectiveness of marketing efforts.* You can see the results of each aspect of promotion and how it affects traffic to the site. You can find out what keywords your customers use to find you and how they respond to your marketing by reviewing your product information.

2. *Figure out where your traffic is coming from.* By knowing where your web visitors come from just before landing on your site, you can determine whether your advertising is working.

3. *Learn what your users like and don't like about your products and messages.* Find out whether it's time to replace or modify those underperforming products or messages featured on pages where visitors tend to bail out. You can assess modifications of an underperforming marketing campaign by changes in visitor activity.

4. *Get to know your customers.* After studying the data coming in and making adjustments to the site, you can learn what your visitors like and what they respond to. Tracking them can tell you what they are looking for when they visit your site.

Where to Find Analytics Tools and How to Apply Them

Web analytics is defined as "the use of data collected from a web site to determine which aspects of the web site work toward the business

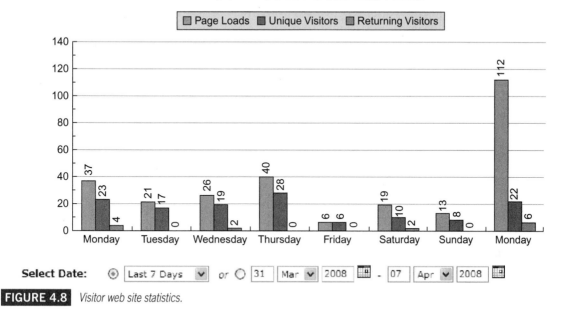

Daily | Weekly | Monthly | Quarterly | Yearly

□ Page Loads ■ Unique Visitors ■ Returning Visitors

Select Date: ⊙ Last 7 Days ▾ or ○ 31 Mar ▾ 2008 ▦ – 07 Apr ▾ 2008 ▦

FIGURE 4.8 *Visitor web site statistics.*

objectives." Many services on the Internet offer web analytics features. After asking a series of questions about how you want to track and compile information, the service will create the code to be inserted into every page of the web site. The code helps the service track activity on the site. The webmaster logs in to the service to view and download the statistics that the system has gathered.

Google Analytics

In early 2005, Google bought the web analytics company Urchin. Later that year, Google revealed its repacked version of Urchin for free: Google Analytics. Google Analytics offers tracking of web visitors and provides the usual tracking statistics plus keyword reports and the ability to measure the effectiveness of AdWords programs. The process requires web owners to insert a small piece of Java code (a piece of computer programming that feeds information to Google from your site) into the head tags of their pages. The statistical output can be viewed on the Google site with the assistance of a dashboard: customizable collection of report summaries.

Quantcast

Quantcast is a new service that provides information on web traffic combined with demographic data to present a clear depiction of a web site's visitors. The service combines sample-based information and analytics of web behavior to present a profile of a web site's traffic, including information on: age, gender, ethnicity, income, education, repeat vs. casual traffic, monthly traffic volume, and a comparison of other sites that your visitors frequent, like, and search for.

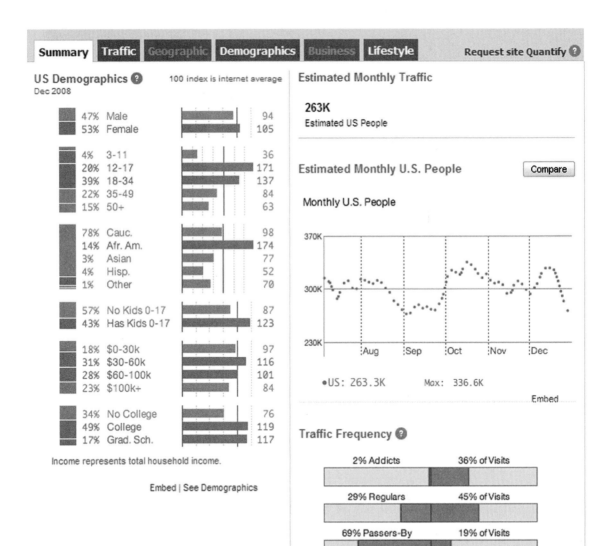

FIGURE 4.9   Quantcast information. (Courtesy of Quantcast)

In the lifestyle analysis, Quantcast analyzes what other sites your fans visit when they are not on your site. This helps develop a profile of what else your visitors are interested in when they are not on your site. For example, visitors to Kelly Clarkson's web site also visit http://bravadousa.com and www.americanidol.com.

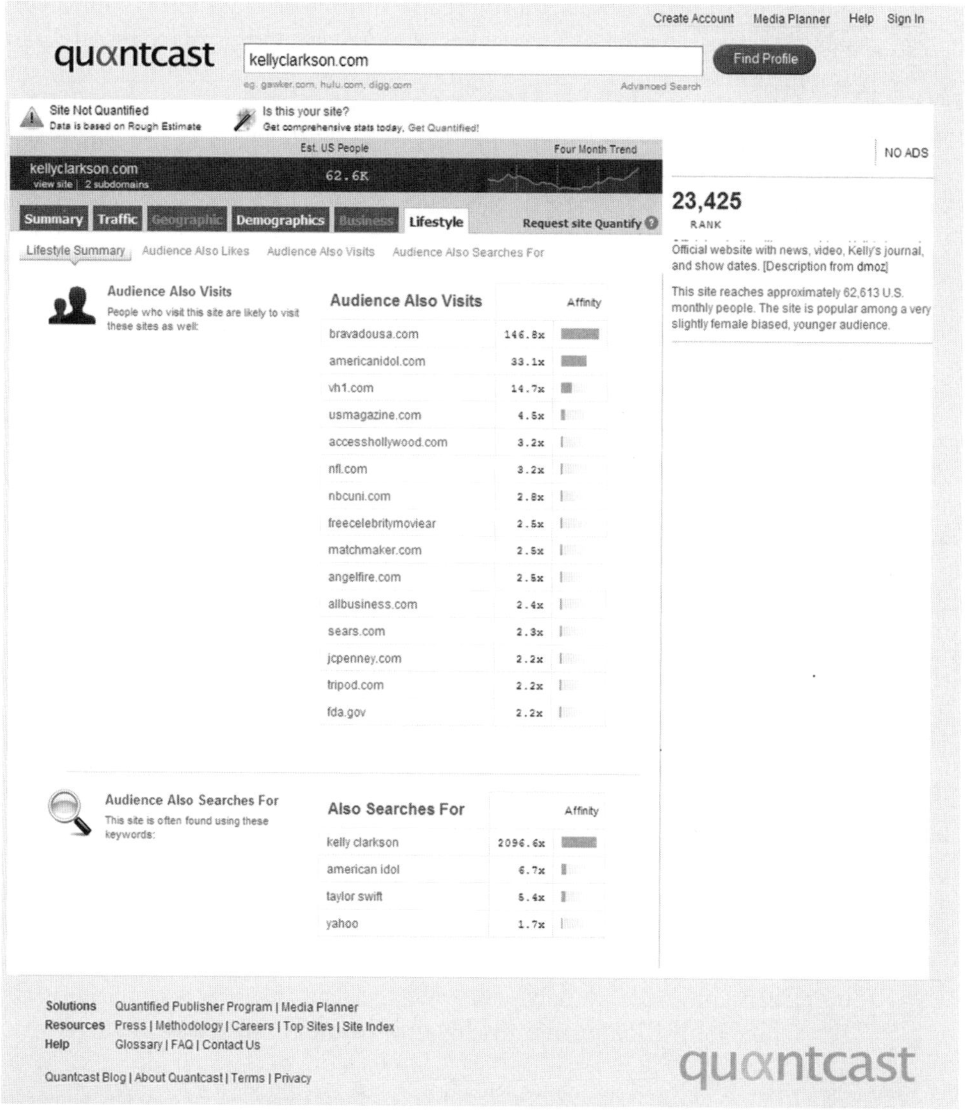

FIGURE 4.10   *Lifestyle information for visitors to Kelly Clarkson's web site. (Courtesy of Quantcast)*

THE USE OF SYNDICATED RESEARCH

Since the introduction of SoundScan and Broadcast Data Systems (BDS), the use of syndicated research has become a valuable tool for making marketing decisions in the record business. Chapter 6 illustrates how SoundScan data can be used as a basis for more in-depth research to detect sales trends and the impact of marketing strategies. Data from BDS can be merged with SoundScan to determine a more precise impact of radio airplay on record sales than was possible 20 years ago. SoundScan and primary research are combined for tour analyses and market profiles, and to persuade radio stations to increase airplay. Data from SoundScan help determine who (what geographic locations) to select for focus groups.[1] The use of SoundScan, coupled with business reply cards, syndicated research from other sources, and occasional focus groups, give marketing departments a better idea of the marketplace and allow for better prediction of marketplace performance of products.

Arbitron provides information on radio listening audience, and much of that information is valuable to the record business. The National Association of Recording Merchandisers (NARM) and RIAA conduct research projects and provide the results to their members. The U.S. Census Bureau provides basic demographic and geographic information that is used to develop marketing strategies. Other syndicated research services provide information about advertising effectiveness in general to help marketers determine which media outlets to use to reach a particular market segment.

Research Sources for Record Labels

Tracking Services

Nielsen SoundScan is arguably the most important company providing research data and information to record labels (see Chapter 7). It is self-described as "an information system that tracks sales of music and music video products throughout the United States and Canada. Sales data from point-of-sale cash registers is collected weekly from over 14,000 retail, mass-merchant and nontraditional (online stores, venues, etc.) outlets" (SoundScan, 2005). Sales results can be compared with marketing efforts for evaluation of the impact of those marketing events. Weekly data from sales are compiled by SoundScan and made available every Wednesday. "Nielsen SoundScan is the sales source for the *Billboard* music charts" (SoundScan, 2005).

---

[1]Ury Greenberg, Linda (February 21, 2005). Personal interview.

In the following example, the effects of artist appearances on *Prairie Home Companion* and the impact of the artist winning the Shure Vocal Competition at the Montreux Jazz Festival are evident in the sales patterns for jazz artist Inga Swearingen.

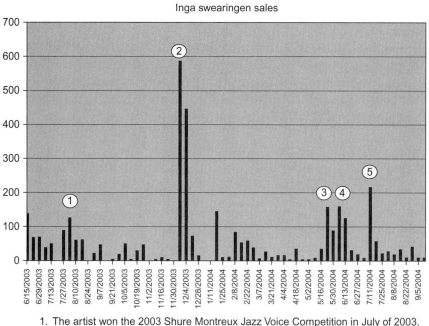

Inga swearingen sales

1. The artist won the 2003 Shure Montreux Jazz Voice Competition in July of 2003.
2. The artist appeared on the National Public Radio show Prairie Home Companion on November 29, 2003.
3. A second appearance on Prairie Home Companion on May 15, 2004, followed by
4. An appearance on May 29, created some sustained sales over several weeks.
5. Another PHC appearance on July 3, 2004 created this final spike in sales.

• And no other marketing or promotional activities were being conducted during this time.

FIGURE 4.11   *Inga Swearingen sales spikes. (Source: SoundScan)*

Nielsen also owns Broadcast Data Systems (BDS). BDS provides airplay tracking for the entertainment industry, using digital pattern recognition technology. Nielsen BDS captures more than 100 million song detections annually on more than 1600 radio stations, satellite radio, and cable music channels in over 140 markets in the United States (BDS, 2009). More information on BDS is presented in Chapter 9.

MediaBase is another company that tracks radio airplay (see Chapter 9). A division of Premiere Radio Networks, MediaBase 24/7 monitors and provides research to nearly 1700 affiliate radio stations in the United States and

Canada on a barter subscription basis. In addition to their subscription-based service, they offer a consumer-oriented service called MediaBase Music. Music fans are invited to rate popular songs on their site www.ratethemusic. com (www.mediabasemusic.com/).

On the indie level, web pages can be set up to offer consumers the chance to rate music after previewing it. This may give the artist and label some indication of the relative popularity of different songs. Several companies offer independent artists the opportunity to upload their music and have it rated by music fans. Artists who register and submit music are able to view ratings and comments on their songs. A piece of programming code can be written into the web page to provide a simple rating feature. Amazon and CD Baby also provide rating features for their products. Often, artists will encourage their fans to leave a favorable review on one of these sites.

Trade Association Research Services

All the major trade organizations provide research for their members. NARM publishes its monthly *Research Briefs*, where it has worked with NPD and Ipsos to present industry/technology information. NARM provides research findings at its annual convention on a variety of current and ever changing industry topics. The RIAA provides annual data on shipments, and contracts with an outside research firm to conduct its annual consumer profile. For this project, nearly 3000 music buyers are surveyed. The RIAA has worked with Taylor Research on a large-scale consumer research survey, with PricewaterhouseCoopers (PWC) on a study of shipments and returns, and NPD Digital for data on what consumers are doing in the digital area. The RIAA also provides research information on illegal downloading. The International Federation of Phonographic Industries (IFPI) collects data from member countries and publishes an annual report called *The Recording Industry in Numbers*. The IFPI also releases periodic reports on digital music and global piracy. The associations tend to conduct issue-oriented research of benefit to all members of the record industry.

Billboard publishes *MarketWatch* in its weekly magazine. MarketWatch provides a weekly synopsis of record sales for both the previous week and year-to-date sales, and compares this with sales figures from 1 year earlier.

Marketing Research Companies

There are several marketing research firms that collect and provide data and analyses on consumer groups. Many of these specialize in Internet consumers, and several also cover the recording industry and technology. Generally, the reports are for sale, and it may be worth purchasing one or two before starting out on a new venture for an artist or a label. These companies are also contracted by the various industry associations to conduct specialized research that is then made available to association members.

Forrester Research is one of the major market research firms focused on the Internet and technology; the company conducts research for the recording industry on all aspects of music and the Internet. Forrester also offers custom research and consulting services to its clients. *Jupitermedia* Corporation is a top provider of original information, images, research, and events for information technology, business, and creative professionals. The associations often hire Jupitermedia to conduct and report on online music consumers. *Edison Media Research* is a leader in political, radio, and music industry research, with clients that include major labels and broadcast groups. *Music Forecasting* does custom research projects on artist imaging and positioning.

The *NPD Group*, recently acquired by *Ipsos*, provides marketing research services through a combination of point-of-sales data and information derived from a consumer panel. NPD's research covers music, movies, software, technologies, video games, and many other product groups. *ComScore* offers consulting and research services to clients in the entertainment and technology industries and conducts audience measurements on web site usage through its *Media Metrix* division. Taylor Nelson Sofres, a U.K. firm, provides both syndicated and custom research of media usage and consumer behavior. Based in France, Ipsos is a global group of researchers providing survey-based research on consumer behavior. *BigChampagne* (owned by Clear Channel) tracks online P2P usage and reports, among other things, the most popular songs on P2P networks. *ResearchMusic* is an online music service that also offers free music market research to registered customers. *OTX* is an online consumer research and consulting firm.

CONCLUSION

Marketing research is a scientific process designed to ascertain information about consumers and their attitudes toward products. It follows a process of problem identification, exploratory research, research design, data collection, analysis, and conclusions or recommendations. Because markets are always a moving target and the recording industry is constantly releasing new and unproven products, marketing research has become a valuable tool enabling record labels to more accurately plan marketing strategies. The recording industry has embraced marketing research to a greater extent in the past 20 years as marketing has grown more competitive and sophisticated.

The Internet has opened up new opportunities to conduct marketing research by allowing for surveys and other forms of data collection to be moved online, and by monitoring web traffic and Internet browsing and buying behavior.

GLOSSARY

coding—The process of identifying and assigning a numerical score or other character symbol to previously edited data.

comma separated values (CSV)—A file format used for the digital storage of data structured in a table of lists form, where each associated item (member) in a group is in association with others also separated by the commas of its set. CSV files are often used for moving tabular data between two different computer programs, for example, between a database program and a spreadsheet program.

concept testing—A means of evaluating ideas by providing an early indicator of the merits of an idea prior to the commitment of resources to development and manufacturing.

custom research—Research that is tailored, proprietary, initiated by a company, and often specific to a particular product developed by that company. Examples include product surveys, focus groups, customer satisfaction surveys, and personal interviews.

data analysis—The application of logic to the data, looking for systematic patterns and details that will answer the research questions.

descriptive research—Research designed to describe characteristics of a population or phenomenon.

experiments—A form of research in which conditions are controlled so that an independent variable(s) can be manipulated to test a hypothesis about a dependent variable.

executive summary—A one- or two-page bulleted-list summary of the highlights of the findings, designed to be read by those people who need to know the main points but do not have time to read through the report to find them.

exploratory research—Initial research conducted to clarify and define the nature of the problem.

focus groups—Unstructured, free-flowing group interviews designed to gain a deeper understanding of consumer feelings, beliefs, and attitudes.

instrument—A data collection form, such as a questionnaire, or other measuring device.

margin of error—The margin of error is a statistic expressing the amount of random sampling error in a survey's results. The larger the margin of error, the less faith one should have that the poll's reported results are close to the "true" figures; that is, the figures for the whole population.

marketing research—The systematic design, collection, analysis, and reporting of data and findings relevant to a specific situation facing a company.

measurement scales—Any series of items on a survey that have been scientifically designed to measure or test a concept.

research report—An oral presentation or written statement of research results, strategic recommendations, and/or other conclusions given to the marketing client.

sample—A small representative subsection of the market segment used as a basis for analysis in research.

sample design—The procedure by which a particular sample is chosen from a population.

sampling frame—The actual set of units from which a sample has been drawn.

secondary research—Going over data that have been previously collected, described, or analyzed for a related project other than this one at hand.

syndicated research—The gathering of continuous or periodic information that is sold in standardized form to all companies involved in the industry. Examples include Arbitron, Nielsen, BDS, and SoundScan.

triangulation—The application and combination of several research methodologies in the study of the same phenomenon.

validity—The ability of a scale to measure what is intended to be measured.

variable—A variable is something that can be changed, such as a characteristic or value.

BIBLIOGRAPHY

American Marketing Association. (2009). wwwmarketingpower.com.

BDS. (2009). www.bdsonline.com/about.html.

Blankenship, A. B., & Breen, G. E. (1993). *State of the art marketing research.* Chicago: NTC Business Books.

Enbysk, M. (n.d.). Five reasons to track web site traffic. www.ecube.ca/5reasonsto trackWebsitetraffic.pdf.

SoundScan. (2005). www.soundscan.com/about.html.

Tutor2U. (2005). Market research—sampling. www.tutor2u.net/business/marketing/research_sampling.asp.

Zikmund, W. G. (1991). *Exploring marketing research.* Orlando: The Dryden Press.

Zulkardi, W. G. (2004). Triangulation in educational research. www.geocities.com/Zulkardi/submit3.html.

CHAPTER 5

Record Label Operations

Amy Macy

LABEL OPERATIONS

In an era when every level of the food chain within the entertainment industry is being scrutinized as to its value, record labels are being squeezed from both sides of the equation. Record sales are diminishing and retailers are looking for more profit in the product. And as technology advances, artists see an opportunity to completely circumvent the "label deal" and sell directly to their fans. The challenge for labels is to create value for both sides of the equation: create various "products" that draw consumers into retail stores on a consistent basis *and* create a loyal fan base to sell product to—which is not as easy as throwing a site onto the World Wide Web and expecting fans to come! To win in today's music business, it takes the creativity and business acumen of a team that is not only savvy about today's technology but also understands long-standing business tactics that can endure strong competition and challenging economic situations.

Every record label is uniquely structured to perform at its best. Oftentimes, the music genre and the "talent pool" of actual label personnel dictate the organization and inner workings of the company. As talent is signed to a label, the "artist" will come in contact with nearly every department in the process of creating a music product for the marketplace.

A typical record label has many departments with very specific duties. Depending on the size of the label, some of these departments may be combined, or even out-sourced, meaning that the task that the department fulfills is hired out to someone not on the label's staff. But the end result

CONTENTS

should be the same—creating a viable music product for the marketplace. In the following structure, there usually is a General Manager/Sr. VP of Marketing who coordinates all the marketing efforts.

GETTING STARTED AS AN ARTIST

As talent is being "found" or developed, the first contact with a record label is usually the A&R department. The artist and repertoire department is always on the hunt for new talent as well as songs for developing and existing artists to record. But before the formal A&R process occurs, the talent has to be signed to the label.

BUSINESS AFFAIRS

The business affairs department is where the lawyers of a label reside. Record company lawyers negotiate in the label's best interest. Most often, new talent will have a manager and lawyer working on his behalf, with the contract in the middle. Clearly, the label wants to protect itself and hopes to reduce risk by maximizing the contract.

Besides being the point person for artist contracts, a label lawyer negotiates and executes many other types of agreements including:

- License of recordings and samples to third parties
- Obtaining the right to use specific album art
- Acting as point person when an artist asks for an accounting or audit of royalties
- Renegotiating artist contracts
- Contractual disputes such as delivery issues
- Conflicts with contract such as guest on another recording
- Vendor contracts and relations

Often, the accounting department's work falls within business affairs because the two are related regarding contractual agreements and financial obligations. The accounting department is the economic force driving all the activity within the record company. It takes money to make money, and the accounting department calculates the budgets for each department as it aligns the forecast of releases. Most record label accounting departments have sophisticated forecasting models that calculate profitability. Each release is analyzed to determine the value of its contribution to the overhead of the company. This analysis is examined in the

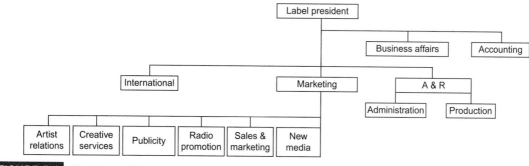

FIGURE 5.1   *Typical record label structure.*

Profit & Loss (P&L) statement, which acts as a predictor equation as to the breakeven of a release and its future value over time. (P&Ls are discussed at great length in Chapter 6.)

Additionally, the accounting department acts as an accounts payable/receivable clearinghouse, managing the day-to-day business of the company.

A&R

The ultimate job of the A&R department is to acquire masters for the label to market. To do so, labels obtain masters in several ways. As in any business, labels need to manage risk. Repackaging and remastering previous recordings is the safest way to produce a master. By creating compilations of known artists with successful sales histories, a master is produced. How it is marketed eventually will determine its success, but the label knows what it has from the start. A successful example of repackaging would be the Beatles *#1* reissue—still the number 1 selling artist on the EMI roster.

Purchasing a master already produced is another way to manage risk. Many artists will work in conjunction with producers to create a recording, with the idea that the master will be marketed and purchased by a label and that the artists will secure a recording contract. Again, the label knows what it's selling because it has the final product in hand. An example is Matchbox Twenty, produced by Matt Scerlatic.

The riskiest of masters is that of signing an unknown artist. Record labels must feel that the talent warrants a contract, with an ear to the creation of the master. Labels take many approaches to this process. But each

> The label's business affairs attorney negotiates agreements between the label and other parties, such as recording artists and independent producers (who may have deals with particular artists whereby the producers own the master recordings). The attorney also negotiates distribution agreements between the label and wholesale distributors that sell CDs to retail stores. Licensing agreements are also key, including contracts with online and mobile music services that offer on-demand streaming and downloads. And from day to day, the attorney advises label executives about legal issues related to marketing, advertising and promotion.
>
> Vincent Peppe
> Director and counsel, Internet and
> New Media Licensing, SESAC

FIGURE 5.2   *Business affairs quote.*

> **A&R**
>
> The A&R process, from securing the talent to finding the musical content can take months. Ideally, masters should be delivered 4 months prior to street date.

It's a seven day a week job. I get into the office around 10:30 am, checking email and voice mail, lining up the day's events will include our weekly A&R meeting, marketing meetings as well as artist and manager meetings. In our weekly A&R, we present new music and artists along with making sure that we prioritize all the shows that we've been invited to view. Easily, we receive 3–15 demos a day, which are officially logged to insure against "unsolicited materials." Depending on the act, I will sometimes reach out to publishers and managers of songwriters, looking for material. I follow the acts that I'm responsible for through the process — from the choosing of the songs, the recording sessions, the photo shoot and the development of the marketing plan. I work weekends too — going to shows and traveling with the artists is all a part of the job.

Louie Bandak
Senior Director of Artist and Repertoire
Capitol Records — Los Angeles

FIGURE 5.3 *Artists and repertoire quote.*

project has similar determining qualities such as which songs should be recorded, how many songs will be included, and who will produce the sessions. Along with the artist, the A&R department designates a producer or producers for a project. Trust is placed in the producer, who needs to be compatible with the artist and hold the same vision that the record label has conceived for the act. Various projects are pivotal on who produces. Some producers are very "hands on" with regards to the creative input, sharing the co-writing role or playing on the master itself. These producers tend to lend credence to the project and assist its marketability. Two examples are Alicia Keys's and Ludacris's debut releases, both produced by Kanye West, who became a notable producer on Jay-Z's albums and who has become an artist in his own right.

Other artists may need less of a heavy hand, but something more like subtle guidance from both the A&R department as well as the producer. Songwriting artists are unique in that both the content as well as the creation are open to direction. And these artists can have strong ideas about their music and how they want it produced. However, A&R representatives can take very active roles in the direction of the project by aiding the songwriting process and nurturing the recording through the entire timeline. It all depends on what the label is looking for in the final product.

ARTIST DEVELOPMENT/RELATIONS

Artist development

Once a contract is signed, an artist is usually assigned an AD representative.

12 months prior to Street Date.

Sometimes called the product development department, this department manages the artist through the maze of the record company and its requirements of the artist. Madelyn Scarpulla describes the product manager as follows: "The product manager in effect is your manager within the label" (Scarpulla, 2002). Artist Development specialists hold the artist's hand, helping him to clarify his niche within the company. Artist Development usually develops strong relationships with the artist and artist manager, with other departments in the company looking to Artist Development as a clearinghouse that helps to prioritize individual department needs with the artist.

Artist Development not only manages the artist through the process, including delivery of the recording, photo and video shoots, and promotional activities, but also looks for additional marketing opportunities that maximize the unique attributes of the act.

CREATIVE SERVICES

Depending on the company, the creative services department can wear many hats. Artist imaging begins with creative services assisting in the development of style and how that style is projected into the marketplace. Special care is taken to help the artist physically reflect her artistry. Image consultants are often hired to assist in the process. "Glam teams" are employed to polish the artist, especially for high-profile events such as photo and video shoots as well as personal appearances.

The creative services department often manages photo and video shoots, setting the arrangements and collaborating on design ideas and concepts with the artist. Once complete, images are selected to be the visual theme of the records and the design process begins. In some cases, creative services contain a full design team that is "in house" and a part of label personnel. Such in-house teams can ensure quality and consistency in imaging of the artist—that the album cover art is the image used on promotional flyers, sales book copy, and advertisements. When there isn't an in-house design crew, design of album cover art and support materials is farmed out to outside designers. Interestingly, the use of subcontractors can enhance unique design qualities beyond the scope of in-house designers, but there can be a lack of cohesive marketing tools if not managed properly.

> **Creative services**
>
> As the A&R process evolves, a photo shoot is needed to represent the content of the recording.
>
> 6 months prior to Street Date.

PUBLICITY

The priority of the publicity department is to secure media exposure for the artists that the label represents. The publicity department is set into motion once an artist is signed. The biography of the artist via an interview is created. Other tools such as photos from the current photo shoot, articles and reviews, discography, and awards and other credits, are collected into one folder, creating a press kit for each artist. These press kits are tools used by the publicity department to aid in securing exposure for their artists and are often sent to both trade and consumer outlets.

> **Publicity**
>
> Creating a press kit for advance awareness occurs shortly after photo shoot.
>
> 5 months prior to Street Date.

Pitching an artist to different media outlets can be a challenge. As an artist tours, the ideal scenario would be that the local paper would review the album and promote the show. Additional activities would be to obtain interviews with the artist in magazines and newspapers that can also be used as incremental content for web sites by these same entities. Booking television shows and other media outlets falls to the publicity department as well. On occasion, artists will hire their publicists to assist in creating higher profile events for the act. These publicists try to work with label publicists to enhance in-house efforts and build on relationships already established.

As artists become more established, many acts hire their own independent publicist to enhance their profile. This additional media punch is usually coordinated between label and indie publicist so that redundancy is avoided as well as efficiency of best efforts among staff and hired gun.

> Day-to-day tasks [of a publicist] include securing coverage, meeting the needs of media (sending press kits, photos, music etc.), controlling the budget, managing independent publicists, communication with managers, booking agents, other label departments etc., reading newspapers & magazines, maintenance of contact database, research and hiring hair & makeup artists for TV & magazine appearances.
>
> Amy Willis
> Media Coordinator
> Sony Music Nashville

FIGURE 5.4 *Publicity quote.*

RADIO PROMOTION

In most record companies, the number 1 agenda for the promotion department is to secure radio airplay. Although the Internet has created a new way for consumers to find new music, surveys continue to show that listeners still learn about new product and artists via the radio. Typically, radio promotion staffs divide up the country into regions, and each promotion representative is responsible for calling on specified radio stations in that region, based on format. Influencing the music director and radio programmer is key in securing a slot on the rotation list of songs played. These communications often take place on the phone, but routinely, radio promotion staffs visit stations, sometimes with the artist in tow, to help introduce new music and secure airplay.

With the consolidation of radio stations continuing, developing an influential relationship with individual stations is getting harder and harder. But there still exists a level of autonomy within each station to create its own playlist as it reflects their listenership. To strengthen the probability of radio airplay, promotion staffs conceive and execute radio-specific marketing activities such as contests, on-air interviews with artists, listener appreciation shows, and much more.

> **Promotion**
>
> Depending on the genre, the first radio single must be released to garner airplay and create demand. Artist visits with radio can enhance airplay.
>
> 3 months prior to Street Date.

SALES AND MARKETING

The sales and marketing department sells product into retail and creates visibility of the product at the consumer level. If all the other departments have done their job correctly, selling the music to retail should be easy because there will be a pent-up demand for the release. But to ensure sales success, the sales and marketing department must create awareness to the gatekeepers.

Many sales departments look to their distributor to be not only their partner but also their first line of customers. The distribution company must be well informed as to new releases and the marketing plan that goes with them so that they can represent the product to retailers. To do so, sales departments educate their distributor by sharing with them detailed marketing plans for upcoming releases.

The second line of customers is retail. Sales departments continue the education process by informing retailers as to new releases and marketing plans. In tandem, record label sales reps, along with the distribution representative, will visit a specific retailer together and on occasion may bring an artist by to visit with the retail buyer. In this visit, the amount of music that is to be purchased by the retailer, along with any specific deal and discount information, is discussed. Co-op advertising is usually secured at this time as well, with pricing and positioning vehicles along with other in-store marketing efforts concluded.

NEW MEDIA

The new media department has their finger in every cookie jar! From receiving clearances via the contract to managing artist web sites, to the development of those web sites, to the creation of "widgets" and banner advertising for Internet marketing in coordination with the marketing and sales department, to the development of tools that may be used as content for online media through magazine and newspaper sites, the new media department can wear various hats. This department can also be responsible for aligning new business arrangements with key digital partners, including telecom entities such as Verizon or AT&T, in creating additional exposure for new releases using downloads as the bait. Depending on the company, this department can be a "catchall" for all things "digital."

> **Sales**
>
> By visiting retail buyers, artists can assist in the setup and sell-through of their record. Solicitation process begins 2 months prior to Street Date.

> I have one of the best jobs on the planet. I work with great people, help develop the careers of many artists, act as the "hub of the wheel" by synchronizing the efforts of sales, promotion, publicity and creative with artists and their managers. I stay in the creative process by personally developing and executing the marketing strategies of several acts including Dave Matthews Band, David Gray, and The Strokes. It's long hours: I'm in the office from 9 am to 8 pm, five days a week, plus shows and dinners after work — not to mention the extensive traveling that goes with the gig. It's not just a lifestyle — my job is my passion.
>
> Hugh Surratt
> Senior Vice President of Marketing
> and Creative Services
> RCA Records — New York

FIGURE 5.5 *Marketing and creative services quote.*

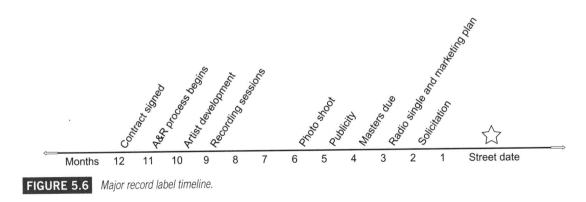

FIGURE 5.6   *Major record label timeline.*

INDEPENDENT LABELS

In an era when album sales continue to decline, the Internet has fragmented the market, and major record labels have tried to maintain their prominence as "creators of superstars," consumers have been looking elsewhere to satiate their burgeoning musical tastes—and they're eating from the independent labels' table. Historically, independent labels have been relatively small in sales stature and were genre- or regionally specific, tending to react quickly to marketplace trends. Some independent labels strike deals of distribution with the major conglomerates, whereas others find their way to consumers through independent distributors—but all are looking to sell records, either literally or virtually to an ever fractionalizing marketplace. And this new age of consumership has brought with it an entirely new dimension for "indie" labels—one that can include the surprise of big artists with big sales along with even bigger opportunities.

Fueled By Ramen has emerged as one of the most successful independent labels so far, with several of their initial acts going on to major label status, such as Jimmy Eat World and Fall Out Boy. Youngster John Janick started his fledgling label from his college dorm room at the University of Florida. Since 1996, he has ridden the viral marketing wave by utilizing the Internet as his label's mouthpiece to cascade popularity of one band with another. His label has been attributed with creating the "360 Deal" where artists are signed to label, tour support, web promotion, and merchandising deals in which Fueled By Ramen and its web site act as the portal for all revenue streams. Panic At the Disco, Gym Class Heroes, and Paramore have all risen to *Billboard* Top 200 chart status via this model. FBR continues to evolutionize its marketing strategies with its fifth most popular YouTube Channel as well as sales of merchandise through teen retailer Hot Topic and placement of recordings on the ever popular video game Rock Band.[1]

---

[1] www.nytimes.com/2008/05/05/business/media/05music.html?pagewanted=1&_r=1.

With what looks like only a dozen staffers, this organization has three folks dedicated to new media with offices in Tampa, Chicago, and New York City.

To look at their organizational chart, **Big Machine Records** in Nashville, Tennessee, is structured similarly to that of any major record label, including in terms of their size—28 strong. But what makes them unique is that they are truly independent, meaning that they are not owned by a major conglomerate but act in their own interest and are in charge of their own destiny—and they seem to know where they're going. With the largest selling act of 2008, Taylor Swift dominated the *Billboard* Top 200 chart with not one, but two albums in the Top 10 and selling over 4 million units. The label, distributed through Universal Music Group Distribution (UMGD), negotiated "free agent" status and can sell directly to consumers—and does so by utilizing cool technologies such as Bandbox. This "free" playback interface allows fans to embed streaming audio and video of their favorite artist with click-through capability to a merchandising center where downloads and other products can be purchased. Big Machine presold Taylor Swift's sophomore album *Fearless* utilizing this new technology. "Her active fan network [helped] spread the (BandBox) widget to all corners of the Internet, essentially providing a Taylor-branded 'end cap' on countless web sites," said

360 deals are contracts that allow a record label to receive a percentage of the earnings from ALL of a band's activities instead of just record sales. Under 360 deals, also called "multiple rights deals," record labels may get a percentage of things that were previously off limits to them, like:

Concert revenue
Merchandise sales
Endorsement deals
Ringtones

In exchange for getting a bigger cut from the artists they represent, the labels say they will commit to promoting the artist for a longer period of time and will actively try and develop new opportunities for them. In essence, the label will function as a pseudo-manager and look after the artist's entire career rather than only focusing on selling records (from http:// musicians.about. com/od/ah/g/ 360deals.htm).

FIGURE 5.7 *BandBox widget of Taylor Swift's album for Big Machine.*

Kelly Rich, VP Sales & Marketing for Big Machine.[2] The administrative staff represents two imprints: Big Machine Records and The Valory Music Company, with a combined count of 11 acts. Three artists have previous platinum status: Reba McEntire, Trisha Yearwood, and Jewel.[3] Big Machine, the label, works on branding itself as well, with the hopes that music consumers will believe in the imprint when the label introduces new artists into the marketplace.[4]

Dualtone Records is "redefining the label as an entertainment company," says Joey Luscinski, Manager of Production; he wakes up with a smile every day, knowing that he works for a cutting edge, independent music entity. Dualtone is meeting artists where the industry has fallen short. Although the company represents acts in the traditional record label manner, this company offers services to artists who want to remain independent themselves. Within a very short 2-week period, Dualtone can have an independent artist released nationwide via Dualtone's download services as well as through their distribution hook-up with independent distributor Alternative Distribution Alliance (ADA). They also offer artist management, radio promotion, film and television placement of recordings, ecommerce and online marketing, and for signed acts they are moving into publishing and booking. "It's all about providing a service, not taking a fee," says General Manager Paul Roper. The idea is to allow the acts to focus on what's important to them and let Dualtone be an expert in areas that will increase the artists' overall wealth. And this model is working. Most recently, Dualtone artist Brett Dennen found himself up late on Conan, Jimmy Kimmel, and Leno—only to then breakfast on CBS's *Early Morning* television show. These engagements, along with opening for John Mayer and other higher profile functions, caused Dualtone to consider a unique partnership with Downtown Records of Gnarls Barkley fame. Together, they hope to take Brett to the next level while sharing in all expenses—and profits. About seven people run Dualtone, each wearing multiple hats and all looking for the next act that will help build their success story.[5]

All of the independent labels noted here have some sort of "multiple rights deal" that allows them to financially benefit in a greater way with success of their artists. But these deals are not limited to the independent label world, with many of the major labels moving toward signings with

[2]Nashville, TN, September 8, 2008. PRNewswire, www.prnewswire.com/mnr/bandbox/34816/.
[3]thevalorymusicco.com/index.cfm?id=5.
[4]Kelly Rich interview, Big Machine Records, February 28, 2009, Nashville, TN.
[5]Joey Luscinski and Paul Roper interviews, Dualtone Records, March 4, 2009, Nashville, TN.

BRETT DENNEN
So Much More

This Oakdale, CA native continues to grow his fan base with a live show consisting of the honest songwriting of a Jack Johnson, the guitar work of a Paul Simon and the attention grabbing enegy of a Ben Harper leaving all in his wake of yearning for more.

TRACK LISTING:
1. Ain't No Reason
2. There Is So Much More
3. Darlin' Do Not Fear
4. Because You Are A Woman
5. She's Mine
6. The One Who Loves You The Most
7. I Asked When
8. When You Feel It
9. So Long Sweet Misery
10. Someday

PRESS: **Rolling Stone** piece confirmed and running with a picture in the fall music preview issue. NPR feature on "Day to Day" confirmed to air around release week. Look for feature coverage in Relix, Honest Tune, USA Today, Entertainment Weekly, New York Times and other major dailies.

RADIO: "Ain't No Reason" is reacting at AAA radio now, expect crossover to AC and Hot AC this fall. Key tastemaker station KCRW has been spinning the album regularly, has had Brett in studio and sponsored a show recently. The station will also be using the album as one of its fund-drive titles this fall. Brett beat out Ray Lamontagne for #1 phones at the station after a recent visit. KEXP, a huge uber-cool station in Seattle with a tremendous range and influence has also recently been spinning the song and had Brett on air live for 30 minutes after his performance at Bumbershoot. KMTT (Seattle), KINK (Portland), WFUV (New York), WXPN (Philadelphia), and KFOG (San Franisco) have all been playing the single and stations already on board include: WCBE (Columbus), WMVY (Cape Cod), KOZT (Ft. Bragg), WNCW (Asheville), KBAC (Sante Fe), KRSH (Santa Rosa), WNCS (Burlington), KHUM (Eureka, CA), WTMD (Baltimore), KTHX (Reno), The Coffeehouse (Sirius), The Spectrum (Sirius) and Hear Music (XM)

DIGITAL/NEW MEDIA: Significant preorder campaign with Brett's fan base. Cross promotions and grassroots marketing campaign with myspace (recently featured on their home page as a featured artist). Ecard available. Google Ad words program, Faccebook. Targeted Text Ad campaign and other online advertising in place. Podcast feature on Brett will be included with tastemaker blogs and online sites. eMusic will feature Brett's "Ain't No Reason" as a Daily Download around street date (this also runs on influential blog pitchforkmedia.com). iTunes is a big supporter and will be featuring Brett around release week.

CONSUMER ADS: RELIX, American Songwriter, Paste, Oxford American, Harp. Three of Brett's songs will be included on the Dualtone Label Sampler that will be handled with the Sep/Oct issues of American Songwriter and for all subscribers and selected newstand for HARP in November.

TOURING: Brett's fan base has grown organically over the past year online and through word of mouth surrounding his electrifying and moving live show. He consistently sells out venues and cities on his first trip through the market which speaks to the power of his fan base and the underground swell surrounding the music. Scott Clayton (My Morning Jacket, John Mayer, Kings Of Leon) from CAA is the booking agent and has solidified Brett as the opening act for the late Summer John Mayer and Sheryl Crow tour, Hotel Cafe tour, and a Fall tour with Ziggy Marley (Jack Johnson and Amos Lee dates are pending). Brett will aslo perform at the three major festivals of the fall: Bumbershoot (Seattle), Austin City Limits and Power to the Peaceful I (San Francisco) with Michael Franti.

"A beautiful and spirited record. Instantly likeable."
~John Mayer

"He's timeless, he's probably twenty-five, but he seems like he'stwenty-five in 1972." ~Rolling Stone

"Brett is on my very short list of artists to watch. So Much More is a joyous dance." ~Chris Douridas, KCRW

"His music has a playfullness in its bounce while the lyrics ring with a truth beyond his years." ~Michael Franti

KEY MARKETS:
San Francisco
Los Angeles
Seattle
Portland
Boulder
Denver
Philadelphia
Boston
New York
Burlington

10.30.06	Freebird Live (w/ Matt Nathanson)	Jacksonville Beach, FL
11.01.06	Studio-A (w/ Matt Nathanson)	Miami, FL
11.02.06	The Social (w/ Matt Nathanson)	Oralando, FL
11.04.06	The Roxy (w/ Matt Nathanson)	Atlanta, GA
11.05.06	The Zydeco (w/ Matt Nathanson)	Birmingham, AL
11.07.06	Gypsy Tea Room (w/Matt Nathanson)	Dallas, TX
11.08.06	Paradise (w/ Xaiver Rudd)	Boston, MA
11.14.06	Sonar (w/ Ziggy Marley)	Baltimore, MD
11.15.06	Stone Pony (w/ Ziggy Marley)	Asbury Park, NJ
11.16.06	Webster Hall (Xaiver Rudd)	New York, NY
11.18.06	Hotel Cafe Tour	Hollywood, CA

8 03020-12402 8

Box Lot: 30
List Price: $13.98
Configuration: CD
UPC: 803020124028
Selection: DTM1240
File under: Pop/Rock
Package type: Jewel Case

Street Date November 21, 2006
5% Discount Through December 5, 2006

FIGURE 5.8 *One-sheet on Brett Dennen.*

new artists that include these elements and more, such as publishing, concert revenues, merchandising, and endorsements. In fact, Warner Brothers Records has gone "on record" by signing all their new acts to "multiple rights deals", each having various elements in place to mutually benefit the relationship.

So, what is considered "success" in the music business today? Some indies have hit it big and compete with majors head on. But there are many smaller independent labels that are making a go at not just surviving, but utilizing this new model of business. In this new model, incremental sales on a smaller level are enough to not only keep the lights on but also to serve as a creative outlet for many more artists who would otherwise not have a venue in which to sing. If selling 50,000 records at 10 bucks a pop makes a half a million dollars and keeps an artist on the road and his small indie label in business for another year—is this success? To the consumer looking for the coffee-house songster whose record may never be heard on mainstream radio—you bet.

BIBLIOGRAPHY

Scarpullo. 2002.

Luscinski, J., & Roper, P. Dualtone Records personal interviews, March 12, 2009.

Rich, K., Big Machine Records personal interview, March 13, 2009.

CHAPTER 6

The Profit and Loss Statement

Amy Macy

PROFITABILITY . . . OR THE LACK THEREOF!

Record companies use profit and loss (P&L) statements both to predict the success of a record prior to its release as well as to analyze a project as it nears the end of its lifecycle. To understand the "math," let's look at all the components of a P&L, recognizing the financial significance of every line item.

SRLP—Suggested Retail List Price

The suggested retail list price (SRLP) is set by the record label and is based on cost of the recording project, the artist's status, genre of music, competitive landscape, and what the market will bear. Although the royalty models are changing, many labels still pay artist royalties based on the SRLP. The SRLP has a correlating wholesale price, which is usually structured by the distribution company.

Card Price

This line item is the wholesale price. The **card price** is the dollar figure that is set by the distribution company that sells the product. For fair trade practices, wholesale card prices are published entities that are the basis of further financial negotiations between seller and buyer. The chapter on retail shows an actual published rate card of wholesale prices.

CONTENTS

Discount

Music product regularly receives a discount, which is set by the label and administered through the distributor. (In some situations, it can be specified by the artist's contract.) In mainstream music, the discount is applied to the wholesale price. In some genres of music such as Christian music, discounts are applied to the retail price, and negotiations are then based on retail prices. But the majority of music is sold on wholesale pricing strategies.

Discounts are based on many variables. If a label has a really hot artist who is a big seller, and demand at the consumer level is high, a discount may not be offered because most retailers will buy the product at full price. To entice retailers to purchase a new artist, record labels will offer discounts, which will increase the margin and profitability at the store level. Fair trade practices require that labels and their distributors offer the same discount on the same release to all retail purchasers. But the discount can be changed based on the marketing elements that the retailer may offer.

For example, a label has a new artist and is offering a 5% discount. This discount will increase the potential profitability of the retailer. If the retailer agrees to include this new artist on a "new artist" endcap for an additional 5%, then the discount will be increased to 10%, adding to the potential profitability of the retailer.

Gross Sales

Gross sales refers to the wholesale price (value) of the product with the reduction of the discount included. It usually reflects the number of records shipped minus returns. Remember that music retailing is basically a consignment business and that stores can return product back to the distributor and label, and receive a credit for this unsold product.

Distribution Fee

Depending on the relationship between the label and distributor, the distribution fee is based on sales after the discount, meaning it is in the distributor's best interest to keep discounts as low as possible to help increase profitability. This fee is a percentage of sales and varies greatly by the distributor. The conglomerates that own both the labels and distributor often charge between 10% and 12%. Independent distributors structure deals with indie labels and artists that range between 18% and 30%, depending on the services being offered.

Gross Sales After Fee

Once again, the gross sales figure is determined on shipped minus returned product, being *net*. Net units are multiplied by the per-unit price, after discounts and distribution fees are deducted.

Return Provision

The industry averages about 20% **returns**, meaning that for every 100 records in the marketplace, 20 will be returned. To protect their business, record labels ensure against returns by "reserving" a percentage of sales. Record labels "reserve" 20% of sales by pocketing these funds in an escrow account. Not until the life of the record has run the majority of its cycle will the reserve be adjusted. In some cases, a record may only have a 5% return in its lifecycle. The record company will then adjust the profitability statement to reflect such a low return, and royalties will then be distributed. In other cases, a record may have a 30% return, which adversely affects the company's overall profitability, because they expected to sell 10% more of a particular release than previously accounted.

Net Sales After Return Reserve

Basically, net sales after return reserve is the computed net sales minus the 20% reserve deduction.

Returns Reserve Opportunity/(Risk)

Labels incur returns on records shipped into the marketplace. For accounting purposes, the profit and loss statement reflects these returns within the overall equation. The returns reserve line item is usually computed at the end of the lifecycle of the release. A label will calculate what actual returns occurred and plug this adjusted number into the P&L, determining the ultimate profitability of the project. But labels will use the P&L or *pro forma* as predictor equations to determine potential profitability of a future project and will plug return reserve standards into the overall equation, helping to evaluate a project's future. Note that return reserves only apply to physical sales and do not apply to digital sales.

Gross Physical Sales

Gross physical sales is the adjusted sales number after deducting discounts, distribution fees, and returns of actual product in the marketplace.

Gross Digital Units/Albums and Singles

Digital sales continue to grow as a significant contributor to the bottom line. According to the Nielsen 2008 year-end music industry report, digital music sales account for 32% of all music purchases (units) in the United States. Although hard to predict, labels are working on models that reflect consumer behavior by genre so that this line item can better serve the *pro forma* equation, helping to predict project potential. Album and single per-unit dollar figures include royalties that have yet to be distributed.

Undetermined is the effect that the pricing strategy of iTunes will have on the digital equation. In April 2009, the three-tiered pricing strategy of $.69, $.99, and $1.29 singles was implemented. Most labels will release new artist singles, at the very least, at the $.99 level, so that predicting income should have some relativity. What will be interesting is the enhanced $1.29 version that can be included in an album sale. iTunes has committed to not selling an album to consumers for over $9.99, but how will that translate as income to the label when the album sold can have up to three enhanced singles on it? And what about the many albums that contain more than 10 tracks? At this point, iTunes will give consumers wanting the entire album credit for singles purchased from that title, and they allow for a price "break" for the remaining tracks in an effort to **upsell** the entire work. It's a noble gesture on the part of iTunes to sell entire CDs, but the financial mystery remains.

Total Net Sales

Total net sales is the adjusted sales number after deducting discounts, distribution fees, and returns of both physical and digital sales.

Cost of Sales

Cost of sales are costs associated with making the actual product, including the pressing of the disc, printing of paper inserts, marketing stickers on the outside of the product, all-in royalties and mechanical royalties, and inventory obsolescence. In accounting terms, these are variable costs because the amount varies based on the number of units produced.

Many of these costs are negotiable, such as the mechanical rate. Labels often receive a reduced mechanical rate on artists who are also the songwriters. Older songs as well as reissued material can often receive lower mechanical rates based on the age and inactivity of the copyrights.

Dealing with the inventory of product that does not sell into the marketplace costs money. So, labels build into the cost of goods an amount that will fund the management of returned and obsolete inventory. It takes manpower and resources to "scrap" a pile of CDs. This includes moving the inventory off the warehouse floor, pulling the inserts and CDs from the jewel cases (which are recycled for new releases), and breaking/melting the actual CDs into pellets, which can then be used to make new CDs.

The one-time fee of preparing the digital masters for online retailers must also be accounted for as included in the cost of sales.

Gross Margin Before Recording Costs

Gross margin before recording costs deducts the cost of sales from the total net sales.

RECORDING COSTS

Know that the recording costs are initially funded by the record label. But built into most artist contracts, recording costs are recoupable, meaning that once the release starts to make money, the record company will pay itself back prior to the artist receiving royalties.

Also included in recording costs are advances. **Advances** are monies fronted to the artist to assist with living expenses. It takes time to record a record, which doesn't allow an artist to make money elsewhere. Advances are recoupable.

Gross Margin

Gross margin reflects gross sales minus discounts, distribution fees, returns, cost of sales, and recording costs. These are revenues made prior to marketing expenses.

MARKETING COSTS

To launch an artist's career in today's climate takes a lot of money. How a company manages its marketing costs can determine the success of an album . . . and artist. Beyond the making of the record, marketing costs include the imaging of an artist, advertising at both trade and consumer publications, publicity, radio promotion, and retail positioning in the stores.

Account Advertising

An expensive marketing cost is the positioning of a record in the retail environment, known as *co-op advertising*. In the grocery business, this is called a *slotting allowance*. Did you ever notice that premium, name-brand items are placed in the most prominent positions within a grocery store? The same occurs in record retail. The pricing and positioning (P&P) of a record, meaning a "sale," or reduced price, along with prime real estate placement in retail stores, can cost record labels hundreds of thousands of dollars.

Other types of account advertising include print advertising via the store's Sunday circulars, endcap positioning, listening stations, event marketing such as artist in-store visits, point-of-purchase materials placement guarantees, and more. Chapter 12 discusses the many aspects of the retail environment.

Advertising

Most schools of advertising include a lesson on internal versus external advertising. Trade advertising is considered an internal promotional activity. Trade advertising creates awareness of a new product to the decision makers of that industry by using strong imaging of the artist and release along with relevant facts such as sales and radio success, tour information, sales data, and upcoming press events.

In the record business, decision makers include music buyers for retail stores, radio stations and their programmers, talent bookers for television shows, reviewers for newspaper and consumer magazines, and talent buyers for venues. Some prominent trade magazines for the music industry include *Billboard*, *Amusement Business*, *Pollstar*, and *Variety*, to name a few.

Consumer advertising is considered external advertising because it is targeting the "end consumer." Consumer advertising also creates awareness, but to end consumers who will purchase music. The most popular consumer advertising today is viral marketing via the Internet. This tool allows for "test driving" the music, along with an endless array of imaging with immediate "**click 'n' buy**" opportunities. Although there are several online marketing strategies through banner placements on key web sites, the most popular consumer advertising has come through funding the preparation of turn-key "**widgets**" that allow for fans to embed and forward them on their personal web sites through social networking.

Video Production

Negotiated into the recording contract, artists are usually responsible for 50% of the cost of video production. Although the record company will fund the video shoot, the company will expect to receive recoupable pay of 100% of the overall costs from record sales and 50% recoupable from video/DVD sales.

Artist Promotion

This line item usually covers the costs associated with introducing and promoting the artist to radio, still a primary source for learning about new music. Labels often take artists to visit with radio stations, including on-air interviews, dinners with music programmers, and Listener Appreciation events. The record company incurs this cost and it is usually not recoupable.

Independent Promotion and Publicity

Depending on the artist's stature and current competitive climate, record companies will hire the services of independent promotion companies and independent publicists. In addition to the label's efforts, these independent agents should enhance the label's strategy by assisting in creating exposure for their artists via additional radio airplay and media. This line item is not recoupable.

Media Travel

Record labels will fund the costs associated with travel for an artist who is doing a media event. Media travel is usually an isolated event where an artist is doing a television talk show or an awards event—again, an expense to the label and not recoupable.

Album Art

The imaging of an artist can take time . . . and money. Most artists receive some type of grooming, if not just a polishing of what the artist already represents. Professionals trained at artist imaging are hired to create a "look" that is unique and defining. Such professionals as hair and make-up artists, clothing specialists, even movement/dance professionals are often required to give an artist a specific shine. As a part of the team, photographers are hired to shoot cover artwork as well as press images for publicity use. And then a designer is hired to create the overall album art concept, from artist image use to album title treatment, booklet layout, and so forth. All of these efforts are expenses to the label and not usually recoupable.

No-Charge Records

A marketing tool often used by labels is the actual CD. No-charge records are those CDs that are used for promotional use such as giveaways on the radio or in-store play copies for record retailers. The value of these records and their use must be accounted for and is not recoupable.

CONTRIBUTION TO OVERHEAD

From the gross margin, a label subtracts the marketing costs including account advertising, trade and consumer advertising, video production, artist promotion, independent promotion and publicity, media travel, album art, and no-charge records, to determine the release's contribution to a record company's overhead.

Note the percentages given within the spreadsheet. The total net sales percentage reflects 100% of money generated by the sale of the project. Each line-item subheading also reflects a percentage, causing each department to consider how much is being spent as a reflection of the project as a whole. Although the contribution to overhead may be a large number, its percentage of the whole determines how effectively and efficiently the project was managed.

Contribution to overhead: example of two different projects

Example	Project A	Project B
Total net sales	$3,500,000	$5,000,000
Contribution to overhead	$1,350,000	$1,500,000
Percentage	38.6%	30.0%

Looking at the numbers, Project B made more money for the company, but as for efficiency, Project A was a more effective release because the company did a better job at managing its expenses.

WAYS AT LOOKING AT PROJECT EFFICIENCY AND EFFECTIVENESS

Record labels look at the "numbers" in many ways to determine how well they are performing. By "spinning" these figures, a company should analyze where spending is less effective, thus causing the overall project to be less profitable.

Return Percentage

$$\text{Gross Ships} - \text{Cumulative Returns} = \text{Net Shipment}$$

$$\frac{\text{Cumulative Returns}}{\text{Gross Ships}} = \text{Return Percentage}$$

Example:

500,000 Units − 35,000 Returns = 465,000 Units Net Shipments

$$\frac{35,000 \text{ Units}}{500,000 \text{ Units}} = 7\% \text{ Return}$$

Again, industry standards reflect an approximate 20% return percentage. Generally, a record that returns more than 20% is not performing at the market average. A couple of decisions could have affected this percentage. The record company oversold the project and caused the returns, or the record company did not promote and have a hit with the project, thus causing returns. In any case, the record company must determine what occurred to ensure that it doesn't happen again.

If a project's return percentage is below the industry average, such as a lifetime return average of 5%, the project could be considered a great success, minimizing return costs as well as manufacturing costs. But the company should also be sensitive to the fact that they may have undersold the project, thus not realizing the full sales potential of the project too. Again, further analysis is vital to the overall success of future projects and the company.

SOUNDSCAN AND SELL OFF

SoundScan is an information system that tracks sales of music and music video product throughout the United States and Canada. Sales data using UPC bar codes from point-of-sale cash registers are collected weekly from over 14,000 retail, mass merchant, and nontraditional (online store, venues, etc.) outlets. Weekly data are compiled and made available every Wednesday. Now branded with the Nielsen name, SoundScan is the sales source for the *Billboard* music charts, major newspapers, magazines, TV, MTV, and VH1 charts. Chapter 6 is dedicated to SoundScan, and its analysis and impact on the industry.

To evaluate inventory, record companies can use a simple equation to know how many units remain in the marketplace:

Net Shipments − SoundScan Sales = Remaining Inventory

$$\frac{\text{SoundScan Sales}}{\text{Net Shipments}} = \textbf{Sell-Off Percentage}$$

Example:

465,000 Net Ships − 437,000 Units SoundScanned = 28,000 Units

$$\frac{437,000 \text{ SoundScan}}{465,000 \text{ Net Ships}} = 94\% \text{ Sell Off}$$

If this project is a steady seller, quietly moving units every week, a record company could use SoundScan weekly sales to determine how many weeks of inventory are left in the marketplace. If this project sold 2000 units each week:

$$\frac{28,000 \text{ Remaining Inventory}}{2000 \text{ SoundScan Units Each Week}} = 14 \text{ Weeks of Inventory Left}$$

Knowing inventory levels and keeping aware of sales are critical to the success of a project. A record company does not want to run out of records, which is called *can't fill*. If it's not on the retailer's shelves, it cannot be purchased.

Percentage of Marketing Costs

Isolating marketing costs and analyzing their effectiveness in selling records are best reflected in the following equations:

Percentage of Marketing Cost to Gross:

$$\frac{\text{Marketing Costs}}{\text{Gross Sales}} \times 100 = \%$$

Percentage of Marketing Cost to Net:

$$\frac{\text{Marketing Costs}}{\text{Net Sales}} \times 100 = \%$$

Example:

$$\frac{\$800,000 \text{ Marketing Costs}}{\$4,238,000 \text{ Gross Sales}} = 18.9\%$$

$$\frac{\$800,000 \text{ Marketing Costs}}{\$3,402,000 \text{ Net Sales}} = 23.5\%$$

A lower percentage reflects a better performing project. Keeping marketing costs in check and knowing when to stop "fueling the fire" is usually a great determiner of seasoned record labels.

Marketing Costs per Gross Unit:

$$\frac{\text{Marketing Costs}}{\text{Gross Shipments}} = \text{Cost per Unit}$$

Marketing Costs per Net Unit:

$$\frac{\text{Marketing Costs}}{\text{Net Shipments}} = \text{Cost per Unit}$$

Example:

$$\frac{\$800{,}000}{500{,}000 \text{ Units Gross Shipped}} = \$1.60$$

$$\frac{\$800{,}000}{465{,}000 \text{ Units Net Shipped}} = \$1.72$$

In all of these equations, the "real" picture is best drawn when using net shipments, because that is the ultimate number of units in the marketplace.

AS A PREDICTOR EQUATION

Record companies can use the profit and loss statement as a predictor of success. Often, labels will "run the numbers" to see how profitable, or not, a potential release could be. By using the spreadsheet and plugging in forecasted numbers, including shipments, cost of sales, recording costs (or the acquisition of a master), and marketing costs, the equation will help a label evaluate and determine whether a project is worth releasing.

Small considerations can dramatically affect the contribution to overhead. List price, discounts, number of pages in the CD booklet, royalties—both artist and mechanical, and the various marketing line items can either make a project profitable or not.

BREAKEVEN POINT

When does a record "**break even**" in covering the costs that it took to make the project, and when does it start to turn a profit? Depending on the equation, record companies look at this value in several ways.

Without marketing costs, a number can be derived simply by dividing the total fixed costs by price, using wholesale dollars.

Breakeven point

$$\text{BE} = \frac{\text{Total Fixed Cost}}{\text{Price} - \text{Total Variable Cost/Units}}$$

Breakeven with costs listed

$$BE = \frac{\text{Recording Cost} + \text{Advances}}{\text{Price} - \text{Variable Cost}}$$

Breakeven example

$$BE = \frac{\$250,000 \text{ Recording Costs} + \$75,000 \text{ Advance}}{\$12.04^a - \left(\$6.05^b + \$1.69^c\right)}$$

[a]Card price of SRLP $18.98.
[b]Based on total cost of sales.
[c]Distribution fee.

BE = 75,581 units without marketing costs.

Cost of sales

Cost of sales			
Manufacturing			0.77
Royalties			
Artist: % based on list price		9.00%	1.7082
Mechanical: no. of tracks based on	0.08	10	0.825
			2.53
Other			0.21
TOTAL COST OF SALES			6.05

List price

List price	18.98		
Card price	12.04	Distribution fee	
Distribution fee	1.69	14.0%	

Adding marketing costs changes the outcome of the equation dramatically:

A label has to derive predicted/budgeted marketing costs to add to the equation. Most seasoned labels have an idea as to how much each activity may cost to launch a record. Using the following dollars, check out the break-even analysis:

Marketing costs

Marketing costs	
Account advertising	100,000
Trade advertising	10,000
Consumer/other advertising	10,000
Video production	75,000
Artist promotion (radio)	100,000
Indie promotion/publicity	5,000
Media travel	2,500
Album art/imaging	25,000
No charge records	10,000
Total marketing costs	337,500

Breakeven example with marketing costs

Example:			
BE =	$250,000 recording costs + $75,000 advance + $337,500 mkt costs		
	$12.04[a] – ($6.05[b] + $1.69[c])		
	[a](card price of SRLP $18.98)	[b](based on total cost of sales)	[c](Distr. Fee)

FIGURE 6.1   *Breakeven point in units.*

$$BE = \frac{\$662,500}{\$5.99} =$$

110,601 units just to recover the costs of making and marketing this record. This does not take into account overhead costs and salaries of the employees getting this job done.

Clearly, the job of the profit and loss statement is multifaceted. It can be used as a predictor of success (or not), as well as an evaluation tool of existing projects. Not a part of the equation is the overhead that it takes to operate the business, such as building expenses, salaries, supplies, and so on. But lessons learned from this type of analysis should aid record labels as to the better allocation of funds and resources.

Account advertising
The most expensive marketing cost is the positioning of a record in the retail environment. The pricing and positioning of a record (P&P), meaning a "sale" or reduced price along with prime real estate placement in store, can cost record labels hundreds of thousands of dollars. Other types of account advertising include print advertising via the store's Sunday circulars, end cap positioning, listening stations, event marketing such as artist instore visits, point of purchase materials placement guarantees, and more.

Advertising
Trade advertising is considered an internal promotional activity. Trade advertising creates awareness of a new product to the decision-makers of that industry by using strong imaging of the artist and release along with relevant facts such a sales and radio success, tour information, sales data, etc.

Video production
Negotiated into the recording contract, artists are usually responsible for 50% of the cost of video production.

Artist promotion
Usually covers the costs associated with introducing and promoting the artist to radio. Labels often take artists to visit with radio stations, including on-air interviews, dinners with music programmers, and Listener Appreciation events. The record company incurs this cost and is usually not recoupable.

Independent promotion and publicity
Record companies will hire the services of independent promotion companies and independent publicists. They assist in creating exposure for their artists via additional radio airplay and media. This line item is not recoupable.

Media travel
Record labels will fund the costs associated with travel for an artist who is doing a media event. This is an expense to the label and not recoupable.

Album art
Photographers are hired to shoot cover artwork as well as press images for publicity use. And then a designer is hired to create the overall album art concept, from artist image use, album title treatment, booklet layout, etc. All of these efforts are expenses to the label and not usually recoupable.

No charge records
A marketing tool often used by labels is the actual CD. No Charge Records are those CDs that are used for promotional use such as give-aways on the radio, or instore play copies for record retailers. The value of these records and their use must be accounted for, and is not recoupable.

Contribution to overhead
From the Gross Margin, a label would subtract the marketing costs including Account Advertising, Trade and Consumer Advertising, Video Production, Artist Promotion, Independent Promotion and Publicity, Media Travel, Album Art, and No Charge Records to determine the release's contribution to the record company's overhead.

Note the percentages given within the spreadsheet. The total net sales percentage reflects 100% of money generated by the sale of the project. Each line item subheading also reflects a percentage, causing each department to consider how much is being spent as a reflection of the project as a whole. Although the Contribution to Overhead may be a large number, its percentage of the whole determines how effective and efficient the project was managed.

FIGURE 6.2 *Profit and loss summary.*

Profit and loss statement
8% discount

	Amt	%	Units	Dollars	
			Gross unit		
CD list price	18.98		325.250		
CD card price	12.04				
Std. discount	0.96	8.0%			
			Net units		
Adj. price to accounts	11.08		276.465		
Gross sales				$3,602,729	
Returns provision (after fee)				–$720,526	20.0%
Sales after reserve	11.08			$2,882,183	
Distribution fee	1.33	12.0%		$345,862	12.0%
Net sales after fee	9.75			$2,536,321	
Sales reserve (opp)/risk			16.263	$180,136	5.0%
Gross physical sales				$2,716,458	
Gross digital units/albums	6.60		118.412	$781,519	
Gross digital units/singles	0.66		1,313.816	$867,119	
Gross digital sales				$1,648,638	
Total net sales (physical and digital)				$4,365,096	
Cost of sales					
Manufacturing (per units shipped)	0.77	7.9%		$250,443	9.2%
Royalties (artist/producer and mechanicals)	2.91	29.9%		$1,328,798	48.9%
Other (sticker of product)	0.21	2.2%		$68,303	2.5%
Returns fee (1.75% of returns)				$854	0.0%
Digital mastering prep for online retailers				$1,000	0.1%
Total cost of sales	3.89	39.9%		$1,649,397	60.7%
Gross margin (before recording costs)	5.86	60.1%		$2,715,699	62.2%
Recording costs					
Recording and advances				$625,000	14.3%
Recording recoupment				–$625.000	–14.3%
Net recording cost				$0	0.0%
Gross margin				$2,715,699	
Marketing costs					
Account advertising				$200,000	4.6%
Trade advertising				$75,000	1.7%
Consumer/other advertising				$100,000	2.3%
Video production				$175,000	4.0%
Artist promotion (radio)				$134,000	3.1%
Indie promotion/publicity				$10,000	0.2%
Media travel				$25,000	0.6%
Album art/imaging				$25,000	0.6%
No charge records				$10,000	0.2%
Total marketing costs				$754,000	17.3%
Contribution to overhead (profit)				$1,961,699	44.9%
Gross shipped to date				325,250	
Cumm return to date				16,263	
Return percentage				5.0%	
SoundScan/physical product only				187,366	
Sell of percentage				60.6%	
% mktg cost to gross				17.3%	
% mktg cost to net				27.8%	
Mkg cost per gross unit				$1.70	
Mktg cost per net unit				$1.91	

FIGURE 6.3 *Profit and loss statement with discount applied.*

Profit and loss statement
0% discount

	Amt	%	Units	Dollars	
			Gross unit		
CD list price	18.98		325.250		
CD card price	12.04				
Std. discount	0.96	0.0%			
			Net units		
Adj. price to accounts	12.04		276.465		
Gross sales				$3,916,010	
Returns provision (after fee)				–$783,202	20.0%
Sales after reserve	12.04			$3,132,808	
Distribution fee	1.44	12.0%		$375,937	12.0%
Net sales after fee	10.60			$2,756,871	
Sales reserve (opp)/risk			16.263	$195,801	5.0%
Gross physical sales				$2,952,672	
Gross digital units/albums	6.60		118.412	$781,519	
Gross digital units/singles	0.66		1,313.816	$867,119	
Gross digital sales				$1,648,638	
Total net sales (physical and digital)				$4,601,309	
Cost of sales					
Manufacturing (per units shipped)	0.77	7.3%		$250,443	8.5%
Royalties (artist/producer and mechanicals)	2.91	27.5%		$1,328,798	45.0%
Other (sticker of product)	0.21	2.0%		$68,303	2.3%
Returns fee (1.75% of returns)				$854	0.0%
Digital mastering prep for online retailers				$1,000	0.1%
Total cost of sales	3.89	36.7%		$1,649,397	55.9%
Gross margin (before recording costs)	6.71	63.3%		$2,951,913	64.2%
Recording costs					
Recording and advances				$625,000	13.6%
Recording recoupment				–$625.000	–13.6%
Net recording cost				$0	0.0%
Gross margin				$2,951,913	
Marketing costs					
Account advertising				$200,000	4.3%
Trade advertising				$75,000	1.6%
Consumer/other advertising				$100,000	2.2%
Video production				$175,000	3.8%
Artist promotion (radio)				$134,000	2.9%
Indie promotion/publicity				$10,000	0.2%
Media travel				$25,000	0.5%
Album art/imaging				$25,000	0.5%
No charge records				$10,000	0.2%
Total marketing costs				$754,000	17.3%
Contribution to overhead (profit)				$2,197,913	47.8%
Gross shipped to date				325,250	
Cumm return to date				16,263	
Return percentage				5.0%	
SoundScan/physical product only				187,366	
Sell of percentage				60.6%	
% mktg cost to gross				16.4%	
% mktg cost to net				25.5%	
Mkg cost per gross unit				$1.70	
Mktg cost per net unit				$1.91	

FIGURE 6.4 Profit and loss statement with no discount.

Profit and loss statement

$1.00% Increased royalty

	Amt	%	Units	Dollars	
			Gross unit		
CD list price	18.98		325.250		
CD card price	12.04				
Std. discount	0.96	8.0%			
			Net units		
Adj. price to accounts	11.08		276.465		
Gross sales				$3,602,729	
Returns provision (after fee)				–$720,526	20.0%
Sales after reserve	11.08			$2,882,183	
Distribution fee	1.33	12.0%		$345,862	12.0%
Net sales after fee	9.75			$2,536,321	
Sales reserve (opp)/risk			16.263	$180,136	5.0%
Gross physical sales				$2,716,458	
Gross digital units/albums	6.60		118.412	$781,519	
Gross digital units/singles	0.66		1,313.816	$867,119	
Gross digital sales				$1,648,638	
Total net sales (physical and digital)				$4,365,096	
Cost of sales					
Manufacturing (per units shipped)	0.77	7.9%		$250,443	9.2%
Royalties (artist/producer and mechanicals)	2.91	40.1%		$1,785,430	65.7%
Other (sticker of product)	0.21	2.2%		$68,303	2.6%
Returns fee (1.75% of returns)				$854	0.0%
Digital mastering prep for online retailers				$1,000	0.1%
Total cost of sales	4.89	50.2%		$2,106,028	77.5%
Gross margin (before recording costs)	4.86	49.8%		$2,259,067	51.8%
Recording costs					
Recording and advances				$625,000	14.3%
Recording recoupment				–$625.000	–14.3%
Net recording cost				$0	0.0%
Gross margin				$2,259,067	
Marketing costs					
Account advertising				$200,000	4.6%
Trade advertising				$75,000	1.7%
Consumer/other advertising				$100,000	2.3%
Video production				$175,000	4.0%
Artist promotion (radio)				$134,000	3.1%
Indie promotion/publicity				$10,000	0.2%
Media travel				$25,000	0.6%
Album art/imaging				$25,000	0.6%
No charge records				$10,000	0.2%
Total marketing costs				$754,000	17.3%
Contribution to overhead (profit)				$1,505,067	34.5%
Gross shipped to date				325,250	
Cumm return to date				16,263	
Return percentage				5.0%	
SoundScan/physical product only				187,366	
Sell of percentage				60.6%	
% mktg cost to gross				17.3%	
% mktg cost to net				33.4%	
Mkg cost per gross unit				$1.70	
Mktg cost per net unit				$1.91	

FIGURE 6.5   *Profit and loss statement with royalty increase.*

A LOOK AT REAL P&L STATEMENTS

This statement was based on a real-life scenario where the new album release was only a month old, but the single had been out for 12 weeks. As you can see, the single has sold over 1.3 million units and the album has sold over 275,000 units, with 118,000 of them being digital. This is a strong indicator of a youth buyer, with over 60% of the album sales being digital. Note that the single has generated more revenue than the digital album sales, so far.

Check out the discount at 8%. This applies to physical sales only at this time.

The recording costs were completely recouped, so this line item cancels itself out.

Marketing costs are also shifting as consumers have taken on the role of word-of-mouth marketing via social networks. Not long ago, account advertising for a release this size would have cost nearly twice as much. This is a pop artist—note the expensive video costs and radio promotion costs because more than one format is being worked.

Even though this release is early in its lifecycle, it has made nearly $2 million.

Change one item, and see what happens to the *profit* line. The discount has been reduced to 0%, meaning that retailers did not receive a discount. Those 8% points would mean that this same project would yield nearly $250,000 *more* to the label—if they did *not* discount the product.

Again, change the royalty by increasing by $1.00 to $3.91, and again the bottom line is greatly impacted. This time, the label loses nearly $500,000. The negotiation of every line item matters—down to the penny and the percentage point.

GLOSSARY

advances—Monies fronted to artists to assist them with living expenses.

breakeven point—The level of sales or production at which the total costs and total revenue of a business are equal.

card price—The wholesale price, or dollar figure that is set by the distribution company that sells the product.

click and buy—Allows web visitors to click on a link or object for immediate online purchase.

returns—CDs and other recorded music products that fail to sell at retail and are returned to the distributor for a refund.

sell-off percentage—The percent of inventory shipped to the marketplace that has already been sold.

upsell—A marketing term for the practice of suggesting higher priced products or services to a customer who is considering a purchase.

widget—A portable chunk of code that can be installed and executed within any separate HTML-based web page by an end user without requiring additional editing or knowledge of HTML programming.

CHAPTER 7

SoundScan and the Music Business

Amy Macy and Trudy Lartz
VP Sales and Service, Nielsen SoundScan

TECHNOLOGY AND THE MUSIC BUSINESS

Technological advancements not only launched the music business, but have continually given it a "shot in the arm" just when it needed it most. From stereo to recordable tape to compact discs to digital files, technology continues to evolve the hardware and software of our industry. Not only has technology changed the format of our business, but it has also changed how the actual business gets done. Technological advancements permeate the core of the record industry, creating the scorecards for success.

SOUNDSCAN

Prior to the current charting system, an archaic reporting system based on undocumented sales information was garnered by *Billboard* to produce the sales charts. *Billboard* had a panel of "reporting" retail stores that stated the best-selling record in their store, based on genre. Often, this information was not verified through actual sales data, but was based on what individuals "thought" to be the best seller. Record labels employed retail promotion teams to help influence these reports, hence the sales charts were not always valid depictions of true sales throughout the nation.

One must remember that this reporting structure was prior to the use of bar code scanning systems and the use of point-of-sales data. So, capturing accurate sales data was difficult, even for the retailer. The use of bar codes, or UPC (universal product codes), has greatly affected product management.

CONTENTS

Not only do UPCs assist in inventory status and reorder generation of hot-selling items, the sales information captured allows the retailer to determine the best-selling item by store and by chain, as confirmed through real sales data.

With the introduction of bar codes and efficient computer management of inventory, a new idea was introduced. Mike Shalet, an ex-record label promotion guy, along with Mike Fine, a statistician who had previously worked with major newspapers and magazines with a focus on surveys, conceived a revolutionary concept that would use this new-found technology to derive the top-selling records of the week. And in 1991, SoundScan was born.

SoundScan is an information system that tracks sales of music and music video products throughout the United States and has expanded to international territories as well. Sales data using UPCs from point-of-sale cash registers as well as International Standard Recording Codes (ISRCs) from digital file sales are collected weekly from over 20,000 retail, mass merchant, and nontraditional outlets (online stores, venues, etc.). Weekly data are compiled and made available every Wednesday. Now owned by Nielsen, SoundScan is the sales source for the *Billboard* music album charts (www.soundscan.com).

Although 20,000 retail outlets sounds like a lot of stores, SoundScan does not capture all music retailers. Through analysis, SoundScan knows which retailers it does not capture and, via statistical equations, is able to derive a fairly accurate sales estimate. The record that sells the most earns the number 1 position for the week.[1]

The UPC, the ISRC, and Registration with SoundScan

The UPCs contain a unique sequence of numbers that identifies a product. For music, record companies are designated a 5- to 6-digit number that identifies the label, by the Uniform Code Council. The record company then assigns a 4- to 5-digit product code that identifies the release, including artist and title of the album. The 11th digit designates the configuration of the product: the number "2" designates a CD, "1" is vinyl, and so forth. The last digit is known as the "check" digit. When scanned, a mathematical equation occurs, determining whether the product has been correctly scanned. The "check" digit is the "answer" to that equation, verifying an accurate scan. In the United States, the standard UPC contain 12 digits.

---

[1] www.soundscan.com/about.html

The ISRC (International Standard Recording Code)is the international identification system for sound recordings and music video recordings. Just like the UPC, each ISRC has a unique and permanent sequence of numbers that identifies each specific recording, which can be permanently encoded into a product as its digital fingerprint. Encoded ISRCs provide the means to automatically identify recordings for royalty payments, key to publishers and songwriters alike. Because these numbers are embedded into the product, this coding system is perfect for the electronic distribution of music,

Table 7.1	Other Nielsen services for the entertainment industry

Nielsen BookScan provides weekly point-of-sale data with the highest degree of accuracy and integrity. Functioning as a central clearinghouse for book-industry data, Nielsen BookScan provides comprehensive reports from a wide variety of perspectives.

Nielsen Broadcast Data Systems is the world's leading provider of airplay tracking for the entertainment industry. Employing a patented, digital-pattern-recognition technology, Nielsen BDS captures in excess of 100 million song detections annually on more than 1400 radio stations in 130+ US markets (including Puerto Rico) and 30 Canadian markets.

Nielsen EDI is a global leader in measurement, information, and research solutions for the motion picture industry, providing up-to-the-minute box office information around the globe in real time, on an hourly basis.

Nielsen Entertainment Television and CPG (NETV/CPC). A full-service market research area, specializes in pilot and episode testing of television programs. In addition, NETV provides vast experience in conducting TV spot, concept, and tracking research.

Nielsen Games provides a full spectrum of products/services for game publishers, PC/Console manufacturers, game developers, ad agencies, and advertisers to analyze the gaming habits and media consumption patterns of gamers.

Nielsen MarketNavigator is a robust SQL server-based software platform developed by Nielsen Entertainment. Applications have been designed specifically for home entertainment sales, marketing and distribution executives, analysts, and their retail partners.

Nielsen MobileScan monitors and analyzes weekly mobile entertainment sales.

Nielsen Research Group (NRG) is a market research and strategic consulting company serving the entertainment industry.

Nielsen SoundScan is an information system that tracks sales of music and music video products throughout the United States and Canada, and serves as the sales source for *Billboard* music charts. Sales data from point-of-sale cash registers is collected weekly from 14,000+ retail, mass merchant, and nontraditional (online stores, venues, etc.) outlets.

Nielsen VideoScan seamlessly integrates point-of-sale (POS) data to provide a comprehensive view of the VHS and DVD sell-through business.

Nielsen SpotScan, the most accurate radio advertising data available, provides detailed reports at the market and station level, including exclusive expenditure data for radio, newspaper, and local TV.

Nielsen RingScan monitors ringtones downloaded through mobile devices.

SoundScan International digital track sales across 35 countries outside of the U.S. and Canada.

Nielsen BuzzMetrics measures Internet chatter across millions of blogs and Internet sites, updated hourly. Can also capture positive or negative sentiment as well as verbatim comments.

with the ease of adoption into the international music community that has been reliable and cost-effective.

Unlike the sophisticated algorithm of the UPC, the ISRC is an internal tracking code that is obtained by application through the U.S. ISRC Agency, which acts as a clearinghouse, ensuring no duplication of numbers. The codes are simple, with origin of country, label or artist indentifier, year of creation, and finally song designation number.[2]

A sample code could look like this: US ASM 09 00001

To capture sales data, the record company that has rights to the digital track gives SoundScan the ISRCs or, if no record company, it comes directly from the digital retail provider like iTunes.

www.ifpi.org/content/section_resources/isrc.html

To learn more about UPC and the Uniform Code Council, contact:

Uniform Code Council, Inc.
7887 Washington Village Drive, Suite 300
Dayton, OH 45459
937-435-3870 (telephone)
937-435-7317 (fax)

www.uc-council.org

To learn more about ISRCs and the application process, check out:

International Federation of Phonographic Industries

www.ifpi.org/isrc

Once a product has UPC, the release can be registered with SoundScan so that sales can be captured on a nationwide basis. To register, the following form along with a copy of the release should be sent to SoundScan (see Figure 7.1).

---

[2] www.ifpi.org/content/section_resources/isrc.html

nielsen
SoundScan

ONE NORTH LEXINGTON AVENUE, 14TH FL.
WHITE PLAINS, NY 10601
(914) 684-5500 or (914) 684-5575 (Direct Line)
E-Mail: dbase@soundscan.com

TITLE ADDITION SHEET

To add a title to the SOUNDSCAN database, each field on the title addition sheet must be completed in order for it to be accepted. Please use a separate form for each additional title.

Title: _____ Release Date: _____

Artist: _____

Label information as it applies to this product

Parent Label: _____ Distribution Co. _____

Sub Label: _____ Label Abbr: __ __ __ __

Please enter all digits of the **U.P.C. Code**. (Including **Prefix** and **Suffix**)
To enter identification codes on how your product should be listed please check the example below.

PLEASE PRINT IN ONE CONFIGURATION FOR EACH LINE

U.P.C. Code
EXAMPLE   PRICE   TYPE   Configuration Types

9 9999999999 9   9.99   C
- ---------- -   ____   ____ | ALBUM   SINGLES
- ---------- -   ____   ____ | A = LP 12" ALBUM   E = CD SINGLE
- ---------- -   ____   ____ | B = CASS ALBUM   F = LP 12" SINGLE
- ---------- -   ____   ____ | C = CD ALBUM   G = CASS. SINGLE
                                     | C = DIGITAL ALB   H = DIGITAL SINGLE
- ---------- -   ____   ____ | D = DVD AUDIO   I = MAXI CD SINGLE
                                     |
PLEASE SELECT ONE GENRE WHICH APPLIES| VIDEO
TO THIS PRODUCT | L = DVD

____ 150 - R & B   ____ 520 - SOUNDTRACK
____ 400 - COUNTRY   ____ 184 - WORLD   VIDEO SUBMISSIONS ONLY
____ 300 - JAZZ   ____ 100 - ROCK
____ 186 - LATIN   ____ 620 - COMEDY   ____ 900 MUSICAL PERFORMANCE
____ 102 - METAL   ____ 640 - GOSPEL   ____ 901 SPORTS
____ 200 - CLASSICAL   ____ 630 - CHRISTIAN   ____ 907 MOVIE
____ 152 - RAP   ____ 625 - KARAOKE   ____ 905 OTHER (E.G. EXERCISE, DOCUMENTARY)
____ 180 - REGGAE   ____ 360 - NEW AGE
____ 690 - CHILDREN   ____ 156 - DANCE/ELECTRONIC
____ 470 - BLUES   ____ 178 - SKA

Please enter your name, phone number and fax number in case we have any questions.

Your Name: _____ PH# (_____) -_____ FAX# (_____) -_____
Email: _____

FIGURE 7.1   *SoundScan title addition sheet.*

A LOOK AT THE DATA

The Charts

By compiling and organizing sales data, SoundScan derives many charts that help with analyzing the marketplace. Let's look at a sample of these charts—there are hundreds of charts, subdivided by specific headings. These data are the actual charts that drive the *Billboard* charts.

nielsen
SoundScan

| HOME | TITLE | SUMMARY | CHARTS | MARKETING | PICKUP | SETS | ARCHIVES | ACCOUNT | HELP |

CHART: Top Current Albums   LOAD TO EXCEL   PRINT

Week Ending: 09/16/2007 ▼   Display: % CHG ▼

Weeks On	Label	2W Rank	LW Rank	TW Rank	Artist	Title	TW Sales	% CHG	LW Sales	RTD Sales
1	DEF			1	WEST*KANYE	GRADUATION	956,936	999	785	957,763
1	INT			2	50 CENT	CURTIS	691,304	999	1,047	692,386
1	BNA			3	CHESNEY*KENNY	JUST WHO I AM: POETS & PIRATES	386,685	999	307	387,009
5	DBV	1	1	4	HIGH SCHOOL MUSICAL 2	SOUNDTRACK	133,088	-19	164,690	1,489,515
12	DBV	4	3	5	HANNAH MONTANA 2: MEET MILEY C	SOUNDTRACK	42,320	-10	46,976	1,292,180
52	AAM	5	2	6	FERGIE	DUTCHESS	40,524	-17	48,590	2,679,107
102	ROAD	10	8	7	NICKELBACK	ALL THE RIGHT REASONS	34,940	-4	36,217	5,928,712
9	UNIV	29	14	8	CAILLAT*COLBIE	COCO	31,660	25	25,298	227,168
9	UME	7	7	9	VARIOUS	NOW 25	31,612	-16	37,846	812,952
53	JIVE	32	12	10	TIMBERLAKE*JUSTIN	FUTURESEX/LOVE SOUNDS	30,968	14	27,194	3,603,193
18	WAR	11	10	11	LINKIN PARK	MINUTES TO MIDNIGHT	29,614	4	28,406	1,687,705
3	REUN	2	4	12	CASTING CROWNS	ALTAR & THE DOOR	28,791	-29	40,501	70,470
24	INT	35	26	13	TIMBALAND	TIMBALAND PRESENTS SHOCK VALUE	27,375	43	19,097	641,959
10	NWLR	6	6	14	HAIRSPRAY	SOUNDTRACK	27,000	-30	38,691	586,947
17	OCAM	19	17	15	MAROON 5	IT WON'T BE SOON BEFORE LONG	26,850	13	23,736	1,244,614
1	HOL			16	GREY'S ANATOMY VOL. 3	SOUNDTRACK	26,560	999	52	26,615
7	GEFN	21	20	17	COMMON	FINDING FOREVER	26,423	26	21,021	353,350
15	DEF	24	21	18	RIHANNA	GOOD GIRL GONE BAD	26,326	28	20,525	662,136
11	ATLG	12	15	19	T.I.	T.I. VS TIP	25,704	2	25,164	1,053,146
27	UNIV	13	9	20	WINEHOUSE*AMY	BACK TO BLACK	25,149	-13	28,930	1,178,291
43	RCA	30	22	21	DAUGHTRY	DAUGHTRY	22,815	16	19,669	3,105,617
1	MCAN			22	YEARWOOD*TRISHA	GREATEST HITS	22,290	999	50	22,344
47	BGMA	15	13	23	SWIFT*TAYLOR	TAYLOR SWIFT	21,943	-14	25,506	1,292,595
6	ATLG	22	18	24	PLIES	REAL TESTAMENT	20,658	-11	23,157	228,339
6	HOL	14	16	25	JONAS BROTHERS	JONAS BROTHERS	20,613	-14	24,011	216,009
88	DBV	17	19	26	HIGH SCHOOL MUSICAL	SOUNDTRACK	19,811	-11	22,381	4,195,324
1	BADB			27	BS	DON'T TALK JUST LISTEN	18,534	999	20	18,556
4	WAR	20	39	28	KWELI*TALIB	EARDRUM	18,165	14	15,962	118,522
20	WAR	25	27	29	BUBLE*MICHAEL	CALL ME IRRESPONSIBLE	18,052	-3	18,581	944,348
3	BADB	3	11	30	YUNG JOC	HUSTLENOMICS	17,442	-37	27,521	114,155
22	RCA	36	30	31	LAVIGNE*AVRIL	BEST DAMN THING	17,120	-7	18,319	1,152,778
15	JIVE	33	34	32	T-PAIN	EPIPHANY	16,769	-3	17,287	650,660
13	ARNV	37	35	33	PAISLEY*BRAD	5TH GEAR	16,196	-4	16,931	584,606
25	CURB	31	37	34	MCGRAW*TIM	LET IT GO	16,116	-2	16,491	1,092,928
14	FUER	27	23	35	PARAMORE	RIOT!	16,042	-18	19,505	274,838
13	ISL	34	28	36	BON JOVI	LOST HIGHWAY	15,471	-16	18,506	802,078
44	KUSU	23	33	37	AKON	KONVICTED	15,187	-14	17,727	2,658,461
22	HOL	28	31	38	PLAIN WHITE T'S	EVERY SECOND COUNTS	15,113	-15	17,813	518,600
7	EPIC	26	32	39	KINGSTON*SEAN	SEAN KINGSTON	14,992	-16	17,779	219,173
6	JIVE	16	25	40	UGK	UNDERGROUND KINGZ	14,899	-23	19,287	316,749
47	DBV	41	40	41	HANNAH MONTANA	SOUNDTRACK	14,347	-8	15,636	2,864,283
2	EVR		5	42	CHIODOS	BONE PALACE BALLET	14,309	-63	38,696	53,276
3	VRGN	9	24	43	HARPER*BEN & INNOCENT CRIMINAL	LIFELINE	14,197	-27	19,333	74,847
20	INT	107	95	44	FEIST	REMINDER	13,996	120	6,370	215,693
3	CULH	18	38	45	LOVETT*LYLE	IT'S NOT BIG IT'S LARGE	13,738	-15	16,182	55,312
45	MERN	59	45	46	SUGARLAND	ENJOY THE RIDE	13,527	8	12,494	1,231,024
38	LAF	48	43	47	PINK	I'M NOT DEAD	12,852	-2	13,085	977,301
96	ARI	46	44	48	UNDERWOOD*CARRIE	SOME HEARTS	12,113	-7	13,019	5,919,215
1	INT			49	ACROSS THE UNIVERSE (DLX)	SOUNDTRACK	12,078	0		12,078
14	SWDG	52	47	50	KEITH*TOBY	BIG DOG DADDY	11,942	1	11,813	527,051

| 1 | 2 | 3 | 4 | 5 | 6 | 7 | 8 | 9 | 10 | 11 |

This is a sample of the Top 200 Albums chart. Each column is identified by number:

1 – Number of weeks the album has been on the Top 200 Chart

2 – Record label

3 – Chart position 2 weeks ago

4 – Chart position last week

5 – Current chart position

6 – Artist name

7 – Album title

8 – Number of units sold this week

9 – Percentage of change from last week to this week

10 – Number of units sold last week

11 – Total number of units sold since the release of the album

FIGURE 7.2   *Example of a Top Current Albums (200) chart. (Source: Nielsen SoundScan)*

nielsen
SoundScan

| HOME | TITLE | SUMMARY | CHARTS | MARKETING | PICKUP | SETS | ARCHIVES | ACCOUNT | HELP |

CHART: 2007 YTD ALBUMS | LOAD TO EXCEL | PRINT

Week Ending: 09/16/2007 ▼

Rank	Label	Artist	Title	YTD Sales
1	RCA	DAUGHTRY	DAUGHTRY	1,984,584
2	WAR	LINKIN PARK	MINUTES TO MIDNIGHT	1,687,705
3	KUSU	AKON	KONVICTED	1,500,430
4	DBV	HIGH SCHOOL MUSICAL 2	SOUNDTRACK	1,489,515
5	AAM	FERGIE	DUTCHESS	1,472,446
6	BNTE	JONES*NORAH	NOT TOO LATE	1,438,081
7	DBV	HANNAH MONTANA 2: MEET MILEY C	SOUNDTRACK	1,292,180
8	ROAD	NICKELBACK	ALL THE RIGHT REASONS	1,283,356
9	ARI	UNDERWOOD*CARRIE	SOME HEARTS	1,266,576
10	OCAM	MAROON 5	IT WON'T BE SOON BEFORE LONG	1,244,614
11	JIVE	TIMBERLAKE*JUSTIN	FUTURESEX/LOVE SOUNDS	1,226,066
12	UNIV	WINEHOUSE*AMY	BACK TO BLACK	1,178,291
13	INT	THICKE*ROBIN	EVOLUTION OF ROBIN THICKE	1,167,404
14	RCA	LAVIGNE*AVRIL	BEST DAMN THING	1,152,778
15	CURB	MCGRAW*TIM	LET IT GO	1,092,928
16	ISL	FALL OUT BOY	INFINITY ON HIGH	1,069,077
17	ATLG	T.I.	T.I. VS TIP	1,053,146
18	CAP	VARIOUS	NOW 24	1,020,985
19	BGMA	SWIFT*TAYLOR	TAYLOR SWIFT	1,000,160
20	DEF	WEST*KANYE	GRADUATION	957,763
21	WAR	BUBLE*MICHAEL	CALL ME IRRESPONSIBLE	944,348
22	CAP	RAE*CORINNE BAILEY	CORINNE BAILEY RAE	925,440
23	DBV	HANNAH MONTANA	SOUNDTRACK	876,602
24	INT	STEFANI*GWEN	SWEET ESCAPE	875,596
25	COL	BEYONCE	B'DAY	862,600
26	JIVE	KELLY*R.	DOUBLE UP	855,265
27	LYST	RASCAL FLATTS	ME AND MY GANG	830,849
28	GEFN	FURTADO*NELLY	LOOSE	817,476
29	UME	VARIOUS	NOW 25	812,952
30	ISL	BON JOVI	LOST HIGHWAY	802,078
31	DEF	NE-YO	BECAUSE OF YOU	794,109
32	INT	50 CENT	CURTIS	692,386
33	COMW	DREAMGIRLS	SOUNDTRACK	678,006
34	UNIV	HINDER	EXTREME BEHAVIOR	674,581
35	COL	MAYER*JOHN	CONTINUUM	670,810
36	DEF	RIHANNA	GOOD GIRL GONE BAD	662,136
37	RCA	CLARKSON*KELLY	MY DECEMBER	650,886
38	JIVE	T-PAIN	EPIPHANY	650,660
39	INT	TIMBALAND	TIMBALAND PRESENTS SHOCK VALUE	641,959
40	WAR	WHITE STRIPES	ICKY THUMP	587,030
41	NWLR	HAIRSPRAY	SOUNDTRACK	586,947
42	ARNV	PAISLEY*BRAD	5TH GEAR	584,606
43	SBSM	VARIOUS	2007 GRAMMY NOMINEES	567,011
44	HRMS	MCCARTNEY*PAUL	MEMORY ALMOST FULL	528,972
45	SWDG	KEITH*TOBY	BIG DOG DADDY	527,051
46	VRGN	STONE*JOSS	INTRODUCING JOSS STONE	510,517
47	ATLG	MUSIQ SOULCHILD	LUVANMUSIQ	507,037
48	UNIV	LLOYD	STREET LOVE	506,013
49	EPIC	FRAY	HOW TO SAVE A LIFE	503,123
50	DEF	YOUNG JEEZY	INSPIRATION: THUG MOTIVATION	498,714

| 1 | 2 | 3 | 4 | 5 |

This is a sample of the Year-to-Date (YTD) Top 200 Chart. Each column is identified by number:

1 – Rank in sales for the year 2007   3 – Artist   5 – Number of units sold in the year 2007
2 – Label   4 – Title of album

FIGURE 7.3   Example of a Year-to-Date Album chart. (Source: Nielsen SoundScan)

Weeks On	Label	2W Rank	LW Rank	TW Rank	Artist	Title	TW Sales	% CHG	LW Sales	RTD Sales
10	DBV	1	1	1	HIGH SCHOOL MUSICAL 2 CAST	WHAT TIME IS IT	6,265	-28	8,667	308,283
5	CAP	2	3	2	J. HOLIDAY	BED - SINGLE	933	1	925	6,841
4	FCFC	3	7	3	TMI BOYZ	SWERVING	746	93	387	1,932
3	JIVE	59		4	TIMBERLAKE*JUSTIN	LOVESTONED/I THINK SHE KNOWS R	603	999	31	762
1	GEFN			5	NASH*KATE	FOUNDATIONS	586	-31	844	1,430
16	FMFR	8	11	6	CHAOS (CCSERVA)	I GET IT IN	427	34	319	4,162
4	DOMI	23	25	7	ANIMAL COLLECTIVE	PEACEBONE	404	156	158	918
10	TBDM	4	25	8	BETTINA	SHE IS	376	66	226	2,226
17	SPAR	6	6	9	MANDISA	ONLY THE WORLD THE SINGLE	371	-8	402	9,162
19	COL	12	10	10	BEYONCE/SHAKIRA	BEAUTIFUL LIAR	348	6	328	30,065
11	COL	7	5	11	BEYONCE	GET ME BODIED	346	-16	413	11,241
3	WAR	10	4	12	NICKS*STEVIE	STAND BACK	329	-38	527	4,498
4	FCFC	19	37	13	KELZ	YE AIN'T BOUT THAT	322	204	106	839
1	TCRE			14	NIGHTMARE OF YOU	BANG	312	999	14	326
5	FCFC	5	8	15	BIGG FACE	BACK DOWN	311	-18	380	2,433
28	KGMM	10	13	16	UNDERWOOD*TEMAR	INSIDE OUT	303	14	265	7,112
6	INT	13	9	17	50 CENT	AYO TECHNOLOGY	291	-15	341	1,570
26	CDBY	16	18	18	A.G.	LET ME SEE SOMETHING	273	16	235	9,126
9	SUBP	14	15	19	IRON & WINE	BOY WITH A COIN	252	4	242	3,818
17	ELES	15	16	20	AYANNA	OOH WEE SINGLE	242	2	238	4,285
4	FCFC	9	12	20	SMOOT	SAVE A LIFE	242	-21	305	1,261
10	YGEG	54	14	22	SKULL	BOOM DI BOOM DI	224	-14	259	1,713
17	UNIV	17	17	23	SHOP BOYZ	PARTY LIKE A ROC	209	-12	237	8,538
11	ATLG	21	19	24	PLIES	SHAWTY	205	-10	227	2,731
2	AVE		21	25	KUZ	BOSS STATUS	197	-10	218	415
53	MLS	33	37	26	MINDLESS SELF INDULGENCE	SHUT ME UP (REMIXES +3)	184	74	106	11,037
7	MUT	18	22	27	VAN DYK*PAUL	WHITE LIES	180	-10	199	1,572
3	INT	62	33	28	WILL.I.AM	I GOT IT FROM MY...	171	45	118	359
2	FCFC		86	29	LATIMORE*JACOB	SUPER STAR	166	286	43	257
4	BUEY	47	68	29	REEC	LE'S GO	166	186	58	433
137	SUBP	37	27	31	POSTAL SERVICE	WE WILL BECOME SILHOUETTES	162	13	144	66,907
8	CDBY	34	24	32	M.E.PAIGE	IN THIS PLACE	161	0	161	1,467
80	MLS	24	26	33	MINDLESS SELF INDULGENCE	STRAIGHT TO VIDEO: REMIXES	136	-6	145	19,062
15	INT	28	29	34	PIPETTES	YOUR KISSES ARE WASTED ON ME	130	-3	134	3,434
3	CDBY	92	77	35	EL GRECO FEAT TERRAH	GET UP ON IT	112	120	51	203
31	UNIV	70		35	MR. C THE SLIDE MAN	CHA CHA SLIDE	112	211	36	32,870
13	ASTW	40	52	37	CHEMICAL BROTHERS	DO IT AGAIN	110	51	73	2,654
5	CDBY	31	60	38	STINGEE FEAT DJ DRAMA	U KNOW U WANT DAT	105	72	61	440
26	COL	25	29	38	ROWLAND*KELLY	LIKE THIS	105	-22	134	11,963
9	XLIN	35	48	40	M.I.A.	BOYZ EP	103	30	79	1,930
15	BADB	37	43	40	YUNG JOC	COFFEE SHOP - SINGLE	103	5	98	1,842
76	INT	36	32	42	NINE INCH NAILS	EVERYDAY IS EXACTLY THE SAME R	101	-18	123	62,041
2	DFER		2	42	CRUXSHADOWS	BIRTHDAY	101	-90	1,044	1,147
31	COMW	29	31	44	BEYONCE	LISTEN	98	-21	124	17,587
30	BUEY	39	35	45	REEC	GET TO THE MONEY	93	-20	116	5,655
34	CDBY	22	40	46	CRYSTAL DOVE/DICE GAMBLE	UNTIL YOU COME BACK TO ME	91	-13	104	13,108
1	GEF			47	FURTADO*NELLY	DO IT	89	536	14	158
10	STRE	49	34	48	J-MIZZ	STRONG ARM	88	-25	117	1,440
14	ASWB	42	44	48	JONES*MIKE	MY 64	88	-2	90	1,879
1	MWM			50	TRIN-I-TEE 5:7	LISTEN	86	562	13	1,071
15	INT	26	49	50	THE MIDWAY STATE	MET A MAN ON TOP OF THE HILL	86	13	76	1,641

1	2	3	4	5	6	7	8	9	10	11

This is a sample of the Top 200 singles chart. Each column is identified by number:

1 – Number of weeks the single has been on the Top 200 chart
2 – Record label
3 – Chart position 2 weeks ago

4 – Chart position last week
5 – Current chart position
6 – Artist name
7 – Single title

8 – Number of units sold this week
9 – Number of units sold last week
10 – Percentage of change from this week to last week
11 – Total number of units sold since the release of the single

FIGURE 7.4 *Example of a Top Singles chart (Source: Nielsen SoundScan)*

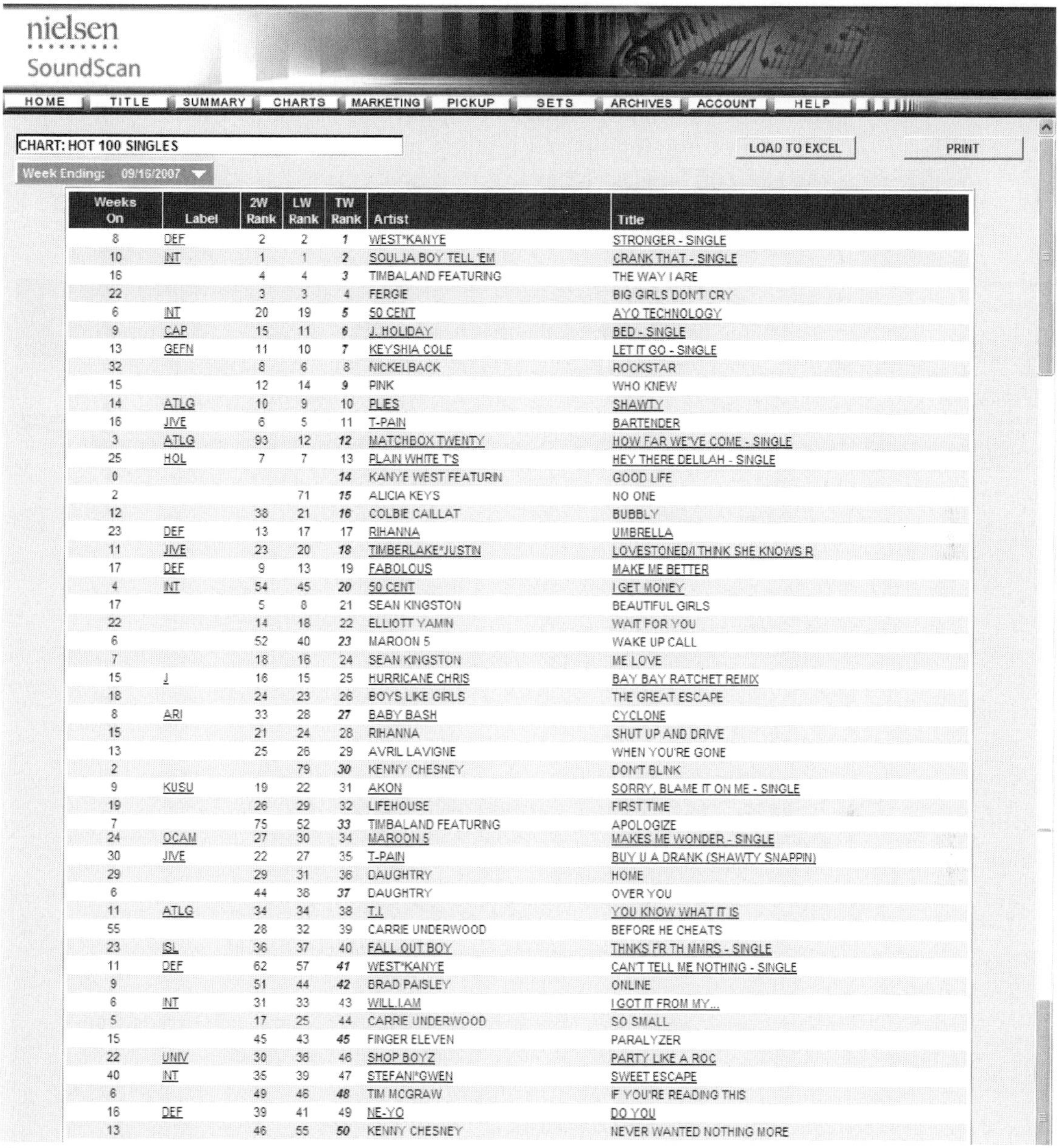

Weeks On	Label	2W Rank	LW Rank	TW Rank	Artist	Title
8	DEF	2	2	1	WEST*KANYE	STRONGER - SINGLE
10	INT	1	1	2	SOULJA BOY TELL 'EM	CRANK THAT - SINGLE
16		4	4	3	TIMBALAND FEATURING	THE WAY I ARE
22		3	3	4	FERGIE	BIG GIRLS DON'T CRY
6	INT	20	19	5	50 CENT	AYO TECHNOLOGY
9	CAP	15	11	6	J. HOLIDAY	BED - SINGLE
13	GEFN	11	10	7	KEYSHIA COLE	LET IT GO - SINGLE
32		8	6	8	NICKELBACK	ROCKSTAR
15		12	14	9	PINK	WHO KNEW
14	ATLG	10	9	10	PLIES	SHAWTY
16	JIVE	6	5	11	T-PAIN	BARTENDER
3	ATLG	93	12	12	MATCHBOX TWENTY	HOW FAR WE'VE COME - SINGLE
25	HOL	7	7	13	PLAIN WHITE T'S	HEY THERE DELILAH - SINGLE
0				14	KANYE WEST FEATURIN	GOOD LIFE
2			71	15	ALICIA KEYS	NO ONE
12		38	21	16	COLBIE CAILLAT	BUBBLY
23	DEF	13	17	17	RIHANNA	UMBRELLA
11	JIVE	23	20	18	TIMBERLAKE*JUSTIN	LOVESTONED/I THINK SHE KNOWS R
17	DEF	9	13	19	FABOLOUS	MAKE ME BETTER
4	INT	54	45	20	50 CENT	I GET MONEY
17		5	8	21	SEAN KINGSTON	BEAUTIFUL GIRLS
22		14	18	22	ELLIOTT YAMIN	WAIT FOR YOU
6		52	40	23	MAROON 5	WAKE UP CALL
7		18	16	24	SEAN KINGSTON	ME LOVE
15	J	16	15	25	HURRICANE CHRIS	BAY BAY RATCHET REMIX
18		24	23	26	BOYS LIKE GIRLS	THE GREAT ESCAPE
8	ARI	33	28	27	BABY BASH	CYCLONE
15		21	24	28	RIHANNA	SHUT UP AND DRIVE
13		25	26	29	AVRIL LAVIGNE	WHEN YOU'RE GONE
2			79	30	KENNY CHESNEY	DON'T BLINK
9	KUSU	19	22	31	AKON	SORRY, BLAME IT ON ME - SINGLE
19		26	29	32	LIFEHOUSE	FIRST TIME
7		75	52	33	TIMBALAND FEATURING	APOLOGIZE
24	OCAM	27	30	34	MAROON 5	MAKES ME WONDER - SINGLE
30	JIVE	22	27	35	T-PAIN	BUY U A DRANK (SHAWTY SNAPPIN)
29		29	31	36	DAUGHTRY	HOME
6		44	38	37	DAUGHTRY	OVER YOU
11	ATLG	34	34	38	T.I.	YOU KNOW WHAT IT IS
55		28	32	39	CARRIE UNDERWOOD	BEFORE HE CHEATS
23	ISL	36	37	40	FALL OUT BOY	THNKS FR TH MMRS - SINGLE
11	DEF	62	57	41	WEST*KANYE	CAN'T TELL ME NOTHING - SINGLE
9		51	44	42	BRAD PAISLEY	ONLINE
6	INT	31	33	43	WILL.I.AM	I GOT IT FROM MY...
5		17	25	44	CARRIE UNDERWOOD	SO SMALL
15		45	43	45	FINGER ELEVEN	PARALYZER
22	UNIV	30	36	46	SHOP BOYZ	PARTY LIKE A ROC
40	INT	35	39	47	STEFANI*GWEN	SWEET ESCAPE
6		49	46	48	TIM MCGRAW	IF YOU'RE READING THIS
16	DEF	39	41	49	NE-YO	DO YOU
13		46	55	50	KENNY CHESNEY	NEVER WANTED NOTHING MORE

1	2	4	3	5	6	7

This is a sample of the Top 100 Albums chart. Each column is identified by number:

1 – Number of weeks the single has been on the Hot 100 chart
2 – Record label
3 – Chart position 2 weeks ago
4 – Chart position last week
5 – Current chart position
6 – Artist
7 – Song title

FIGURE 7.5 *Example of Hot 100 Singles chart. (Source: Nielsen SoundScan)*

The Hot 100 Singles chart is a combination of single sales and radio airplay, as calculated by *Billboard*. Utilizing the sales of downloaded tracks combined with the sales of the available singles at retail, *Billboard* spins the data with audience information captured by BDS that listens to all formats of radio—from Top 40 and hip-hop to country, Latin, and rock—to create the Hot 100 Singles chart.

nielsen SoundScan

HOME | TITLE | SUMMARY | CHARTS | MARKETING | PICKUP | SETS | ARCHIVES | ACCOUNT | HELP

CHART: Hot Digital Tracks   LOAD TO EXCEL   PRINT

Week Ending: 12/30/2007   Display: % CHG

Weeks On	Label	2W Rank	LW Rank	TW Rank	Artist	Title	TW Sales	% CHG	LW Sales	RTD Sales
9	ATLG	1	1	1	FLO RIDA	LOW (FEAT T-PAIN) (ALBUM VERS)	467,149	163	177,518	1,551,766
11	JIVE	4	5	2	BROWN*CHRIS	KISS KISS	277,144	233	83,225	1,454,485
24	INT	2	3	3	TIMBALAND FEAT. ONEREPUBLIC	APOLOGIZE	272,211	179	97,532	2,113,219
11	EPIC	5	2	4	BAREILLES*SARA	LOVE SONG(ALBUM VERSION)	261,658	123	117,500	612,000
16	J	3	4	5	KEYS*ALICIA	NO ONE (RADIO EDIT)	255,145	171	94,191	1,654,768
25	INT	10	11	8	SOULJA BOY TELL'EM	CRANK THAT (SOULJA BOY)	244,589	330	58,851	2,713,920
35	WNDU	7	6	7	FINGER ELEVEN	PARALYZER	240,490	190	83,013	1,351,746
24	UNIV	9	10	8	CAILLAT*COLBIE	BUBBLY	220,525	214	70,121	1,731,242
14	JIVE	8	9	9	SPARKS*JORDIN	TATTOO(MAIN VERSION)	219,837	200	73,350	859,883
13	AAM	6	7	10	FERGIE	CLUMSY	211,709	171	78,245	987,277
20	COL	13	8	11	JEAN*WYCLEF FEATURING AKON, LI	SWEETEST GIRL (DOLLAR BILL)(AL	187,896	153	74,203	712,130
23	ARI	15	17	12	BABY BASH FEATURING T-PAIN	CYCLONE	155,437	229	47,285	1,049,082
11	EPIC	12	12	13	KINGSTON*SEAN	TAKE YOU THERE	148,141	180	52,922	483,051
41	ROAD	27	27	14	NICKELBACK	ROCKSTAR	131,178	302	32,593	1,912,985
24	EPIC	17	18	15	GOOD CHARLOTTE	DANCE FLOOR ANTHEM	129,726	184	45,756	728,517
13	EPIC	14	15	16	BEDINGFIELD*NATASHA FEATURING	LOVE LIKE THIS	128,483	167	48,026	505,543
8	JIVE	16	14	17	BROWN*CHRIS	WITH YOU	123,909	152	49,168	289,884
31	BGMA	25	23	18	SWIFT*TAYLOR	OUR SONG	120,777	243	35,182	666,986
17	ATLG	21	22	19	MATCHBOX TWENTY	HOW FAR WE'VE COME (RADIO VERS	119,558	227	36,550	1,122,310
9	JIVE	19	13	20	SPEARS*BRITNEY	PIECE OF ME	114,869	131	49,824	362,192
48	BGMA	29	29	21	SWIFT*TAYLOR	TEARDROPS ON MY GUITAR	105,631	233	31,700	952,886
10	WAR	20	20	22	LINKIN PARK	SHADOW OF THE DAY (ALBUM VERS)	103,829	151	41,302	331,241
26	FUER	31	31	23	PARAMORE	MISERY BUSINESS (ALBUM VERSION)	103,647	238	30,632	681,326
16	DEF	22	28	24	RIHANNA	HATE THAT I LOVE YOU	102,213	222	31,776	760,534
14	ARI	18	21	25	SANTANA FEATURING CHAD KROEGER	INTO THE NIGHT(ALBUM VERSION)	98,587	145	40,280	489,902
21	HOL	64	52	26	JONAS BROTHERS	S.O.S.	96,223	376	20,646	694,545
37	AAM	36	40	27	FERGIE	BIG GIRLS DON'T CRY	96,904	277	25,724	2,452,888
23	DEF	30	38	28	WEST*KANYE	STRONGER	88,735	232	26,690	1,788,232
18	DEF	32	25	29	RIHANNA	DON'T STOP THE MUSIC	87,990	158	34,126	301,941
22	RCA	35	43	30	DAUGHTRY	OVER YOU	85,531	247	24,628	704,714
16	DEF	33	37	31	WEST*KANYE FEAT. T-PAIN	GOOD LIFE	84,929	216	26,882	857,792
4	COL	58	35	32	BOYS LIKE GIRLS	HERO/HEROINE(TOM LORD-ALGE MIX	84,211	203	27,754	135,676
33	INT	28	33	33	TIMBALAND	WAY I ARE	82,978	185	29,082	1,869,785
22	OCAM	39	39	34	MAROON 5	WAKE UP CALL	79,247	208	25,735	700,974
30	DEF	56	46	35	RIHANNA	UMBRELLA	79,217	252	22,503	1,617,765
15	ATLG	41	41	36	PLIES FEATURING AKON	HYPNOTIZED (FEATURING AKON) (E	78,879	212	25,296	383,394
9	DBV	47	36	37	CYRUS*MILEY	SEE YOU AGAIN	77,997	185	27,402	223,547
11	FUER	48	42	38	PARAMORE	CRUSHCRUSHCRUSH (ALBUM VERSION	78,297	203	25,202	243,383
43	FEAR	66	58	39	PLAIN WHITE T'S	HEY THERE DELILAH	75,528	293	19,220	2,006,743
8	EVSM	40	30	40	BUCKCHERRY	SORRY (ALBUM VERSION)	72,985	138	30,679	222,105
39	WAR	72	56	41	LINKIN PARK	WHAT I'VE DONE (ALBUM VERSION)	72,026	265	19,740	1,370,562

1 – Number of weeks on the Hot Digital Tracks chart
2 – Label
3 – Rank two weeks ago
4 – Rank last week
5 – Rank this week
6 – Artist
7 – Title
8 – This week's sales numbers
9 – Percent change in sales from previous week
10 – Last week's sales numbers
11 – Sales numbers since the release of the track

FIGURE 7.6   *Example of Hot Digital Tracks chart. (Source: Nielsen SoundScan)*

The Hot Digital Tracks report acts as a barometer as to the next big single or the next up-and-coming act that will break through onto the marketplace. Labels are retooling their business models by looking at the single (and this chart) as a money-making entity, with the leveraging of ringtones, ringbacks, commercial uses, and the larger sales volume that legal downloading has afforded.

The Digital Album chart is an exclusive album chart dedicated to downloaded sales of full-length purchases.

nielsen
SoundScan

| HOME | TITLE | SUMMARY | CHARTS | MARKETING | PICKUP | SETS | ARCHIVES | ACCOUNT | HELP | |

CHART: Digital Albums

LOAD TO EXCEL   PRINT

Week Ending: 12/30/2007 ▼   Display: % CHG   ▼

Weeks On	Label	2W Rank	LW Rank	TW Rank	Artist	Title	TW Sales	% CHG	LW Sales	RTD Sales
3	RHI	11	8	1	JUNO	SOUNDTRACK	25,555	270	6,910	37,197
3	RAZ	10	5	2	ALVIN & THE CHIPMUNKS	SOUNDTRACK	18,686	87	9,974	35,709
8	EPIC	6	6	3	BAREILLES*SARA	LITTLE VOICE	16,028	81	8,847	100,719
2	WAR		7	4	SWEENY TODD THE DEMON BARBER O	SOUNDTRACK	13,447	80	7,457	20,904
2	GEFN		1	5	BLIGE*MARY J.	GROWING PAINS	11,945	-49	23,509	35,454
6	INT	4	10	6	ONEREPUBLIC	DREAMING OUT LOUD	11,792	75	6,723	62,092
2	ATLG		2	7	FIASCO*LUPE	LUPE FIASCO'S THE COOL	11,149	-38	16,006	29,155
2	OCAM		4	8	MAROON 5	B-SIDE COLLECTION	11,100	11	10,009	21,109
7	J	5	11	9	KEYS*ALICIA	AS I AM	10,023	51	6,654	125,864
3	DBV	32	23	10	HIGH SCHOOL MUSICAL 2	SOUNDTRACK	9,457	204	3,112	175,684
29	FUER	27	22	11	PARAMORE	RIOT!	8,712	178	3,136	88,314
24	UNIV	15	18	12	CAILLAT*COLBIE	COCO	8,049	99	4,038	207,052
2	COL		17	13	ONCE	SOUNDTRACK	7,812	98	3,945	69,392
7	ATLG	8	20	14	LED ZEPPELIN	MOTHERSHIP	6,989	89	3,689	66,617
2	DBV		28	15	HANNAH MONTANA 2: MEET MILEY C	SOUNDTRACK	6,914	165	2,607	72,868
12	WAR	1	3	16	GROBAN*JOSH	NOEL	6,822	-56	15,551	113,072
33	INT	12	14	17	FEIST	REMINDER	6,685	44	4,643	119,526
14	DEF		48	18	WEST*KANYE	GRADUATION	6,059	205	1,984	237,487
1	NWLR			19	HAIRSPRAY	SOUNDTRACK	6,046	233	1,815	136,312
1	OCAM			20	FLYLEAF	FLYLEAF	6,006	270	1,624	49,925
1	OCAM			21	MAROON 5	IT WON'T BE SOON BEFORE LONG	5,878	235	1,755	252,345
6	BGMA	46	40	22	SWIFT*TAYLOR	TAYLOR SWIFT	5,748	165	2,173	84,224
12	RCA	36	29	23	FOO FIGHTERS	ECHOES SILENCE PATIENCE & GRAC	5,736	124	2,563	90,694
10	ARI	45	41	24	UNDERWOOD*CARRIE	CARNIVAL RIDE	5,498	154	2,161	92,684
1	ATLG			25	P.S. I LOVE YOU	SOUNDTRACK	5,413	280	1,426	8,530
6	DBV	26	38	26	ENCHANTED	SOUNDTRACK	5,084	128	2,233	30,175
3	WAR	49	49	27	LINKIN PARK	MINUTES TO MIDNIGHT	5,035	154	1,983	189,002
1	HOL			28	JONAS BROTHERS	JONAS BROTHERS	4,965	185	1,742	46,856
6	OCCR	39		29	COOK*DANE	ROUGH AROUND THE EDGES-LIVE FR	4,928	166	1,856	44,174
9	JIVE	30	27	30	SPEARS*BRITNEY	BLACKOUT	4,733	75	2,698	95,453
28	INT	28	32	31	TIMBALAND	TIMBALAND PRESENTS SHOCK VALUE	4,622	92	2,406	150,611
1	RCA			32	DAUGHTRY	DAUGHTRY	4,581	154	1,807	263,197
2	COL			33	BOYS LIKE GIRLS	BOYS LIKE GIRLS	4,510	234	1,350	86,773
4	UNIV	29	24	34	WINEHOUSE*AMY	BACK TO BLACK	4,422	49	2,962	241,453
8	JIVE	41	44	35	BROWN*CHRIS	EXCLUSIVE	4,363	111	2,064	32,238
1	DBV			36	HANNAH MONTANA	SOUNDTRACK	4,126	160	1,587	71,012
10	ROUN	34	25	37	PLANT/KRAUSS	RAISING SAND	4,061	42	2,850	48,361
2	INT		39	38	U2	U218 SINGLES	4,051	88	2,176	58,482
1	AAM			39	FERGIE	DUTCHESS	4,020	124	1,798	128,342
2	ATLG		9	40	JAHEIM	MAKINGS OF A MAN	3,704	-48	6,875	10,579
1	JIVE			41	TIMBERLAKE*JUSTIN	FUTURESEX/LOVE SOUNDS	3,643	158	1,411	278,563

FIGURE 7.7   *Example of Digital Albums chart, (Source: Nielsen SoundScan)*

Archives

SoundScan archives the previous years' sales data, going back through 1994. Figure 7.9 is a YTD albums report that shows week-ending sales of January 5, 2003.

136 CHAPTER 7: SoundScan and the Music Business

Copyright © 2007 Nielsen SoundScan, a division of Nielsen Entertainment, LLC. All rights reserved.

SoundScan can reduce the sales data to create charts within specific DMAs. This chart reflects sales in the New York DMA, stating sales of the best-selling albums for the week ending Sunday, December 30, 2007, including a full week of sales:

1 – Weeks on chart
2 – Record label
3 – Rank 2 weeks prior
4 – Rank last week

5 – Current ranking
6 – Artist
7 – Album title
8 – Sales this week

9 – % change from previous week
10 – Sales last week
11 – Sales since release in specific DMA

FIGURE 7.8 Example of Top Albums DMA chart for New York City. (Source: Nielsen SoundScan)

```
CHART: 2003 YTD ALBUMS                          Page   1 of 4
                                                Week Ending 01/05/03
                                                2003 YTD
          Label  Artist                Title              Sales(est)
          -----  ------                -----              ----------
    1)    INT    8 MILE                SOUNDTRACK              119284
    2)    BNTE   JONES*NORAH           COME AWAY WITH ME       118354
    3)    ARI    LAVIGNE*AVRIL         LET GO                  113091
    4)    COL    DIXIE CHICKS          HOME                     96994
    5)    RCA    AGUILERA*CHRISTINA    STRIPPED                 95390
    6)    EPIC   LOPEZ*JENNIFER        THIS IS ME...THEN        94856
    7)    JIVE   TIMBERLAKE*JUSTIN     JUSTIFIED                86270
    8)    MERN   TWAIN*SHANIA          UP!                      83064
    9)    UNBG   AALIYAH               I CARE 4 U               80173
   10)    UME    VARIOUS               NOW 11                   78001
   11)    EEG    ELLIOTT*MISSY MISDEMEANOR  UNDER CONSTRUCTION  75670
   12)    COL    NAS                   GOD'S SON                72846
   13)    INT    EMINEM                EMINEM SHOW              70268
   14)    INT    TUPAC                 BETTER DAYZ              67747
   15)    UNIV   NELLY                 NELLYVILLE               66825
   16)    VRGN   ROLLING STONES        FORTY LICKS              64464
   17)    DEF    JA RULE               LAST TEMPTATION          61339
   18)    WAR    GROBAN*JOSH           JOSH GROBAN              60390
   19)    RCA    PRESLEY*ELVIS         ELVIS 30 NO. 1 HITS      57892
   20)    CURB   MCGRAW*TIM            TIM MCGRAW & THE DANCEHALL  55374
   21)    EPIC   GOOD CHARLOTTE        YOUNG & THE HOPELESS     52465
   22)    DEF    JAY-Z                 BLUEPRINT 2: GIFT & THE CUR  52116
   23)    J      STEWART*ROD           GREAT AMERICA SONG BOOK  50857
   24)    ARI    SANTANA               SHAMAN                   50419
   25)    COL    SYSTEM OF A DOWN      STEAL THIS ALBUM         46554
   26)    WAR    HILL*FAITH            CRY                      46542
   27)    COL    MAYER*JOHN            ROOM FOR SQUARES         46221
   28)    UNIV   3 DOORS DOWN          AWAY FROM THE SUN        45627
   29)    EPIC   B2K                   PANDEMONIUM!             45090
   30)    ISL    CAREY*MARIAH          CHARMBRACELET            43216
   31)    CAP    MCCARTNEY*PAUL        BACK IN THE U.S. LIVE 2002  43127
   32)    GEFN   NIRVANA               NIRVANA                  39240
   33)    PRR    SNOOP DOGG            PAID THA COST TO BE DA BOSS  39177
   34)    LAF    PINK                  MISSUNDAZTOOD            38282
   35)    DEF    VA-SOURCE PRESENTS    VOL. 6-HIP HOP HITS      37014
   36)    EPIN   AUDIOSLAVE            AUDIOSLAVE               36420
   37)    ATLG   MATCHBOX TWENTY       MORE THAN YOU THINK YOU ARE  35380
   38)    LAVA   KID ROCK              COCKY                    34590
   39)    CAP    COLDPLAY              RUSH OF BLOOD TO THE HEAD  33920
   40)    ISL    JOHN*ELTON            GREATEST HITS 1970-2002  33400
   41)    ARI    HOUSTON*WHITNEY       JUST WHITNEY             32840
   42)    DRMW   KEITH*TOBY            UNLEASHED                30749
   43)    WAR    LORD OF THE RINGS:TWO TOWER  SOUNDTRACK        29961
   44)    GEFN   PUDDLE OF MUDD        COME CLEAN               29821
   45)    J      TYRESE                I WANNA GO THERE         28496
   46)    BNA    CHESNEY*KENNY         NO SHOES NO SHIRT NO PROBLE  27829
   47)    UNCM   BABY AKA THE #1 STUNNA  BIRDMAN                27685
   48)    ISL    SUM 41                DOES THIS LOOK INFECTED?  26996
   49)    LYST   RASCAL FLATTS         MELT                     26843
   50)    EPIC   CHEVELLE              WONDER WHAT'S NEXT        25808
```

FIGURE 7.9   *Example of archive report YTD Sales 2003. (Source: Nielsen SoundScan)*

MARKETING REPORTS

SoundScan creates various marketing charts that analyze the marketplace by segmenting sales into many categories.

When reading the YTD—Sales by Format Genre Album report, add (000) to the end of the units to accurately depict album sales by genre. Note that the CD format dominates sales, but that digital is emerging. Additionally, certain releases can be counted twice because they may be considered to be in more than one genre.

What are not calculated in this report are the genre percentages as they relate to the total. This information is in a different SoundScan report.

nielsen
SoundScan

| HOME | TITLE | SUMMARY | CHARTS | MARKETING | PICKUP | SETS | ARCHIVES | ACCOUNT | HELP |

Marketing Report: YTD - Sales By Format/Genre Album   **LOAD TO EXCEL**

Week Ending: 12/30/2007 ▼

Units Sold (000's) Format: Album	Total YTD	CD YTD	LP YTD	Cassette YTD	DVD YTD	Digital YTD
Total	500,396	449,018	988	270	93	50,004
Alternative	88,604	74,547	347	10	11	13,687
Classic	18,044	16,767	1	1	5	1,266
Country	62,698	59,839	10	25	4	2,815
Hard Music	52,951	47,858	91	17	5	4,979
Jazz	14,047	12,453	14	5	4	1,570
R&B	95,555	89,372	114	72	3	5,993
Rap	41,662	38,592	108	23	1	2,937
Soundtrack	24,570	21,232	11	4	0	3,323
Latin	31,853	31,297	1	16	23	515
Gospel	5,910	5,670	1	9	1	229
New Age	3,210	2,855	1	1	0	352
Rock	149,414	129,247	584	37	36	19,507
Electronic	6,938	5,503	68	1	0	1,361

FIGURE 7.10   *Example marketing report: YTD sales by format/album. (Source: Nielsen SoundScan)*

Label Market Share Report

The Label Share Marketing Report shows the percentage of business by distribution companies, as well as indies, for the year ending December 30, 2007. Each distributor, shown as Level 1, has "owned" labels that are part of the conglomerate and "distributed" labels that have contracted the distributor to place their records into the marketplace, which are noted as

nielsen
SoundScan

| HOME | TITLE | SUMMARY | CHARTS | MARKETING | PICKUP | SETS | ARCHIVES | ACCOUNT | HELP |

Marketing Report: Label Share - Units - Year to Date

LOAD TO EXCEL

Week Ending: 12/30/2007 ▼   Period: Year To Date ▼   Level: All Levels ▼

Format: Album	Level	Total (%)	Cass (%)	CD (%)	LP (%)	DVD (%)	Digital (%)
Warner Bros	2	6.93	7.08	6.93	4.73	16.94	6.87
Warner-ADA	3	.06	.00	.05	.37	.00	.14
Atlantic Records	2	5.95	4.46	6.05	1.92	3.87	5.08
ADA-Atlantic	3	.19	.00	.19	.72	.00	.24
ADA-Fueled by Ram	3	.11	.00	.10	.00	.00	.16
Bad Boy Records	3	.30	.02	.31	.27	.00	.17
Roadrunner Record	3	.73	.05	.76	.04	.00	.46
WEA/Fueled by Ram	3	.15	.00	.15	.06	.00	.22
WSM/Rhino	2	.96	3.65	1.03	.11	1.33	.41
CURB	2	1.09	.10	1.17	.00	.00	.43
WEA Latina	2	.27	.09	.29	.00	.10	.10
Word/Curb	2	.25	.07	.26	.00	.01	.18
V2 Records	2	.08	.01	.06	.46	.00	.18
Others	2	.79	.14	.71	.23	.35	1.51
Total WEA	1	16.31	15.61	16.51	7.45	22.59	14.77
Columbia	2	5.47	1.63	5.52	3.12	.00	5.15
Epic	2	2.55	.68	2.49	1.02	.00	3.09
Sony Classical	2	.49	.10	.47	.02	.00	.66
SonyBMG Norte	2	1.37	.48	1.50	.00	.00	.28
SMCMG	2	.60	8.46	.66	.03	.00	.01
Others	2	.75	.17	.73	.11	2.17	.88
SMD Subtotal		11.22	11.53	11.37	4.30	2.17	10.08
The RCA Music Group	2	4.38	.82	4.40	.61	3.03	4.25
Arista Records	3	.90	.05	.94	.00	.83	.53
J Records	3	.96	.23	1.00	.12	.14	.66
RCA Records	3	1.98	.17	1.94	.34	.07	2.36
RCA Victor	3	.11	.07	.10	.01	.00	.18
Arista SMG	3	.26	.14	.26	.02	2.00	.28
RCA SMG	3	.17	.15	.17	.12	.00	.24
Sony BMG Nashville	2	2.73	.57	2.86	.01	.00	1.64
Arista Nashville	3	.57	.03	.61	.00	.00	.21
BNA Records	3	.40	.07	.42	.00	.00	.25
Columbia Nashvill	3	.78	.22	.82	.01	.00	.41
RCA Nashville	3	.30	.07	.31	.00	.00	.13
Provident Label G	3	.28	.05	.26	.00	.00	.39
BMG Classics	2	.03	.02	.03	.00	.00	.05
BMG SMG	2	.10	.03	.10	.02	.46	.08
Zomba Group	2	2.77	1.35	2.88	.89	.53	1.87
Jive	3	1.78	.28	1.85	.32	.36	1.22
La Face	3	.36	.08	.38	.11	.17	.26
So So Def	3	.02	.01	.02	.00	.00	.01
Star Trak	3	.01	.00	.01	.01	.00	.01
Verity	3	.36	.30	.39	.00	.00	.12
Volcano	3	.20	.25	.20	.43	.00	.20
Zomba SMG	3	.03	.43	.03	.01	.00	.05
Kinetic Rec	2	.00	.00	.00	.00	.00	.00
Razor & Tie	2	.53	.02	.54	.17	.00	.42
Robbins Ent.	2	.03	.00	.03	.00	.00	.03
Sanctuary/CMC	2	.03	.04	.03	.04	.00	.06
Windup Rec	2	.46	.00	.46	.00	.00	.42
Others	2	.06	.31	.04	.02	.00	.24
BMG Subtotal		11.11	3.17	11.36	1.76	4.02	9.06
Total Sony/BMG	1	22.33	14.70	22.73	6.06	6.19	19.14

A

FIGURE 7.11   *Example marketing report: label shares—units. (Source: Nielsen SoundScan)*

Interscope/A&M	2	4.96	1.72	4.87	1.90	2.17	5.90
A&M/Octone	3	.05	.00	.05	.03	.00	.09
Geffen	2	3.00	4.73	2.97	1.13	2.08	3.24
Island Def Jam Music	2	4.54	3.53	4.57	4.18	1.24	4.21
Def Jam/Def Soul	3	1.84	1.28	1.90	1.26	.03	1.35
Island	3	2.42	2.23	2.41	1.37	1.18	2.49
Roadrunner	3	.09	.01	.09	.06	.00	.06
Lost Highway	3	.18	.00	.17	1.49	.03	.29
Mercury	3	.01	.00	.01	.00	.00	.01
Distributed Labls	3	.00	.00	.00	.00	.00	.00
Universal Records Gr	2	3.76	4.81	3.77	1.11	1.41	3.74
UMG-Nashville	2	1.73	1.08	1.85	.02	.61	.72
MCA Nashville	3	1.15	.65	1.23	.00	.04	.38
Mercury Nashville	3	.52	.41	.55	.02	.57	.26
Rounder	3	.06	.02	.06	.01	.00	.08
Big Machine	2	.45	.00	.48	.00	.00	.17
Show Dog Nashville	2	.22	.00	.24	.00	.00	.06
The Verve Group	2	.50	.06	.48	.55	2.37	.66
Verve	3	.42	.03	.41	.16	2.31	.53
GRP	3	.07	.03	.06	.39	.06	.14
Classics	2	1.54	.12	1.52	.32	2.75	1.78
Velour Records	3	.01	.00	.00	.05	.00	.02
Rounder Records	3	.39	.03	.40	.24	.00	.31
Universal Latino	2	.58	.24	.62	.00	.00	.18
Bohemia	3	.00	.00	.00	.00	.00	.00
Protel	3	.00	.00	.00	.00	.00	.00
Platano	3	.00	.00	.00	.00	.00	.00
Latino	3	.00	.00	.00	.00	.00	.00
Karen	3	.00	.01	.00	.00	.03	.01
Rodven	3	.08	.07	.09	.00	.00	.01
RMM	3	.00	.04	.00	.00	.00	.00
Siente Music	3	.03	.00	.04	.00	.00	.00
Venemusic	3	.05	.00	.05	.00	.00	.00
Machete Music	2	.35	.00	.38	.00	.00	.07
Lideres	2	.01	.01	.01	.00	.00	.00
OLE	2	.00	.00	.00	.00	.00	.00
Univision Music Grp	2	2.16	.94	2.40	.00	.00	.07
Fonovisa Records	3	.00	.88	.89	.00	.00	.03
Disa	3	.82	.04	.91	.00	.00	.01
Emusica	2	.01	.00	.01	.00	.00	.00
Viva Discos	2	.00	.00	.00	.00	.00	.00
Hollywood Records	2	1.76	.06	1.80	.02	.00	1.47
Hollywood	3	1.04	.05	1.03	.02	.00	1.11
Mammoth	3	.00	.00	.00	.00	.00	.01
Lyric Street	3	.72	.01	.76	.00	.00	.35
Disney/Buena Vista	2	2.75	1.19	2.91	.00	.61	1.45
H.O.L.A.	2	.00	.01	.00	.00	.00	.00
Concord Records	2	.97	.09	.96	.47	1.05	1.07
Peak Records	3	.05	.00	.05	.00	.00	.04
Fantasy Records	3	.38	.05	.37	.42	.55	.54
Telarc	3	.16	.04	.16	.00	.50	.18
Hear Music	3	.22	.00	.24	.04	.00	.12
CURB	2	.04	.00	.04	.00	.00	.04
Curb/MCA Nashvil	3	.00	.00	.00	.00	.00	.01
Curb/MCA Los Ange	3	.00	.00	.00	.00	.00	.01
Curb/Universal	3	.00	.00	.00	.00	.00	.00
Curb/Mercury Nash	3	.00	.00	.00	.00	.00	.00
Curb/Lost Highway	3	.03	.00	.03	.00	.00	.02
Thump	2	.05	.06	.05	.03	.00	.01
Universal Music Ente	2	.74	10.37	.76	.07	.00	.49
Hip-O	3	.50	.21	.52	.06	.00	.33
Now	3	.01	.00	.01	.00	.00	.01
Special Products	3	.23	10.10	.23	.01	.00	.15
Varese	2	.16	.00	.14	.00	.00	.36
Trinity Records	2	.00	.00	.00	.00	.00	.00
Mock & Roll	2	.00	.00	.00	.00	.00	.00
Beyond Music	2	.00	1.02	.00	.00	.00	.00
Ark 21	2	.01	.00	.01	.00	.00	.02
Bungalo Records	2	.04	.00	.04	.00	.00	.02
Northsound	2	.00	.09	.00	.00	.00	.00
VI Music	2	.00	.00	.00	.00	.00	.00
VP Records	2	.07	.03	.07	.50	.41	.07
Others	2	1.50	.53	1.48	2.32	.28	1.90
Total UMGD	**1**	**31.90**	**30.68**	**32.41**	**12.61**	**14.98**	**27.69**

B

FIGURE 7.11 *(Continued)*

Blue Note Label Grou	2	1.38	.66	1.41	.18	.32	1.17
Capitol Nashville	2	.71	.97	.75	.00	.00	.34
Capitol Music Group	2	4.82	7.59	4.60	4.68	5.32	4.79
Capitol Records	3	3.37	7.15	3.40	3.50	1.87	3.03
Virgin	3	1.25	.44	1.20	1.18	3.44	1.76
Caroline Music	2	.55	.03	.48	5.91	.25	1.13
Christian Music Grou	2	.84	.16	.60	.04	.00	1.21
EMI Latin	2	.41	1.60	.45	.00	.00	.10
Aries Records	3	.02	.00	.02	.00	.00	.01
Max Mex Records	3	.00	.00	.00	.00	.00	.00
UBO Latin	3	.01	.00	.01	.00	.00	.00
Venevision	3	.01	.00	.01	.00	.00	.00
EMI America Records	2	.02	.00	.02	.00	.00	.02
Petrol Records	3	.01	.00	.01	.00	.00	.00
Unidentified	2	.22	.08	.14	2.52	1.84	.93
Others	2	.21	.26	.19	2.69	.00	.36
Total EMM	1	8.97	11.56	8.83	16.02	7.72	10.05
Madacy	2	2.05	.23	2.26	.00	.00	.18
Tommy Boy	2	.02	.07	.02	.01	.00	.01
St. Clair	2	.23	.09	.26	.00	1.74	.00
Laserlight/Delta	2	.14	.12	.15	.00	5.03	.01
KTEL/qwil	2	.01	.14	.00	.00	.00	.01
Broken Bow	2	.19	.01	.21	.00	.00	.06
American Gramaphone	2	.02	.00	.02	.00	.00	.00
TVT	2	.15	.08	.15	.26	.00	.11
Rounder	2	.01	.01	.01	.00	.00	.02
Direct Source	2	.29	.00	.32	.00	.00	.00
Rykodisc	2	.08	.09	.08	.07	.00	.04
Warlock/Quality	2	.02	.05	.02	.00	.00	.01
Telarc	2	.00	.02	.00	.00	.00	.00
Caroline Records	2	.01	.00	.01	.22	.00	.01
Starbucks	2	.29	.01	.33	.00	.00	.01
Sub pop	2	.26	.00	.20	4.48	.00	.63
Malaco/Muscle Shoals	2	.03	.78	.03	.00	.00	.01
Alligator	2	.04	.05	.04	.00	.00	.04
Saddle Creek Records	2	.08	.00	.07	1.75	.00	.15
BCI Music	2	.21	.00	.24	.00	.86	.00
Sanctuary	2	.14	.02	.13	.07	.00	.26
Matador	2	.08	.00	.08	1.78	.00	.23
Freddie Records	2	.11	.09	.12	.00	24.82	.00
VP Records	2	.00	.01	.00	.08	.00	.00
Image Entertainment	2	.10	.00	.09	.00	.87	.16
Metal Blade	2	.14	.02	.14	.01	.00	.12
Vanguard	2	.08	.00	.07	.03	.00	.18
Oh boy	2	.02	.01	.02	.03	.00	.03
Side One Dummy	2	.10	.00	.09	.66	.00	.11
Lakeshore	2	.06	.00	.04	.00	.00	.27
Earache	2	.01	.00	.01	.02	.00	.00
Fat Wreck Chords	2	.05	.00	.05	.50	.00	.09
Lil Joe	2	.01	.03	.01	.03	.00	.01
Kung Fu Records	2	.00	.08	.00	.00	.00	.00
Sugar Hill	2	.06	.06	.05	.00	.00	.10
Turn Up The Music	2	.05	.04	.05	.00	.00	.01
Merge	2	.18	.00	.13	2.44	.00	.63
Century Media	2	.06	.01	.06	.02	.00	.04
IM Records	2	.06	.00	.07	.00	.00	.00
CMH	2	.06	.00	.06	.00	.00	.03
Ultra	2	.08	.00	.07	.00	.00	.12
Epitaph Records	2	.36	.02	.33	2.11	.12	.66
Merge	2	.18	.00	.13	2.44	.00	.61
NAXOS	2	.14	.06	.13	.00	.93	.22
Astralwerks	2	.04	.00	.03	.34	.00	.10
Platinum Disk	2	.04	.00	.04	.00	.00	.00
Collectables Records	2	.07	.04	.08	.00	.00	.00
Domino	2	.03	.00	.02	.77	.00	.09
Cleopatra	2	.09	.00	.09	.07	.00	.04
Hopeless Records	2	.06	.00	.05	.04	.00	.12
Nuclear Blast	2	.03	.00	.03	.00	.23	.04
Original Sound Rec	2	.01	.45	.01	.00	.00	.00
Victory Records	2	.22	.00	.22	.09	.00	.24
Nitro Records	2	.00	.00	.00	.00	.00	.00
New West	2	.10	.00	.10	.06	.00	.13
Mute	2	.02	.00	.02	.11	.00	.04
Ruthless	2	.03	.00	.04	.00	.00	.00

C

FIGURE 7.11 *(Continued)*

Label						
Matador	2	.08	.00	.06	1.78	.00	.23
Freddie Records	2	.11	.09	.12	.00	24.82	.00
VP Records	2	.00	.01	.00	.06	.00	.00
Image Entertainment	2	.10	.00	.09	.00	.87	.16
Metal Blade	2	.14	.02	.14	.01	.00	.12
Vanguard	2	.08	.00	.07	.03	.00	.18
Oh boy	2	.02	.01	.02	.03	.00	.03
Side One Dummy	2	.10	.00	.09	.66	.00	.11
Lakeshore	2	.06	.00	.04	.00	.00	.27
Earache	2	.01	.00	.01	.02	.00	.00
Fat Wreck Chords	2	.05	.00	.05	.50	.00	.09
Lil Joe	2	.01	.03	.01	.03	.00	.01
Kung Fu Records	2	.00	.08	.00	.00	.00	.00
Sugar Hill	2	.06	.06	.05	.00	.00	.10
Turn Up The Music	2	.05	.04	.05	.00	.00	.01
Merge	2	.18	.00	.13	2.44	.00	.63
Century Media	2	.06	.01	.06	.02	.00	.04
IM Records	2	.06	.00	.07	.00	.00	.00
CMH	2	.06	.00	.06	.00	.00	.03
Ultra	2	.08	.00	.07	.00	.00	.12
Epitaph Records	2	.36	.02	.33	2.11	.12	.66
Merge	2	.18	.00	.13	2.44	.00	.61
NAXOS	2	.14	.06	.13	.00	.93	.22
Astralwerks	2	.04	.00	.03	.34	.00	.10
Platinum Disk	2	.04	.00	.04	.00	.00	.00
Collectables Records	2	.07	.04	.08	.00	.00	.00
Domino	2	.03	.00	.02	.77	.00	.09
Cleopatra	2	.09	.00	.09	.07	.00	.04
Hopeless Records	2	.06	.00	.05	.04	.00	.12
Nuclear Blast	2	.03	.00	.03	.00	.23	.04
Original Sound Rec	2	.01	.45	.01	.00	.00	.00
Victory Records	2	.22	.00	.22	.09	.00	.24
Nitro Records	2	.00	.00	.00	.00	.00	.00
New West	2	.10	.00	.10	.06	.00	.13
Mute	2	.02	.00	.02	.11	.00	.04
Ruthless	2	.03	.00	.04	.00	.00	.00
Comedy Central Recor	2	.23	.00	.20	.00	.00	.55
Beggars Banquet	2	.06	.00	.04	.54	.00	.21
Touch & Go	2	.04	.00	.03	1.78	.00	.09
Six Degrees	2	.00	.00	.00	.00	.00	.00
Blix Street	2	.02	.00	.02	.01	.00	.00
Psychopathic	2	.06	.00	.06	.00	.00	.02
Ferret	2	.07	.00	.08	.00	.00	.06
Trustkill	2	.04	.00	.04	.01	.00	.04
Genius Products	2	.04	.00	.05	.00	.00	.00
Eagle Records/Vision	2	.04	.00	.04	.00	.00	.04
System/Double Blind	2	.01	.00	.04	.04	.00	.01
V2 Records	2	.01	.00	.01	.00	.00	.04
Koch Entertainment	2	1.98	.57	1.99	4.50	3.75	1.87
Koch Records	3	.59	.31	.60	.72	.00	.53
Shanachie	3	.07	.02	.07	.00	.00	.08
Kinkysweet	3	.01	.00	.01	.00	.00	.00
Compendia	3	.00	.01	.00	.00	.00	.00
Musicrama	3	.08	.00	.09	.00	.20	.01
Navarre	3	.30	.03	.30	.34	2.49	.28
Cleopatra	3	.05	.00	.05	.07	.00	.02
Equity Music Grp	3	.06	.00	.07	.00	.00	.04
Koch Other	3	.82	.19	.81	3.36	1.06	.89
Others	2	10.91	24.15	9.90	32.51	10.17	19.43
Total Others	1	20.49	27.46	19.52	57.86	48.52	28.35

D

FIGURE 7.11 (Continued)

Level 2. Label groups have sublabels that are noted as Level 3. Each label is part of the cumulative percentage.

From a competitive standpoint, these data allow distributors and their labels to evaluate their performance as it compares to others. These data are compiled weekly, monthly, and year to date.

Because of the merger between SONY and BMG, their business is broken out, with a combined percentage at the bottom of their information. *Note:* The data for label market share are not current. The current version of this chart reflects the fact that Sony has since purchased BMG's holdings and is now known as Sony Music Entertainment, Inc.

Marketing Report: YTD Percent Sales by DMA/Genre

The YTD Percentage of Sales by DMA/Genre report calculates the percentage of music sold in a specific DMA by genre. DMA stands for designated market area, with the regions being derived by Nielsen, the company that calculates television ratings.

Example: New York DMA purchased 6.77% of all records sold. For every 100 records sold, nearly 7 are purchased in the New York DMA.

Additionally, New York DMA purchased 10.80% of all classical records sold, making New York a ripe market for classical music. Again, for every 100 classical records sold, nearly 11 records are purchased in the New York DMA.

This report helps marketers to look at DMAs as a whole. Labels can compare the percentage of their artists' sales to that of the overall market, seeing whether an artist's sales have over- or underperformed (see Figure 7.12).

National Sales Summary Report

This report denotes sales by store type, geographic region, and population density. Note that chain stores sold the most, with Mass Merchants not far behind. Chain Stores include electronic superstores as well as retail chains.

Although cities have higher population density per square mile, rural areas out-purchase the city geographic marketplace, with the suburbs being the largest subsection of music purchasers.

Nielsen SoundScan Marketing Report   Page 1 of 4

Marketing Report: YTD - % Sales By DMA/Genre Album

Week Ending: 12/30/2007   DMA Type: Full

Units Sold (000's) DMA	Total YTD	Alternative YTD	Classical YTD	Country YTD	Hard Music YTD	Jazz YTD	R & B YTD	Rap YTD
Total	100.00	100.00	100.00	100.00	100.00	100.00	100.00	100.00
New York, NY	6.77	6.50	10.80	2.64	5.40	10.74	7.13	6.00
Los Angeles, CA	5.92	6.45	5.21	2.75	5.52	6.40	6.02	6.29
Chicago, IL	3.26	3.35	4.30	2.01	2.79	3.79	3.44	3.33
Philadelphia, PA	2.61	2.67	3.10	1.72	2.44	3.15	3.19	2.70
SF-Okland-San Jose	2.56	2.51	3.84	1.27	1.83	4.01	2.45	2.34
Boston, MA	2.30	2.71	3.25	1.64	2.42	3.20	1.78	1.91
Dallas-Ft. Worth, TX	2.01	1.92	1.72	2.37	1.74	1.59	1.91	1.84
Detroit, MI	1.44	1.48	1.87	1.16	1.43	1.86	1.93	1.68
Washington, DC	2.58	2.23	3.25	1.91	1.88	3.62	3.60	2.98
Houston, TX	1.77	1.46	1.38	1.77	1.55	1.36	1.98	1.95
Cleveland, OH	1.14	1.14	1.27	1.24	1.20	1.17	1.29	1.25
Atlanta, GA	2.15	1.75	1.66	2.01	1.57	2.04	3.31	3.08
Minneapolis-St. Paul, MN	1.50	1.76	1.68	1.66	1.57	1.53	1.15	1.23
Tampa-St.Petersburg, FL	1.19	1.10	1.20	1.26	1.11	1.15	1.11	1.11
Seattle-Tacoma, WA	1.93	2.33	2.12	1.68	1.94	2.40	1.63	1.79
Miami, FL	1.32	.91	1.66	.49	.86	1.60	1.64	1.80
Pittsburgh, PA	.82	.87	.88	1.03	.95	.83	.71	.88
St. Louis, MO	.94	1.02	.88	1.09	1.01	1.08	.87	.72
Denver, CO	1.62	1.91	1.53	1.52	1.82	1.52	1.28	1.56
Phoenix, AZ	1.61	1.70	1.43	1.52	1.80	1.32	1.51	1.77
Sacramento-Stockton, CA	1.20	1.20	1.17	1.08	1.21	1.22	1.19	1.21
Baltimore, MD	1.30	1.31	1.06	.88	1.14	1.74	2.05	1.72
Hartford-New Haven, CT	.83	.84	1.07	.68	.86	1.00	.81	.79
San Diego, CA	1.05	1.21	1.05	.61	.97	1.15	.94	1.00
Orlndo-Daytona Bch-Mlbrne	1.28	1.32	1.18	1.12	1.30	1.21	1.23	1.27
Indianapolis, IN	.89	.94	.78	1.21	.92	.78	.75	.73
Portland, OR	1.30	1.41	1.84	1.21	1.25	2.01	.78	.80
Milwaukee, WI	.69	.74	.84	.71	.67	.68	.64	.61
Kansas City, KS-MO	.79	.80	.71	1.03	.79	.69	.82	.88
Cincinnati, OH	.73	.81	.70	.93	.80	.69	.65	.62
Charlotte, NC	.91	.81	.73	1.07	.87	.74	1.10	1.03
Nashville, TN	.89	.86	.56	1.43	.93	.64	.87	.89
Raleigh-Durham, NC	.93	.82	.74	.97	.75	.85	1.17	1.03
Columbus, OH	.76	.83	.73	.91	.92	.69	.70	.68
Greenville-Sprtnbrg-Ashvll	.54	.53	.46	.94	.67	.48	.61	.61
Roanoke-Lynchburg, VA	.35	.33	.26	.61	.40	.23	.33	.32
Lexington, KY	.37	.36	.18	.67	.43	.20	.27	.27
Austin, TX	.66	.73	.55	.66	.55	.60	.47	.51
Rochester, NY	.33	.36	.53	.32	.30	.44	.27	.24
Omaha, NE	.36	.41	.31	.45	.41	.31	.28	.30
Portland-PolandSpring, ME	.36	.44	.37	.41	.48	.38	.22	.25
Sprngfld-Decatr-Chmpgn,IL	.29	.29	.28	.47	.35	.24	.22	.22
Pdch-CpGrdu-Hrsbg-Mrion,KY-IL	.24	.23	.13	.49	.31	.10	.18	.18
Spokane, WA	.39	.49	.33	.55	.57	.29	.26	.32
Davnprt-Rcklslnd-Molin,IL	.23	.24	.21	.37	.29	.18	.17	.19
Tucson, AZ	.41	.43	.43	.35	.45	.39	.40	.53
Hntsville-Decatr-Flornc,AL	.28	.27	.15	.45	.34	.16	.30	.31
CdarRpds-Wtrloo-Dubuqu,IA	.26	.29	.25	.40	.33	.19	.18	.21
Columbia, SC	.33	.25	.19	.32	.28	.26	.62	.59
Springfield, MO	.32	.33	.22	.63	.40	.18	.21	.23
Chattanooga, TN	.28	.27	.15	.48	.34	.14	.24	.25
Southbend-Elkhart, IN	.25	.26	.23	.31	.32	.19	.21	.24
Jackson, MS	.28	.21	.17	.32	.23	.21	.45	.41
Brstl-Kngsprt-JhnsnCty,TN	.20	.19	.11	.44	.28	.08	.14	.13
Johnstown-Altoona, PA	.18	.21	.16	.30	.29	.10	.11	.10
Youngstown, OH	.19	.20	.17	.26	.24	.17	.19	.18
Madison, WI	.32	.34	.45	.40	.33	.31	.21	.23
Las Vegas, NV	.66	.66	.51	.46	.68	.57	.75	.81
Brlngtn-Plattsbrgh, VT-NY	.31	.35	.36	.41	.33	.36	.19	.21
Evansville, IN	.20	.20	.13	.40	.27	.10	.14	.15
Baton Rouge, LA	.31	.27	.16	.27	.30	.23	.59	.62
Lincoln-Hastings-Kearney	.21	.24	.18	.35	.27	.13	.15	.18
Ft. Myers-Naples, FL	.29	.23	.30	.29	.26	.26	.24	.29
Waco-Temple-Bryan, TX	.31	.28	.17	.48	.33	.15	.36	.42
Springfield, MA	.27	.30	.29	.20	.31	.30	.26	.31
ColoradoSprngs-Pueblo, CO	.36	.40	.33	.42	.45	.29	.34	.43
Hawaii	.52	.41	.58	.25	.34	.68	.45	.42
Others	13.44	12.66	9.78	18.47	15.10	8.80	12.07	13.74

FIGURE 7.12   *Sample marketing report: YTD—% sales by DMA/genre album (Source: Nielsen SoundScan).*

nielsen
SoundScan

Orlndo-Daytona Bch-Mlbrne	1.27	1.27	1.36	1.53	1.34	1.34	1.29	1.28	1.30	1.75
Indianapolis, IN	.94	1.00	.97	.52	.60	.82	.81	.79	.96	.68
Portland, OR	1.20	1.04	1.11	.80	.47	1.77	1.44	1.48	1.41	1.62
Milwaukee, WI	.71	.76	.79	.38	.62	.92	.64	.63	.72	.62
Kansas City, KS-MO	.81	.86	.89	.51	.79	1.01	.76	.76	.79	.54
Cincinnati, OH	.74	.79	.83	.23	.70	.78	.72	.72	.80	.53
Charlotte, NC	.95	1.01	.90	.97	1.70	.82	.85	.84	.82	.63
Nashville, TN	.93	.99	.85	.69	.83	.65	.82	.80	.89	.56
Raleigh-Durham, NC	.97	.96	.86	1.13	2.32	.81	.87	.85	.81	.75
Columbus, OH	.77	.82	.78	.24	.69	.78	.73	.73	.85	.66
Greenvlle-Sprtnbrg-Ashvll	.64	.64	.57	.57	1.09	.55	.62	.62	.62	.45
New Orleans, LA	.56	.57	.52	.64	1.13	.39	.58	.59	.48	.43
Grnd Rpds-Klmzo-Bttle Crk	.51	.55	.56	.26	.25	.69	.50	.49	.57	.32
Buffalo, NY	.59	.61	.60	.11	.66	.63	.64	.66	.74	.55
Memphis, TN	.52	.55	.48	.36	1.56	.36	.49	.50	.40	.28
Oklahoma City, OK	.55	.58	.54	.43	.36	.53	.54	.53	.57	.34
Salt Lake City, UT	1.17	1.18	1.49	1.03	.57	1.65	1.20	1.18	1.33	1.10
Nrflk-Prtsmth-NwptNws-Hmp	.74	.78	.65	.37	1.80	.62	.65	.64	.64	.50
San Antonio, TX	.74	.69	.74	1.59	.46	.64	.72	.69	.67	.70
Providence-New Bedford,RI	.54	.56	.63	.22	.17	.57	.60	.62	.64	.65
Harrsbrg-Yrk-Lncstr-Lbnon	.56	.60	.58	.24	.30	.74	.53	.53	.61	.35
Louisville, KY	.65	.63	.59	.27	.42	.44	.63	.64	.74	.64
Birmingham, AL	.48	.49	.44	.41	.75	.33	.43	.43	.44	.28
Charleston-Huntington, WV	.34	.38	.34	.03	.13	.20	.30	.30	.34	.12
Greensbro-WnstnSalm-HiPnt	.51	.53	.47	.54	1.01	.45	.43	.42	.41	.27
WstPlmBch-FrtPerc-VeroBch	.52	.51	.51	.95	.56	.63	.53	.53	.46	.56
Albuquerque, NM	.68	.64	.57	.86	.18	.79	.60	.80	.76	.66
Dayton, OH	.40	.45	.41	.07	.37	.41	.37	.36	.43	.22
Albany-Schnctady-Troy, NY	.44	.46	.50	.10	.17	.54	.49	.50	.53	.32
Wilkes-Barre-Scranton, PA	.41	.44	.43	.09	.10	.37	.41	.41	.52	.24
Mobile-Pensacola, AL-FL	.50	.52	.43	.40	.93	.39	.50	.50	.49	.36
Jacksonville, FL	.59	.61	.56	.42	1.19	.50	.57	.57	.59	.56
Little Rock, AR	.41	.43	.39	.30	.86	.31	.41	.40	.38	.20
Tulsa, OK	.39	.42	.39	.27	.24	.35	.39	.38	.41	.24
Flint-Saginaw-BayCity, MI	.35	.39	.37	.04	.47	.42	.33	.33	.37	.15
Richmond, VA	.67	.69	.56	.37	1.78	.52	.62	.62	.52	.43
Wichita-Hutchison, KS	.34	.35	.32	.38	.19	.46	.32	.31	.34	.18
Fresno-Visalla, CA	.45	.45	.40	1.08	.19	.35	.38	.36	.37	.27
Toledo, OH	.35	.39	.39	.12	.20	.51	.34	.33	.39	.19
Knoxville, TN	.47	.48	.45	.22	.24	.37	.47	.47	.52	.34
Springfield, MA	.26	.25	.26	.21	.13	.29	.28	.28	.30	.33
ColoradoSprngs-Pueblo, CO	.36	.36	.35	.20	.14	.49	.36	.35	.40	.36
Hawaii	.47	.37	.46	.13	.19	.82	.62	.65	.40	.59
Others	13.55	13.39	12.23	19.51	13.23	12.46	13.24	13.00	13.09	8.65

nielsen
SoundScan

FIGURE 7.12 (Continued)

nielsen SoundScan

HOME | TITLE | SUMMARY | CHARTS | MARKETING | PICKUP | SETS | ARCHIVES | ACCOUNT | HELP

Marketing Report: National Sales By Strata   LOAD TO EXCEL | PRINT

Week Ending: 12/30/2007 ▼   Format: Alb/Sng ▼   Display: % TOT ▼

Units Sold (000's)	Total				Chain				Indep				Mass M				Non T			
Format: All	TW	% TOT	LW	YTD	TW	% TOT	LW	YTD	TW	% TOT	LW	YTD	TW	% TOT	LW	YTD	TW	% TOT	LW	YTD
Total	14,560		25,617	502,722	6,082		9,471	179,668	799		1,183	31,374	4,987		11,967	203,065	2,692		2,995	88,615
Geographic Region																				
North East	803	6	1,384	25,592	452	7	689	12,306	25	3	25	692	148	3	465	6,802	178	7	204	5,792
Middle Atlantic	1,909	13	3,401	64,723	945	16	1,514	27,470	163	20	184	5,170	391	8	1,206	17,900	411	15	497	14,182
E.North Central	2,054	14	3,997	69,936	895	15	1,423	24,497	65	8	101	3,143	754	15	2,081	31,409	339	13	392	10,887
W.North Central	927	6	1,700	31,703	339	6	538	9,706	30	4	49	1,338	418	8	959	16,146	140	5	155	4,513
South Atlantic	2,845	20	5,139	100,828	1,208	20	1,915	36,703	96	12	143	5,284	1,048	21	2,536	43,057	494	18	545	15,803
South Central	2,373	16	4,176	82,477	891	15	1,382	26,557	115	14	169	4,581	1,033	21	2,280	41,009	334	12	345	10,331
Mountain	1,149	8	1,901	40,813	461	8	690	14,648	45	6	77	1,834	429	9	915	17,267	214	8	219	7,064
Pacific	2,499	17	3,917	86,851	891	15	1,318	27,782	260	33	435	9,352	766	15	1,526	29,475	582	22	638	20,043
Geographic Place																				
City	2,260	16	3,502	85,102	975	16	1,491	33,639	342	43	464	13,425	430	9	914	17,074	514	19	632	20,964
Suburb	8,505	58	14,850	280,752	4,128	68	6,486	118,778	417	52	628	15,854	2,605	52	6,240	102,627	1,354	50	1,497	43,493
Rural	3,795	26	7,265	136,868	979	16	1,494	27,251	40	5	92	2,094	1,951	39	4,813	83,364	824	31	866	24,158

FIGURE 7.13 *Example summary report: national sales by strata. (Source: Nielsen SoundScan)*

A LOOK AT SOUNDSCAN TITLE REPORTS (ALICIA KEYS/*AS I AM*)

In the current interface, SoundScan provides comprehensive look-up charts for individual artists. After requesting a specific artist, using the last name first (for example, Keys*Alicia), all releases for this artist will be listed, with release-to-date (RTD) information included. Each title has a drop-down menu where the researcher can look at specific title reports.

For artist's specific album titles, reports are generated that are similar to national marketing reports. J Records inaugural artist Alicia Keys has had unprecedented success. Her 2006 hit *As I Am* has scanned over 2.5 million units. The title report reflects store and geographical information that is key to record label, artist management, booking agent, and retailer alike.

Mass merchants dominate sales with 52% of the overall take-home, with chains doing 33% percent of Alica Keys's business. Interestingly, 12% of Alicia's sales were nontraditional, with half of those being digital album sales. Remember that this information is based on 2007 data, and entire album downloads were just becoming popular.

Suburb sales are 60% of purchases, where the other two geographic "places" split 20/20 between city and rural.

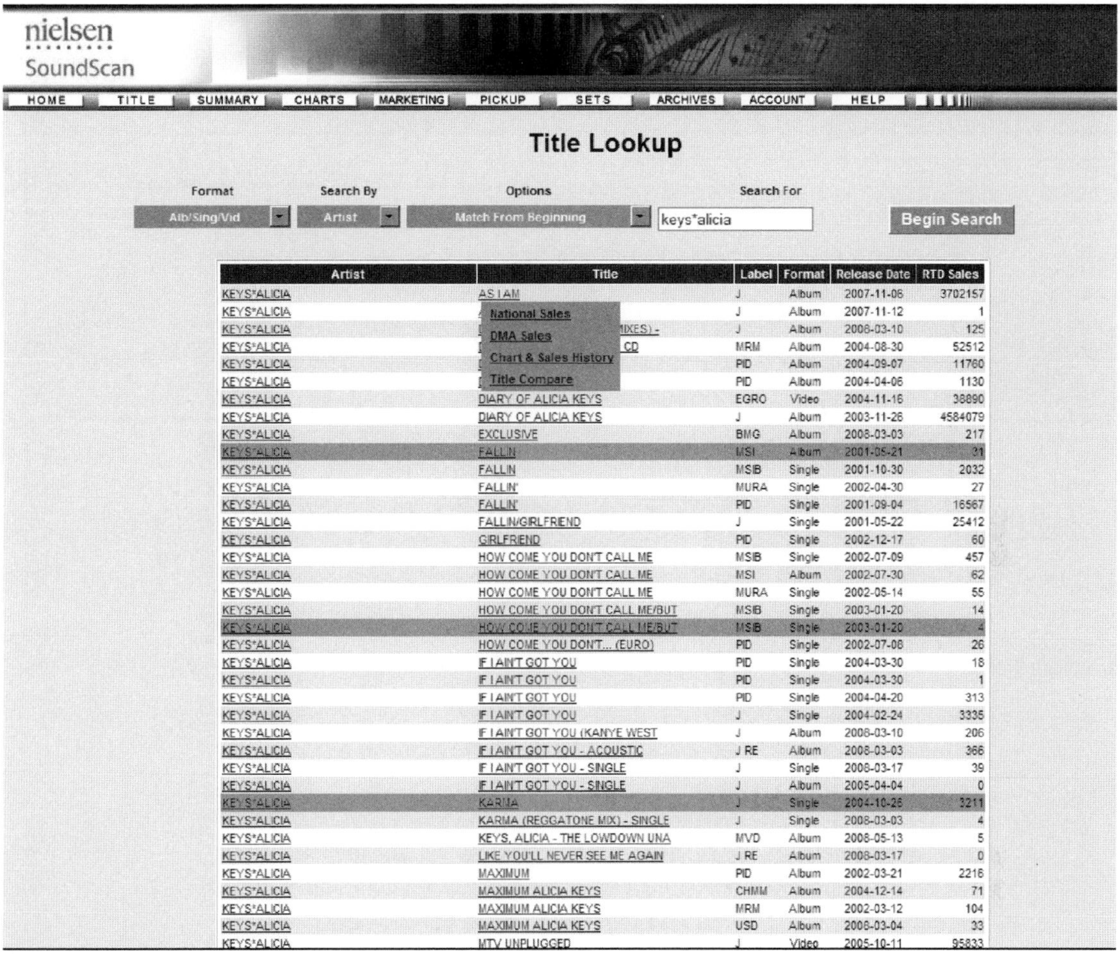

Artist	Title	Label	Format	Release Date	RTD Sales
KEYS*ALICIA	AS I AM	J	Album	2007-11-06	3702157
KEYS*ALICIA	[illegible]	J	Album	2007-11-12	1
KEYS*ALICIA	[illegible] MIXES) -	J	Album	2008-03-10	125
KEYS*ALICIA	[illegible] CD	MRM	Album	2004-08-30	52512
KEYS*ALICIA	[illegible]	PID	Album	2004-09-07	11760
KEYS*ALICIA	[illegible]	PID	Album	2004-04-06	1130
KEYS*ALICIA	DIARY OF ALICIA KEYS	EGRO	Video	2004-11-16	38890
KEYS*ALICIA	DIARY OF ALICIA KEYS	J	Album	2003-11-26	4584079
KEYS*ALICIA	EXCLUSIVE	BMG	Album	2008-03-03	217
KEYS*ALICIA	FALLIN	MSI	Album	2001-05-21	31
KEYS*ALICIA	FALLIN	MSIB	Single	2001-10-30	2032
KEYS*ALICIA	FALLIN'	MURA	Single	2002-04-30	27
KEYS*ALICIA	FALLIN'	PID	Single	2001-09-04	16567
KEYS*ALICIA	FALLIN/GIRLFRIEND	J	Single	2001-05-22	25412
KEYS*ALICIA	GIRLFRIEND	PID	Single	2002-12-17	60
KEYS*ALICIA	HOW COME YOU DON'T CALL ME	MSIB	Single	2002-07-09	457
KEYS*ALICIA	HOW COME YOU DON'T CALL ME	MSI	Album	2002-07-30	62
KEYS*ALICIA	HOW COME YOU DON'T CALL ME	MURA	Single	2002-05-14	55
KEYS*ALICIA	HOW COME YOU DON'T CALL ME/BUT	MSIB	Single	2003-01-20	14
KEYS*ALICIA	HOW COME YOU DON'T CALL ME/BUT	MSIB	Single	2003-01-20	4
KEYS*ALICIA	HOW COME YOU DON'T... (EURO)	PID	Single	2002-07-08	26
KEYS*ALICIA	IF I AIN'T GOT YOU	PID	Single	2004-03-30	18
KEYS*ALICIA	IF I AIN'T GOT YOU	PID	Single	2004-03-30	1
KEYS*ALICIA	IF I AIN'T GOT YOU	PID	Single	2004-04-20	313
KEYS*ALICIA	IF I AIN'T GOT YOU	J	Single	2004-02-24	3335
KEYS*ALICIA	IF I AIN'T GOT YOU (KANYE WEST	J	Album	2008-03-10	206
KEYS*ALICIA	IF I AIN'T GOT YOU - ACOUSTIC	J RE	Album	2008-03-03	366
KEYS*ALICIA	IF I AIN'T GOT YOU - SINGLE	J	Single	2008-03-17	39
KEYS*ALICIA	IF I AIN'T GOT YOU - SINGLE	J	Album	2005-04-04	0
KEYS*ALICIA	KARMA	J	Single	2004-10-26	3211
KEYS*ALICIA	KARMA (REGGATONE MIX) - SINGLE	J	Single	2008-03-03	4
KEYS*ALICIA	KEYS, ALICIA - THE LOWDOWN UNA	MVD	Album	2008-05-13	5
KEYS*ALICIA	LIKE YOU'LL NEVER SEE ME AGAIN	J RE	Album	2008-03-17	0
KEYS*ALICIA	MAXIMUM	PID	Album	2002-03-21	2216
KEYS*ALICIA	MAXIMUM ALICIA KEYS	CHMM	Album	2004-12-14	71
KEYS*ALICIA	MAXIMUM ALICIA KEYS	MRM	Album	2002-03-12	104
KEYS*ALICIA	MAXIMUM ALICIA KEYS	USD	Album	2008-03-04	33
KEYS*ALICIA	MTV UNPLUGGED	J	Video	2005-10-11	95833

FIGURE 7.14 *Title lookup (Source: Nielsen SoundScan)*

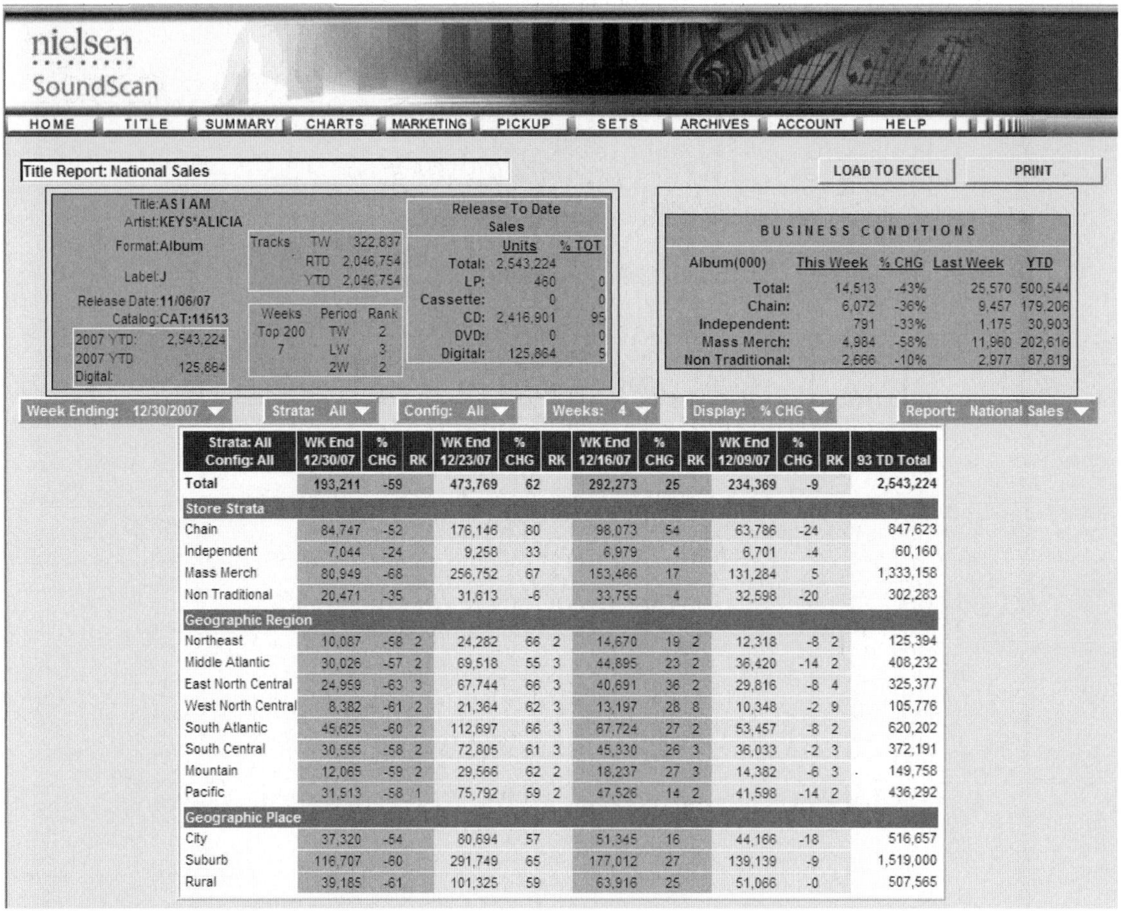

FIGURE 7.15 | *Example title report: national sales* Alicia Keys, As I Am. *(Source: Nielsen SoundScan)*

Title Report: DMA

DMA Title Reports shows sales by DMA. Again, by doing some fast math, marketers can analyze the data to show what markets are the best performing, based on population, overall sales of the market, genre specific sales, and so forth. Marketers, who include label personnel, managers, and booking agents as well as advertisers, look to this type of analysis to create a blueprint—a map as to where to market, tour, create promotions, and so on. The desired result would be increased records sales but could also enhance the artist's profile with increased ticket sales and bigger product endorsements. Note that of the 2.5 million albums sold, 125,864 were digitally purchased at the end of 2007.

nielsen
SoundScan

HOME | TITLE | SUMMARY | CHARTS | MARKETING | PICKUP | SETS | ARCHIVES | ACCOUNT | HELP

Title Report: DMA Sales

LOAD TO EXCEL

GRAPH
PRINT

Title: AS I AM
Artist: KEYS*ALICIA
Format: Album
Label: J
Release Date: 11/06/07
Catalog CAT:11613
2007 YTD 2,543,224
2007 YTD 125,864
Digital:

Tracks TW 322,837
RTD 2,046,754
YTD 2,046,754

Weeks Period Rank
Top 200 TW 2
7 LW 3
2W 2

Release To Date Sales
	Units	% TOT
Total:	2,543,224	
LP:	460	0
Cassette:	0	0
CD:	2,416,901	95
DVD:	0	0
Digital:	125,864	5

BUSINESS CONDITIONS
Album(000)	This Week	% CHG Last Week	YTD
Total:	14,513	-43%	25,570 900,544
Chain:	6,072	-36%	9,457 179,206
Independent:	791	-33%	1,175 30,903
Mass Merch:	4,984	-58%	11,980 202,616
Non Traditional:	2,666	-10%	2,977 87,819

Week Ending: 12/30/2007 ▼   Strata: All ▼   Config: All ▼   Weeks: 4 ▼   Display: % CHG ▼   Report: DMA Sales ▼

DMA	WK End 12/30/07	% CHG	RK	WK End 12/23/07	% CHG	RK	WK End 12/16/07	% CHG	RK	WK End 12/09/07	% CHG	RK	92 TD Total
Total	193,209	-59		473,769	62		292,278	25		234,371	-9		2,543,231
New York, NY	17,848	-53	2	38,094	54	2	24,742	15	2	21,488	-15	2	246,796
Los Angeles, CA	11,690	-58	1	27,828	85	3	16,855	10	2	15,254	-18	2	172,255
Chicago, IL	6,703	-64	3	18,516	77	3	10,471	34	2	7,818	-14	2	93,154
Philadelphia, PA	6,781	-61	3	17,379	59	3	10,901	32	2	8,285	-17	2	98,482
SF-Okland-San Jose	4,868	-61	2	12,405	88	3	7,404	13	2	6,568	-20	2	72,727
Boston, MA	4,190	-59	2	10,182	82	2	6,298	24	2	5,064	-10	2	52,130
Dallas-Ft. Worth, TX	4,302	-58	2	10,289	65	3	6,251	24	2	5,031	-2	3	51,071
Detroit, MI	3,151	-64	3	8,656	65	3	5,255	37	2	3,835	-11	2	45,374
Washington, DC	6,093	-60	2	15,334	87	3	9,172	21	2	7,574	-17	2	95,399
Houston, TX	4,325	-62	2	11,381	87	3	6,835	32	2	5,194	1	3	55,528
Cleveland, OH	2,572	-58	3	6,162	57	3	3,923	63	2	2,408	-17	4	28,940
Atlanta, GA	6,381	-58	2	15,351	79	2	8,595	23	2	7,014	-10	2	89,205
Minneapolis-St. Paul, MN	1,834	-65	2	5,180	68	2	3,076	30	6	2,388	-8	5	26,024
Tampa-St. Petersburg, FL	2,418	-60	3	6,064	63	3	3,731	40	2	2,661	-2	3	29,734
Seattle-Tacoma, WA	3,131	-59	2	7,572	50	2	5,041	24	2	4,071	-7	2	38,968
Miami, FL	3,682	-57	2	8,504	62	2	5,245	17	2	4,474	-7	2	46,972
Pittsburgh, PA	1,552	-61	2	3,958	50	2	2,629	44	3	1,826	-10	3	18,677
St. Louis, MO	1,586	-61	3	4,017	69	3	2,375	23	6	1,931	5	6	19,540
Denver, CO	2,043	-61	2	5,240	69	3	3,107	31	4	2,363	-14	3	26,402
Phoenix, AZ	3,079	-61	1	7,945	73	2	4,581	24	2	3,709	-5	3	38,544
Sacramento-Stockton, CA	2,432	-61	1	6,287	64	2	3,839	18	2	3,259	-11	2	33,547
Baltimore, MD	3,151	-58	2	7,501	47	3	5,101	36	2	3,760	-16	2	47,178
Hartford-New Haven, CT	2,013	-61	3	5,138	73	3	2,977	8	2	2,754	-10	2	27,832
San Diego, CA	2,327	-58	1	5,335	55	2	3,432	9	2	3,140	-13	2	30,817
Orlndo-Daytona Bch-Mlbrne	2,388	-61	3	6,116	72	3	3,547	31	2	2,704	1	3	29,774
Indianapolis, IN	1,403	-65	2	3,993	62	3	2,464	37	4	1,797	-8	5	18,750
Portland, OR	1,593	-56	2	3,613	47	2	2,455	38	3	1,782	-1	3	17,958
Milwaukee, WI	1,131	-65	2	3,274	75	3	1,869	13	3	1,652	-2	3	16,757
Kansas City, KS-MO	1,417	-58	3	3,204	61	4	1,990	37	6	1,454	-12	5	16,555
Cincinnati, OH	1,209	-61	2	3,381	72	2	1,967	22	4	1,606	-0	4	15,885
Charlotte, NC	2,143	-64	2	5,873	93	3	3,036	24	2	2,458	-6	2	27,845
Nashville, TN	1,464	-60	2	3,897	83	4	2,270	24	4	1,832	-3	4	19,437
Raleigh-Durham, NC	2,448	-58	2	5,828	83	3	3,182	22	2	2,606	-10	2	30,175
Columbus, OH	1,492	-61	2	3,793	73	3	2,197	37	2	1,609	-1	2	17,341
Greenville-Sprtnbrg-Ashvll	957	-61	2	2,458	61	4	1,526	24	3	1,229	3	3	12,327
New Orleans, LA	1,317	-53	2	2,819	41	3	1,999	19	2	1,683	-15	2	18,769
Grnd Rpds-Klmzo-Bttle Crk	862	-62	2	2,282	63	2	1,402	49	4	942	-6	6	9,846
Buffalo, NY	1,061	-58	3	2,526	51	3	1,671	29	2	1,296	10	4	13,515
Memphis, TN	1,406	-55	2	3,144	62	2	1,942	30	2	1,494	-18	2	16,522
Oklahoma City, OK	1,070	-47	1	2,000	65	5	1,209	10	6	1,104	6	6	10,310
Salt Lake City, UT	1,264	-59	3	3,092	56	6	1,986	30	9	1,533	-2	10	15,842
Nrflk-Prtsmth-NwptNws-Hmp	1,826	-62	2	4,745	67	3	2,839	37	2	2,073	-17	2	28,146
San Antonio, TX	1,807	-56	1	4,123	65	2	2,505	31	2	1,905	5	2	19,608
Providence-New Bedford, RI	1,504	-58	1	3,613	82	2	1,983	20	2	1,655	-5	2	16,936
Harrsbrg-Yrk-Lncstr-Lbnon	945	-58	2	2,277	46	2	1,558	32	4	1,181	-13	4	11,703
Louisville, KY	823	-65	2	2,366	66	4	1,427	34	5	1,067	-6	4	11,221
Birmingham, AL	920	-57	2	2,142	78	4	1,204	32	3	909	-7	3	10,366
Charlesbro-Huntington, WV	372	-63	6	1,002	45	10	689	27	11	543	3	10	4,680
Greensbro-WnstnSalm-HiPnt	1,051	-64	2	2,931	63	3	1,794	30	3	1,379	9	3	14,073
WstPlmBch-FrtPerc-VeroBch	1,174	-59	1	2,871	54	3	1,860	32	2	1,411	-2	2	14,633
Albuquerque, NM	1,293	-58	1	2,948	53	2	1,925	24	2	1,550	-1	3	15,198
Dayton, OH	708	-61	2	1,805	78	3	1,012	30	5	781	1	6	8,012
Albany-Schnctady-Troy, NY	842	-61	2	2,183	54	2	1,420	31	2	1,082	-11	2	10,584
Wilkes-Barre-Scranton, PA	685	-59	2	1,652	52	3	1,086	36	4	798	-17	4	8,087

A

FIGURE 7.16 *Example title report: DMA sales. (Source: Nielsen SoundScan)*

City												Total
Mobile-Pensacola, AL-FL	1,058	-53	2	2,246	54	3	1,460	40	2	1,041	-12	3	12,014
Jacksonville, FL	1,138	-62	2	2,962	68	3	1,763	19	2	1,483	6	2	15,482
Little Rock, AR	629	-58	2	1,485	53	5	973	42	6	685	-8	7	7,462
Tulsa, OK	543	-61	6	1,403	106	7	681	19	7	572	-18	10	6,235
Flint-Saginaw-BayCity, MI	747	-58	2	1,780	66	2	1,071	-45	3	737	-6	4	7,988
Richmond, VA	1,514	-59	2	3,663	61	4	2,280	25	2	1,818	-14	2	22,824
Wichita-Hutchison, KS	431	-59	5	1,058	78	7	593	24	9	480	4	9	4,770
Fresno-Visalia, CA	1,073	-60	1	2,702	64	2	1,646	16	2	1,425	-11	2	13,711
Toledo, OH	631	-59	2	1,558	40	2	1,112	44	3	771	6	4	7,340
Knoxville, TN	520	-64	5	1,430	70	5	840	18	8	711	24	6	6,286
Shrvport-Txrcana, AR-LA-TX	545	-56	2	1,249	59	3	785	32	3	594	-3	3	6,260
Des Moines, IA	410	-60	5	1,026	49	6	690	31	8	525	9	9	4,856
Green Bay-Appleton, WI	294	-69	5	943	52	6	622	14	9	548	15	9	4,223
Syracuse, NY	629	-60	2	1,569	51	2	1,036	51	2	687	-1	4	7,243
Roanoke-Lynchburg, VA	562	-65	1	1,590	53	4	1,040	26	3	823	0	3	7,726
Lexington, KY	403	-67	7	1,208	68	7	720	18	8	608	-2	8	5,753
Austin, TX	1,004	-63	1	2,715	73	3	1,566	41	4	1,111	-8	4	12,586
Rochester, NY	641	-62	2	1,704	63	3	1,047	32	3	793	-1	2	7,992
Omaha, NE	505	-64	3	1,421	65	5	863	30	6	663	-9	5	7,373
Portland-PolandSpring, ME	522	-51	2	1,072	68	3	637	18	4	539	13	4	4,882
Sprngfld-Decatr-Chmpgn, IL	396	-62	2	1,050	56	5	875	42	8	474	-7	9	4,862
Pdch-CpGrdu-Hrsbg-Mrion, KY-IL	234	-69	8	758	47	7	515	44	9	357	-5	10	3,477
Spokane, WA	436	-59	3	1,085	47	3	726	22	6	594	9	6	5,130
Davnprt-Rcklsnd-Molin, IL	311	-64	4	861	62	5	530	32	8	401	22	9	3,748
Tucson, AZ	732	-57	2	1,693	59	2	1,068	16	2	919	6	2	8,831
Hntsvlle-Decatr-Flornc, AL	464	-53	2	996	41	7	707	28	6	551	6	5	5,331
CdarRpds-Wtrloo-Dubuqu, IA	334	-61	3	863	52	7	569	26	8	451	-1	9	4,033
Columbia, SC	945	-57	2	2,191	72	2	1,276	28	2	993	-9	2	12,085
Springfield, MO	327	-50	7	660	44	12	459	41	11	326	-12	13	3,289
Chattanooga, TN	296	-67	9	898	68	9	536	19	8	451	18	6	4,114
Southbend-Elkhart, IN	367	-65	2	1,046	62	3	645	34	5	481	5	4	4,596
Jackson, MS	708	-52	2	1,470	59	2	924	24	2	744	-3	2	7,674
Brstl-Kngsprt-JhnsnCty, TN	197	-65	6	556	58	9	352	24	12	285	20	10	2,503
Johnstown-Altoona, PA	162	-71	7	554	80	7	308	22	11	252	-5	9	2,366
Youngstown, OH	386	-60	2	955	49	2	642	43	2	449	-14	3	4,470
Madison, WI	392	-64	3	1,089	61	4	675	40	5	481	-4	5	4,937
Las Vegas, NV	1,477	-55	1	3,256	51	2	2,151	27	2	1,694	-15	2	19,014
Brlngtn-Plattsbrgh, VT-NY	319	-55	5	710	37	7	519	35	5	384	-15	6	3,811
Evansville, IN	187	-71	10	636	59	10	400	24	12	323	11	11	2,775
Baton Rouge, LA	691	-56	2	1,557	37	3	1,140	23	3	930	-13	2	10,182
Lincoln-Hastings-Kearney	295	-56	5	673	98	7	340	-3	10	349	6	9	3,151
Ft. Myers-Naples, FL	445	-61	2	1,154	55	2	743	32	2	962	-1	3	5,821
Waco-Temple-Bryan, TX	624	-51	2	1,261	55	2	814	17	3	695	-7	3	7,136
Springfield, MA	691	-53	1	1,472	62	2	908	16	2	782	-8	2	7,863
Colorado Sprngs-Pueblo, CO	517	-62	2	1,366	62	3	842	29	4	652	-7	4	6,898
Hawaii, HI	892	-50	2	1,796	9	3	1,853	20	2	1,377	-18	3	12,915
Other	19,810	-59		48,324	56		30,882	24		24,858	-0		244,702

B

FIGURE 7.16 (Continued)

Table 7.2	Example title report: DMA sales with comparison column added

Title Report: DMA Sales

Title: CRAZY EX-GIRLFRIEND	Tracks TW: 20721	2007 YTD: 304999	RTD Units	% TOT
Artist: LAMBERT*MIRANDA	Tracks RTD: 249069	YTD Dig: 20150	Total: 304999	0
Format: Album	Tracks YTD: 249069		LP: 0	0
Label: CLNV	Wks Top 200: 35		Cassette: 0	0
Release Date: 2007-05-01	TW Rank: 176		CD: 284849	93
Catalog:CAT: 78932	LW Rank: 167		DVD: 0	0
	2W Rank: 186		Digital: 20150	7

Week Ending: 12/30/2007

DMA Total	92 TD Total 304,876	M Lambert % Total	Country % Total	% difference
New York, NY	6085	2.00%	2.64%	−0.64%
Los Angeles, CA	7588	2.49%	2.75%	−0.26%
Chicago, IL	6876	2.26%	2.01%	0.25%
Philadelphia, PA	4706	1.54%	1.72%	−0.18%
SF-Okland-San Jose, CA	3230	1.06%	1.27%	−0.21%
Boston, MA	3806	1.25%	1.64%	−0.39%
Dallas-Ft. Worth, TX	10,098	3.31%	2.37%	0.94%
Detroit, MI	3714	1.22%	1.16%	0.06%
Washington, DC	6300	2.07%	1.91%	0.16%
Houston, TX	4877	1.60%	1.77%	−0.17%
Cleveland, OH	3546	1.16%	1.24%	−0.08%
Atlanta, GA	7250	2.38%	2.01%	0.37%
Minneapolis-St. Paul, MN	6374	2.09%	1.66%	0.43%
Tampa-St. Petersburg, FL	3046	1.00%	1.26%	−0.26%
Seattle-Tacoma, WA	5434	1.78%	1.68%	−0.10%
Miami, FL	1049	0.34%	0.49%	−0.15%
Pittsburgh, PA	2994	0.98%	1.03%	−0.05%
St. Louis, MO	3095	1.02%	1.09%	−0.07%
Denver, CO	3760	1.23%	1.52%	−0.29%
Phoenix, AZ	3371	1.11%	1.52%	−0.41%

Title Report DMA Sales Calculations

By manipulating the data, marketers can examine how well an artist is performing. This example of artist Miranda Lambert's debut album *Crazy Ex-Girlfriend* detects markets where Miranda is overperforming and underperforming. By calculating average sales for the market and comparing them to the national DMA genre percentage, Miranda's sales performance can be evaluated. A simple math equation, subtracting the national average from the artist's average, will create a positive or negative value. Being positive, Chicago, Dallas/Ft. Worth, Detroit, and DC are all DMAs that are overperforming in the Top 20 markets. The negative markets denote that Miranda is not selling at the market average. An analysis of these DMAs can help create a strategy to enhance sales in these markets.

Title Report: Index DMA Current

The Index Report reveals how well an artist is performing in a specific DMA. To read this report, SoundScan statistically evaluates each DMA by genre and gives the DMA a par score of 100, meaning that the genre sales average index score in that specific market equals 100. If an artist index number is over 100, the artist is overperforming in that DMA. If the index is under 100, the artist is underperforming in the market.

The index number "This Week" in New York is 146, showing that sales performed well compared to par, but the overall index for the record since its release in NY is 153, making New York a strong market overall for Keys's *As I Am*; however, for the week, sales did not do as well as previously. .

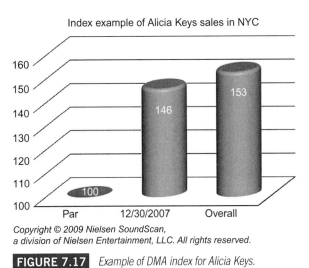

Index example of Alicia Keys sales in NYC

FIGURE 7.17   *Example of DMA index for Alicia Keys.*

Title Report: Artist History

This report is a great overview for every artist, listing all releases with pertinent information such as label, release date, and weekly and overall sales. When online, most reports are hyperlinked so that the viewer can instantly connect to the detail of the data.

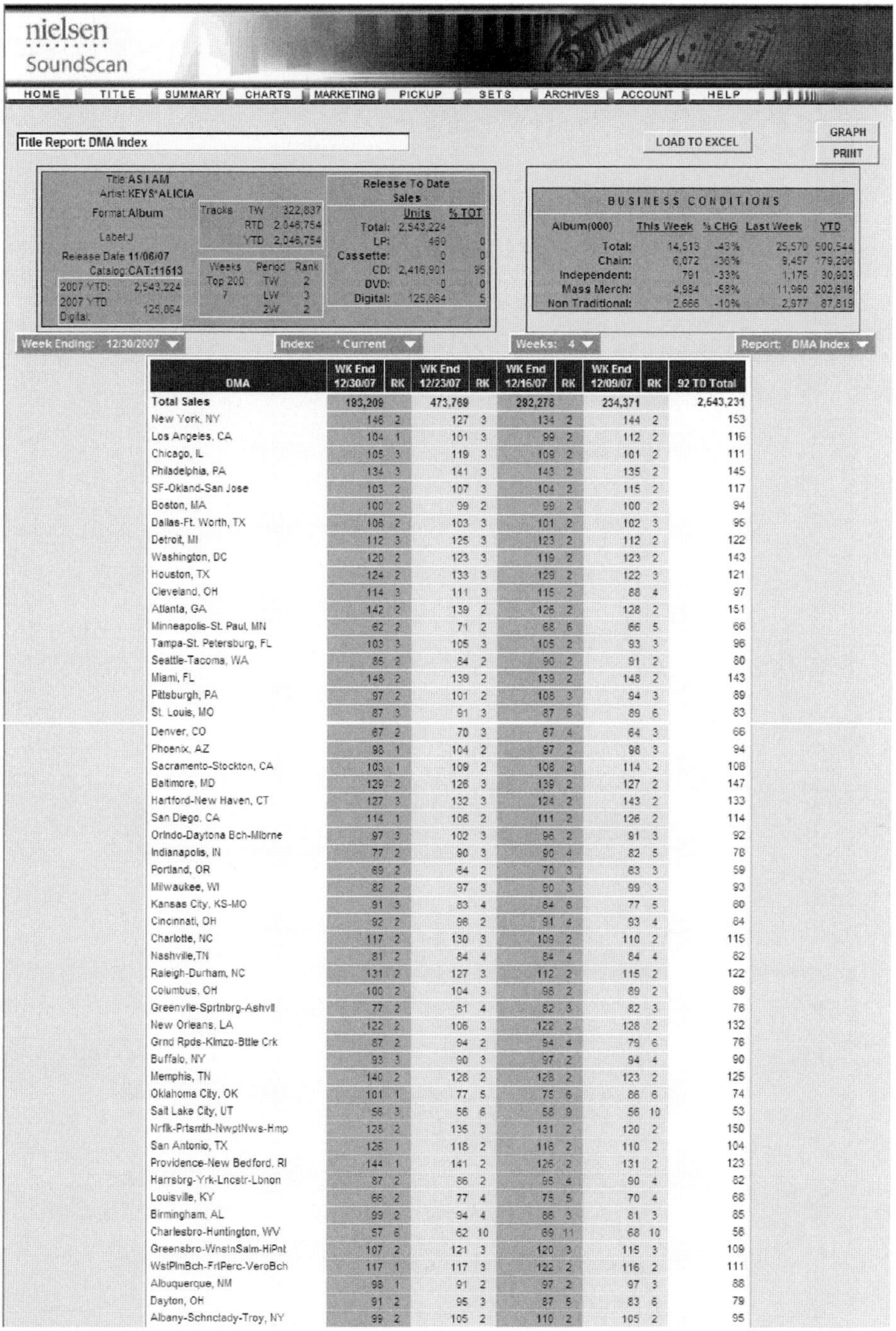

nielsen SoundScan

HOME | TITLE | SUMMARY | CHARTS | MARKETING | PICKUP | SETS | ARCHIVES | ACCOUNT | HELP

Title Report: DMA Index

LOAD TO EXCEL | GRAPH | PRINT

Title: AS I AM
Artist: KEYS*ALICIA
Format: Album
Label: J
Release Date 11/06/07
Catalog: CAT:11613

2007 YTD:	2,543,224
2007 YTD Digital:	125,864

Tracks		
	TW	322,837
	RTD	2,046,754
	YTD	2,046,754

Weeks Top 200	Period	Rank
7	TW	2
	LW	3
	2W	2

Release To Date Sales

	Units	% TOT
Total:	2,543,224	
LP:	460	0
Cassette:	0	0
CD:	2,416,901	95
DVD:	0	0
Digital:	125,864	5

BUSINESS CONDITIONS

Album(000)	This Week	% CHG	Last Week	YTD
Total:	14,513	-43%	25,570	500,544
Chain:	6,072	-36%	9,457	179,206
Independent:	791	-33%	1,175	30,903
Mass Merch:	4,984	-58%	11,960	202,816
Non Traditional:	2,666	-10%	2,977	87,819

Week Ending: 12/30/2007 | Index: * Current | Weeks: 4 | Report: DMA Index

DMA	WK End 12/30/07	RK	WK End 12/23/07	RK	WK End 12/16/07	RK	WK End 12/09/07	RK	92 TD Total
Total Sales	183,209		473,769		282,278		234,371		2,543,231
New York, NY	146	2	127	3	134	2	144	2	153
Los Angeles, CA	104	1	101	3	99	2	112	2	116
Chicago, IL	105	3	119	3	109	2	101	2	111
Philadelphia, PA	134	3	141	3	143	2	135	2	145
SF-Okland-San Jose	103	2	107	3	104	2	115	2	117
Boston, MA	100	2	99	2	99	2	100	2	94
Dallas-Ft. Worth, TX	106	2	103	3	101	2	102	3	95
Detroit, MI	112	3	125	3	123	2	112	2	122
Washington, DC	120	2	123	3	119	2	123	2	143
Houston, TX	124	2	133	3	129	2	122	3	121
Cleveland, OH	114	3	111	3	115	2	88	4	97
Atlanta, GA	142	2	139	2	126	2	128	2	151
Minneapolis-St. Paul, MN	62	2	71	2	68	6	66	5	66
Tampa-St. Petersburg, FL	103	3	105	3	105	2	93	3	96
Seattle-Tacoma, WA	85	2	84	2	90	2	91	2	80
Miami, FL	148	2	139	2	139	2	148	2	143
Pittsburgh, PA	97	2	101	2	108	3	94	3	89
St. Louis, MO	87	3	91	3	87	6	89	6	83
Denver, CO	67	2	70	3	67	4	64	3	66
Phoenix, AZ	98	1	104	2	97	2	98	3	94
Sacramento-Stockton, CA	103	1	109	2	108	2	114	2	108
Baltimore, MD	129	2	126	3	139	2	127	2	147
Hartford-New Haven, CT	127	3	132	3	124	2	143	2	133
San Diego, CA	114	1	106	2	111	2	126	2	114
Orlndo-Daytona Bch-Mlbrne	97	3	102	3	96	2	91	3	92
Indianapolis, IN	77	2	90	3	90	4	82	5	78
Portland, OR	69	2	64	2	70	3	63	3	59
Milwaukee, WI	82	2	97	3	90	3	99	3	93
Kansas City, KS-MO	91	3	83	4	84	6	77	5	80
Cincinnati, OH	92	2	96	2	91	4	93	4	84
Charlotte, NC	117	2	130	3	109	2	110	2	115
Nashville,TN	81	2	84	4	84	4	84	4	82
Raleigh-Durham, NC	131	2	127	3	112	2	115	2	122
Columbus, OH	100	2	104	3	98	2	89	2	89
Greenvlle-Sprtnbrg-Ashvll	77	2	81	4	82	3	82	3	76
New Orleans, LA	122	2	106	3	122	2	128	2	132
Grnd Rpds-Klmzo-Bttle Crk	87	2	94	2	94	4	79	6	76
Buffalo, NY	93	3	90	3	97	2	94	4	90
Memphis, TN	140	2	128	2	128	2	123	2	125
Oklahoma City, OK	101	1	77	5	75	6	86	6	74
Salt Lake City, UT	56	3	56	6	58	9	56	10	53
Nrflk-Prtsmth-NwptNws-Hmp	128	2	135	3	131	2	120	2	150
San Antonio, TX	126	1	118	2	116	2	110	2	104
Providence-New Bedford, RI	144	1	141	2	126	2	131	2	123
Harrsbrg-Yrk-Lncstr-Lbnon	87	2	86	2	95	4	90	4	82
Louisville, KY	66	2	77	4	75	5	70	4	68
Birmingham, AL	99	2	94	4	88	3	81	3	85
Charlesbro-Huntington, WV	57	6	62	10	69	11	68	10	56
Greensbro-WnstnSalm-HiPnt	107	2	121	3	120	3	115	3	109
WstPlmBch-FrtPerc-VeroBch	117	1	117	3	122	2	116	2	111
Albuquerque, NM	98	1	91	2	97	2	97	3	88
Dayton, OH	91	2	95	3	87	5	83	6	79
Albany-Schnctady-Troy, NY	99	2	105	2	110	2	105	2	95

A

FIGURE 7.18 *Example title report: DMA index Alicia Keys,* As I Am. *(Source: Nielsen SoundScan)*

Market	WK End 12/30/07	RK	WK End 12/23/07	RK	WK End 12/16/07	RK	WK End 12/09/07	RK	92 TD Total
Mobile-Pensacola, AL-FL	110	2	95	3	100	2	89	3	94
Jacksonville, FL	100	2	106	3	102	2	107	2	103
Little Rock, AR	79	2	76	5	61	6	71	7	72
Tulsa, OK	72	6	76	7	80	7	83	10	63
Flint-Saginaw-BayCity, MI	110	2	107	2	105	3	90	4	90
Richmond, VA	117	2	115	4	116	2	116	2	134
Wichita-Hutchison, KS	66	5	66	7	60	9	60	9	55
Fresno-Visalia, CA	123	1	127	2	125	2	135	2	120
Toledo, OH	93	2	94	2	109	3	94	4	82
Knoxville, TN	57	5	64	5	61	5	65	6	53
Shrvport-Txrcana, AR-LA-TX	94	2	88	3	90	3	84	3	82
Des Moines, IA	64	5	66	6	72	8	68	9	58
Green Bay-Appleton, WI	51	5	66	6	71	9	78	9	55
Syracuse, NY	99	2	100	2	107	2	89	4	86
Roanoke-Lynchburg, VA	81	1	93	4	99	3	98	3	84
Lexington, KY	56	7	69	7	67	6	70	8	61
Austin, TX	79	1	87	3	81	4	72	4	75
Rochester, NY	101	2	109	3	109	3	103	2	95
Omaha, NE	71	3	81	5	80	6	76	5	78
Portland-PolandSpring, ME	82	2	89	3	66	4	70	4	58
Sprngfld-Decatr-Chmpgn, IL	71	2	76	5	80	6	70	9	66
Pdch-CpGrdu-Hrsbg-Mrion, KY-IL	50	8	67	7	73	9	63	10	57
Spokane, WA	61	3	61	3	67	6	68	6	55
Davnprt-Rcklsnd-Moln, IL	87	4	76	5	76	8	71	9	61
Tucson, AZ	92	2	87	3	89	2	96	2	85
Hntsvlle-Decatr-Flornc, AL	83	2	72	7	63	6	81	5	72
CdarRpds-Wtrloo-Dubuqu, IA	66	3	70	7	75	8	74	9	61
Columbia, SC	144	2	136	2	128	2	125	2	140
Springfield, MO	53	7	44	12	49	11	43	13	40
Chattanooge, TN	55	9	68	9	65	6	69	6	58
Southbend-Elkhart, IN	73	2	85	3	85	5	79	4	70
Jackson, MS	136	2	115	2	117	2	118	2	112
Brstl-Kngsprt-JhnsnCty, TN	49	6	56	9	57	12	58	10	47
Johnstown-Altoona, PA	44	7	62	7	55	11	57	9	49
Youngstown, OH	100	2	101	2	110	2	96	3	88
Madison, WI	65	3	74	4	74	5	66	5	63
Las Vegas, NV	114	1	103	2	110	2	108	2	112
Brlngtn-Plattsbrgh, VT-NY	57	5	52	7	61	5	56	6	52
Evansville, IN	46	10	84	10	65	12	66	11	52
Baton Rouge, LA	112	2	103	3	122	3	124	2	125
Lincoln-Hastings-Kearney	89	5	85	7	53	10	68	9	56
Ft. Myers-Naples, FL	79	2	84	2	88	2	83	3	79
Waco-Temple-Bryan, TX	101	2	83	2	87	3	93	3	88
Springfield, MA	128	1	119	2	119	2	128	2	119
Colorado Sprngs-Pueblo, CO	74	2	80	3	80	4	77	4	75
Hawaii, HI	98	2	81	3	120	2	125	3	106
Other	73		73		75		76		89

Sub DMA	WK End 12/30/07	RK	WK End 12/23/07	RK	WK End 12/16/07	RK	WK End 12/09/07	RK	92 TD Total
New York, NY									
New York - Manhattan	127		93		109		109		126
Bronx_Queens-BrK-SI NY	216		130		153		182		208
Nassau-Suffolk NY	181		162		156		169		169
Northern Suburbs NY	151		144		145		156		156
N. J. -Pa. NY	122		126		128		136		141
Los Angeles, CA									
Los Angeles - Los Angeles	110		107		108		122		129
San Fernando Valley LA	107		101		92		110		120
Ventura LA	82		74		75		80		81
Orange County LA	86		91		90		97		99
Riverside LA	106		100		96		105		103
San Bernardino LA	109		107		104		124		120
Other Los Angeles LA	131		111		84		107		119
Chicago, IL									
Chicago - Northern Suburbs	89		102		97		89		92
Chicago CH	104		96		92		93		113
Southern Suburbs CH	129		150		130		113		128
Indiana CH	91		120		109		108		104
Philadelphia, PA									
Philadelphia - Northern Suburbs PH	116		125		128		119		124
Philadelphia PH	166		164		161		162		214
Southern Subs-Delaware PH	130		129		140		121		134
N.J. Suburbs PH	137		149		147		146		149
SF-Oakland-San Jose									
San Francisco - North Area SF	89		94		101		105		101
San Francisco-Oakland SF	107		111		104		116		123
South Bay-San Jose SF	107		114		110		123		120

B

FIGURE 7.18 (Continued)

nielsen
SoundScan

HOME | TITLE | SUMMARY | CHARTS | MARKETING | PICKUP | SETS | ARCHIVES | ACCOUNT | HELP

Title Report: Artist History

Title	Label	Release Date	Format	TW Sales	YTD Sales	2008	2007	RTD Sales
AS I AM	J	2007-11-06	Album	2,999	23,802	1,122,642	2,543,224	3,689,768
AS I AM	J	2007-11-12	Album	0	0	1	0	1
DIARY (DANCE VAULT REMIXES) -	J	2008-03-10	Album	5	15	98	0	114
DIARY OF ALICIA + BONUS CD	MRM	2004-08-30	Album	2	9	279	2,362	52,503
DIARY OF ALICIA KEYS	J	2003-11-26	Album	952	6,873	89,951	54,516	4,577,822
DIARY OF ALICIA KEYS	EGRO	2004-11-16	Video	12	59	873	1,187	32,839
DIARY OF ALICIA KEYS	PID	2004-04-06	Album	0	1	3	27	1,129
DIARY OF ALICIA KEYS	PID	2004-09-07	Album	23	208	1,914	2,095	11,624
EXCLUSIVE	BMG	2008-03-03	Album	6	32	179	0	211
FALLIN	MSIB	2001-10-30	Single	0	2	1	0	2,032
FALLIN	MSI	2001-05-21	Album	0	0	0	0	31
FALLIN	PID	2001-09-04	Single	0	0	1	1	16,567
FALLIN	MURA	2002-04-30	Single	0	0	0	0	27
FALLIN/GIRLFRIEND	J	2001-05-22	Single	0	0	1	0	25,412
GIRLFRIEND	PID	2002-12-17	Single	0	0	0	0	68
HOW COME YOU DON'T CALL ME	MSIB	2002-07-09	Single	0	0	0	0	457
HOW COME YOU DON'T CALL ME	MSI	2002-07-30	Album	0	0	2	1	62
HOW COME YOU DON'T CALL ME	MURA	2002-05-14	Single	0	0	0	0	55
HOW COME YOU DON'T CALL ME/BUT	MSIB	2003-01-20	Single	0	0	0	0	14
HOW COME YOU DON'T CALL ME/BUT	MSIB	2003-01-20	Single	0	0	0	0	4
HOW COME YOU DON'T... (EURO)	PID	2002-07-08	Single	0	0	0	0	26
F I AIN'T GOT YOU	J	2004-02-24	Single	0	0	1	8	3,335
F I AIN'T GOT YOU	PID	2004-04-20	Single	0	0	0	0	313
F I AIN'T GOT YOU	PID	2004-03-30	Single	0	0	0	0	1
F I AIN'T GOT YOU	PID	2004-03-30	Single	0	0	0	0	18
F I AIN'T GOT YOU (KANYE WEST	J	2008-03-10	Album	8	30	184	0	184
F I AIN'T GOT YOU - ACOUSTIC	J RE	2008-03-03	Album	8	57	297	0	354
F I AIN'T GOT YOU - SINGLE	J RE	2008-03-17	Single	0	4	27	0	31
F I AIN'T GOT YOU - SINGLE	J	2005-04-04	Album	0	0	0	0	0
KARMA	J	2004-10-26	Single	0	2	10	8	3,210
KARMA (REGGATONE MIX) - SINGLE	J	2008-03-03	Single	0	0	4	0	4
KEYS, ALICIA - THE LOWDOWN UNA	MVD	2008-05-13	Album	0	3	1	0	4
LIKE YOU'LL NEVER SEE ME AGAIN	J RE	2008-03-17	Album	0	0	0	0	0
MAXIMUM	PID	2002-03-21	Album	0	0	0	1	2,216
MAXIMUM ALICIA KEYS	USD	2008-03-04	Album	0	1	32	0	33
MAXIMUM ALICIA KEYS	MRM	2002-03-12	Album	0	0	0	1	104
MAXIMUM ALICIA KEYS	CHMM	2004-12-14	Album	0	0	5	11	71
MTV UNPLUGGED	J	2005-10-11	Video	64	615	9,790	9,811	95,403
NO ONE	J	2007-10-30	Single	0	4	377	228	609
NO ONE - EP	J RE	2008-03-17	Single	0	0	0	0	0
SING-A-LONG	BCIM	2004-04-13	Album	0	1	7	9	660
SONGS IN A MINOR	MSIB	2002-10-15	Album	0	3	37	29	892
SONGS IN A MINOR	J	2001-06-12	Album	1,228	9,983	77,972	55,830	6,191,605
SONGS IN A MINOR	PID	2003-06-19	Album	0	0	2	13	96
SONGS IN A MINOR	MRM	2005-07-26	Album	4	15	169	1,206	7,237
SONGS IN A MINOR/ASIAN	PID	2002-06-10	Album	0	0	14	0	167
SONGS IN A MINOR/JAPAN	PID	2002-03-04	Album	0	0	0	0	18
SONGS IN A MINOR/UK	PID	2002-03-08	Album	1	3	15	273	802
TEENAGE LOVE AFFAIR	J	2008-04-22	Single	1	5	130	0	135
THE DIARY OF ALICIA KEYS	J RE	2008-03-03	Album	0	0	0	0	0
UNLOCKED /THE OFFICIAL	J	2006-11-21	Video	73	177	144	589	890
UNBREAKABLE	J	2005-10-11	Single	0	0	3	16	1,118
UNBREAKABLE - SINGLE	J RE	2008-03-17	Single	3	14	90	0	104
UNPLUGGED	J	2005-10-11	Album	799	5,783	36,486	39,106	918,391
UNTITLED	PID	2003-11-04	Album	0	0	0	0	20
UNTITLED	PID	2003-11-04	Album	0	0	0	0	2
WOMAN WORTH (X4)	MSI	2002-02-26	Single	0	0	0	0	591
WOMAN'S WORTH	MSI	2000-10-31	Single	0	0	0	0	37
WOMAN'S WORTH	J	2001-09-25	Single	0	0	3	0	7,034
WOMAN'S WORTH	PID	2002-03-01	Single	0	0	0	0	12
WOMAN'S WORTH	PID	2002-01-29	Single	1	4	60	72	2,885
WOMAN'S WORTH	MURA	2002-05-14	Single	0	0	0	0	4
WOMAN'S WORTH (X5)	MSI	2002-04-30	Single	0	0	0	0	577
YOU DON'T KNOW MY NAME	PID	2003-11-25	Single	1	3	46	66	2,303
YOU DON'T KNOW MY NAME	PID	2004-03-23	Single	0	0	0	0	3
YOU DON'T KNOW MY NAME	J	2003-10-07	Single	0	0	1	8	7,593
Total			Album	6,031	46,421	1,318,558	2,698,704	15,456,121
			Single	6	38	755	410	74,351
			Video	149	851	10,607	11,347	135,132

FIGURE 7.19 *Example title report: artist history. (Source: Nielsen SoundScan). Copyright © 2009 Nielsen SoundScan, a division of Nielsen Entertainment, LLC. All rights reserved.*

Chart History

The chart history report documents sales from the first week of release to current day. Each week sales along with ranking of title on the Top 200 chart is presented. Additional information such as ranking at retail and mass merchants, as well as the specific genre is also included. These data drive the lifecycle sales discussed earlier in this text.

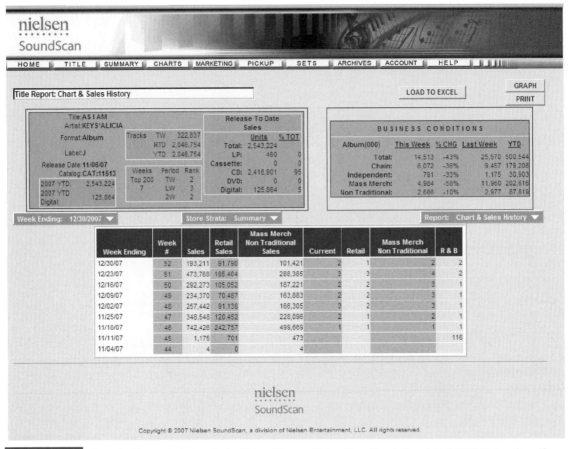

FIGURE 7.20　*Example title report: chart and sales history. (Source: Nielsen SoundScan). Copyright © 2007 Nielsen SoundScan, a division of Nielsen Entertainment, LLC. All rights reserved.*

A DEEPER LOOK AT SOUNDSCAN

SoundScan data give both summaries as well as in-depth analysis as to overall sales of music, a particular genre of music, a particular artist, a particular market, and many more aspects of the business as it pertains to sales of music. By manipulating SoundScan beyond the scope of their predetermined charts, marketers can better understand the marketplace and its drivers.

Seasonality and Record Sales

Like most products, there is seasonality to the sales of music. Every year, sales trends show a similar pattern, with sales spikes at Valentine's Day, Easter, a lull through summer months, and then a steady rise through the fall going into the holiday selling season. Using the weekly sales charts of the Top 200, this overlay of years of weekly sales crystallized seasonal sales trends.

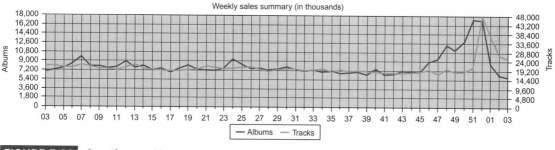

FIGURE 7.21   *SoundScan weekly album sales.*

As students of the business, record company executives analyze when the best time of year would be to release a new artist versus that of a platinum act. Strictly based on seasonality alone, most would say to release a new act shortly after the New Year to take advantage of the spring sales spikes and summer touring. And if the act has radio success, the fall selling season would be healthy for the current release already in the marketplace. As for the superstar act, a fourth-quarter release would be perfect timing to capitalize on shoppers while minimizing long-term advertising dollars.

Lifecycles

As discussed earlier in this text, there are four stages within the traditional product lifecycle: Introduction, Growth, Maturity, and Decline. As a product is adopted into use, and as others learn of its availability, its sales will grow. Eventually, the product will hit maturity, level off in sales, and decline. Either its maker will "reinvent" the product as "new and improved" and evolve it in some manner, or it will no longer exist.

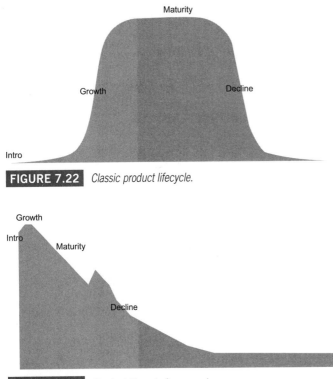

FIGURE 7.22 *Classic product lifecycle.*

FIGURE 7.23 *Product lifecycle for new releases.*

Classic Product Lifecycles

Product lifecycles occur in music too. Although there is an occasional exception to the rule, such as a second or third single from a release being the song that drives sales, most album releases have a similar sales pattern. Once an artist is established, sales patterns rarely vary, which is why the first few months of a release are so critical to the success of a record. Historically, the sales success of a release has depended on how many units were shipped initially into retail, but with the burgeoning digital market, there is a never-ending supply of a particular title on hand ready for delivery. Yet how many units are initially sold is usually dependent on the pent-up demand felt from the marketplace, including radio airplay, publicity, touring, press, and other marketing events. Recognize that "brick 'n' mortar" retailers buying music also look at the track record of the artist and his previous sales as well, remembering that as consumers shift from physical to digital consumption, the current trend is approximately 60% to 65% CD, with the remaining being a file format purchase.

Looking at these examples, most of the releases show a similar pattern in sales. Note that they are different artists, genres, sales plateaus, times of year, and yet the similarity in sales trend is unmistakable. An established artist's sales trend shows an undeniable peak in sales early in the lifecycle that tapers off within the first 6 weeks and 3 months. This is why most record labels pack their marketing strategies into this small window of time.

The exceptions to the rule are the artists that are "breaking." Check out the first record of these now superstar acts. Most of their initial releases had a "slow boil" effect, meaning that sales did not catch on at street date, but later, as consumers became knowledgeable about the artist.

Example: Jack Johnson's initial independent release took nearly a full year to make it to its peak position on the *Billboard* Top 200. *Brushfire*

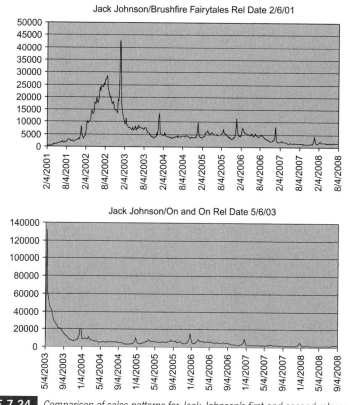

FIGURE 7.24 *Comparison of sales patterns for Jack Johnson's first and second releases. (Source: Nielsen SoundScan)*

Fairytales lumbered into the number 34 position, but this "slow boil" made for a nice long stew for Johnson as his sophomore major label release came out of the box strong. *On & On* sold over 134,000 units its initial week on the charts, debuting in the number 3 position. Johnson toured with Ben Harper while sharpening his songwriting and performance chops. With a background in film making, which precipitated his need to create music for his works, his subsequent releases including the soundtrack to the highly successful *Curious George* movie have made him an artist to keep watching.

Released in 2001, *Room for Squares* was John Mayer's "official" first album release. The hit single *No Such Thing* peaked at radio, depending on the format, in mid-2002, with the bigger hit *Your Body Is a Wonderland* driving sales even higher. Note the "slow boil" effect with the first release.

Any Given Thursday is a live album release that was to bridge sales demand while Mayer recorded the next album. Although a classic "established" artist sales profile occurs, the actual volume was much lower than that of Mayer's initial release. *Heavier Things* continued the sales trend of an established act, though showing a quick die-off in sales, mostly because of the lack of a big single at radio. *Continuum* sold well, with over 2.3 million units in the market, but Mayer's summer 2008 release has not performed as well, with only 300,000+ units sold to date. To plot his cumulative solo album sales, there is a trend that vaguely shadows that of a classic product lifecycle. It is not uncommon for most artists to live a "product" lifecycle—it's just imperative that artists plan for such an occurrence.

DMAs and Market Efficiencies

The designated market area is A. C. Nielsen's geographic market design, which defines each television market. DMAs are composed of counties (and possibly also split counties) and are updated annually by the A. C. Nielsen Company based on historical television viewing patterns. Every county or split county in the United States is assigned exclusively to one DMA.

Radio audience estimates for DMAs are published in the radio market reports of all standard radio markets whose metros are located within the DMA *and* whose names are contained in the DMA name. For example, radio audience estimates for the San Francisco-Oakland-San Jose DMA are reported in both the San Francisco and the San Jose radio market reports; however, radio audience estimates for the New York DMA are reported in the New York report, but not in the Nassau-Suffolk report (Katz Media Group radio resource area).

The following data reflects the percentage of music sold in each of the DMAs. SoundScan lists the DMAs in order of population, with New York being the most populated area surveyed, Los Angeles being the second largest populated area surveyed, and so on, with some variances based on radio or television markets. In *Radio & Records (R&R)*, populations are based on 12+ age category, people that live within the surveyed area that are 12 years or more in age. When looking at the markets, record labels should consider efficiency of the advertising dollars and marketing efforts.

Each genre of music contains a unique profile. For example, the percentage of music sold in the top DMAs for jazz equals 31.29%. Think about it—for every 100 jazz records sold, 31 of them are sold in these top 6 DMA markets. As a marketing department, advertising in these DMAs should have a big bang for the buck, considering the efficiency of targeting to the buyers in these markets. In contrast, to sell 33 records out of 100 in the country genre,

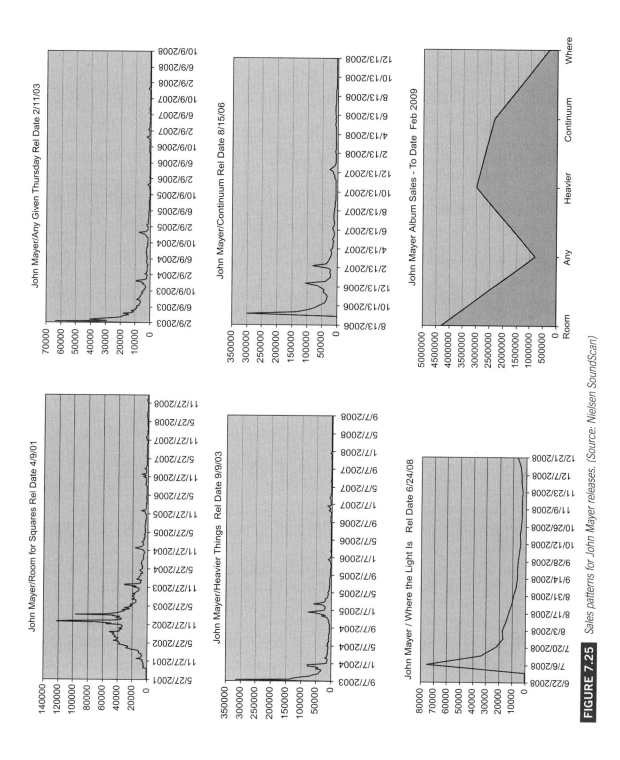

FIGURE 7.25 Sales patterns for John Mayer releases. (Source: Nielsen SoundScan)

the top 20 DMA markets are needed to achieve that percentage. Instantly, to reach buyers of country music, country record labels must spread their marketing dollars and efforts thinner, or *smarter*, to effectively reach the same percentage of buyers of the genre.

Best-Selling Markets vs. Strongest Markets

It sounds confusing—why wouldn't the best-selling markets be the *strongest* markets when looking at genre sales and/or title reports of a specific release? Let's look at how record companies can manipulate SoundScan data to be *smarter* marketers.

These data are a new sort with the same numbers. By ranking the DMAs by percentage of sales, marketing experts can now view the best-selling

Table 7.3	Comparison of the number of DMAs needed to get 30% of the market for jazz and country genres (Source: Nielsen SoundScan)		
	DMA	**Jazz**	**Country**
1	New York	10.74	2.64
2	Los Angeles	6.4	2.75
3	Chicago	3.79	2.01
4	Philadelphia	3.15	1.72
5	SF–Oakland	4.01	1.27
6	Boston	3.2	1.64
		31.29	
7	Dallas	1.59	2.37
8	Detroit	1.86	1.16
9	Washington	3.62	1.91
10	Houston	1.36	1.77
11	Cleveland	1.17	1.24
12	Atlanta	2.04	2.01
13	Minneapolis	1.53	1.66
14	Tampa	1.15	1.26
15	Seattle	2.4	1.68
16	Miami	1.6	0.49
17	Pittsburgh	0.83	1.03
18	St. Louis	1.08	1.09
19	Denver	1.52	1.52
20	Phoenix	1.32	1.52
			32.74

Table 7.4	YTD percent of sales by DMA/Genre with population ratio			
Rank	DMA	Jazz	12+ Population	Ratio sales %/Pop
9	Washington	3.62	4176300	8.67E-07
6	Boston	3.2	3838300	8.34E-07
15	Seattle	2.4	3257200	7.37E-07
4	Philadelphia	3.15	4360200	7.22E-07
1	New York	10.74	15291100	7.02E-07
19	Denver	1.52	2194800	6.93E-07
5	SF–Oakland	4.01	5891900	6.81E-07
11	Cleveland	1.17	1794200	6.52E-07
2	Los Angeles	6.4	10826600	5.91E-07
13	Minneapolis	1.53	2662100	5.75E-07
12	Atlanta	2.04	4085000	4.99E-07
14	Tampa	1.15	2314300	4.97E-07
3	Chicago	3.79	7738000	4.90E-07
8	Detroit	1.86	3888300	4.78E-07
18	St. Louis	1.08	2282700	4.73E-07
16	Miami	1.6	3533000	4.53E-07
20	Phoenix	1.32	3058000	4.32E-07
17	Pittsburgh	0.83	1998800	4.15E-07
7	Dallas	1.59	4838600	3.29E-07
10	Houston	1.36	4469900	3.04E-07

markets in order. (For this book's example, only the top 20 markets are being analyzed.) But are these the *best* markets for jazz sales? To determine the strongest markets, using population data in the equation helps to determine where to place marketing efforts.

Here is another look at the same data. By adding age 12+ population data and doing a simple ratio, the best markets emerge. The age 12+ data come from *Radio and Records* DMA ratings information. The equation: sales percentage/age 12+ population

Where New York and Los Angeles were ranked number 1 and number 2 based on population size, the strongest market for jazz based on percentage of sales to the population of the DMA would be Washington, with Boston closely behind. Although Washington does not sell as much jazz as New York, the propensity of the population to buy jazz in the DC marketplace is over 50% greater, making it a better or stronger market for jazz music.

Table 7.5	The Fray DMA cumulative sales and population ratio		
DMA	**Cume Sales**		**Ratio**
Total	**2252317**	**12+ pop**	**Sales/pop**
New York, NY	142605	15291100	9.33E-03
Los Angeles, CA	120927	10826600	1.12E-02
Chicago, IL	92036	7738000	1.19E-02
Philadelphia, PA	72834	4360200	1.67E-02
SF-Oakland-San Jose, CA	52497	5891900	8.91E-03
Boston, MA	68987	3838300	1.80E-02
Dallas-Ft. Worth, TX	54025	4838600	1.12E-02
Detroit, MI	36890	3888300	9.49E-03
Washington, DC	51252	4176300	1.23E-02
Houston, TX	37313	4469900	8.35E-03
Cleveland, OH	29634	1794200	1.65E-02
Atlanta, GA	47912	4085000	1.17E-02
Minneapolis-St. Paul, MN	50511	2662100	1.90E-02
Tampa-St. Petersburg, FL	22381	2314300	9.67E-03
Seattle-Tacoma, WA	41236	3257200	1.27E-02
Miami, FL	21026	3533000	5.95E-03
Pittsburgh, PA	21110	1998800	1.06E-02
St. Louis, MO	25199	2282700	1.10E-02
Denver, CO	74073	2194800	3.37E-02
Phoenix, AZ	38897	3058000	1.27E-02

This same type of information can be used with title report data. By looking at the sales of an artist's specific record, labels can determine where to place marketing and dollars.

This example is The Fray's *How to Save a Life* album that was released September 13, 2005. Again, ranked by DMA, then ranked by sales, then ranked by sales/population ratio, look at the variance in market strength.

Based on sales per population, Denver emerges as the number one strongest market for sales of The Fray's *How to Save a Life*. By concentrating on markets that have a stronger probability of sales, labels can better manage their marketing dollars through succinct activities that may include radio promotions, in-store events, touring, and so on. The goal is to maximize the market and sell records. This record also falls into the "slow boil" category, with the band being featured on the hit TV series and subsequent compilation of *Grey's Anatomy*. By year's end 2008, *How to Save a Life* had sold over 2 million copies.

Table 7.6	The Fray DMA cumulative sales and population ratio ranked by sales		
DMA	**Cume Sales**		**Ratio**
Total	**2252317**	**12+ pop**	**Sales/pop**
New York, NY	142605	15291100	9.33E-03
Los Angeles, CA	120927	10826600	1.12E-02
Chicago, IL	92036	7738000	1.19E-02
Denver, CO	74073	2194800	3.37E-02
Philadelphia, PA	72834	4360200	1.67E-02
Boston, MA	68987	3838300	1.80E-02
Dallas-Ft. Worth, TX	54025	4838600	1.12E-02
SF-Oakland-San Jose, CA	52497	5891900	8.91E-03
Washington, DC	51252	4176300	1.23E-02
Minneapolis-St. Paul, MN	50511	2662100	1.90E-02
Atlanta, GA	47912	4085000	1.17E-02
Seattle-Tacoma, WA	41236	3257200	1.27E-02
Phoenix, AZ	38897	3058000	1.27E-02
Houston, TX	37313	4469900	8.35E-03
Detroit, MI	36890	3888300	9.49E-03
Cleveland, OH	29634	1794200	1.65E-02
St. Louis, MO	25199	2282700	1.10E-02
Tampa-St. Petersburg, FL	22381	2314300	9.67E-03
Pittsburgh, PA	21110	1998800	1.06E-02
Miami, FL	21026	3533000	5.95E-03

Table 7.7	The Fray DMA cumulative sales and population ratio ranked by sales ratio		
DMA	**Cume sales**		**Ratio**
Total	**2252317**	**12+ pop**	**Sales/pop**
Denver, CO	74073	2194800	3.37E-02
Minneapolis-St. Paul, MN	50511	2662100	1.90E-02
Boston, MA	68987	3838300	1.80E-02
Philadelphia, PA	72834	4360200	1.67E-02
Cleveland, OH	29634	1794200	1.65E-02
Phoenix, AZ	38897	3058000	1.27E-02
Seattle-Tacoma, WA	41236	3257200	1.27E-02
Washington, DC	51252	4176300	1.23E-02
Chicago, IL	92036	7738000	1.19E-02
Atlanta, GA	47912	4085000	1.17E-02
Los Angeles, CA	120927	10826600	1.12E-02
Dallas-Ft. Worth, TX	54025	4838600	1.12E-02
St. Louis, MO	25199	2282700	1.10E-02
Pittsburgh, PA	21110	1998800	1.06E-02
Tampa-St. Petersburg, FL	22381	2314300	9.67E-03
Detroit, MI	36890	3888300	9.49E-03
New York, NY	142605	15291100	9.33E-03
SF-Oakland-San Jose, CA	52497	5891900	8.91E-03
Houston, TX	37313	4469900	8.35E-03
Miami, FL	21026	3533000	5.95E-03

CHAPTER 8

How Radio Works

Paul Allen

RADIO

Despite the talk on blogs and other online chatter about the pending demise of traditional commercial radio, radio continues to be an immense influence on music purchase decisions that consumers make. Arbitron provides radio audience measurement services for the broadcast and advertising industries, and in its 2008 report the company notes that 90% of people in the United States listen to commercial radio every week. On average, these same consumers listen to 18 hours of radio each week, down only 75 minutes over the previous 6 years; on a daily basis, they listen to one or more radio stations for more than 2½ hours each day (Arbitron, 2008a). Additionally, the PEW Internet and Life Project reported in May, 2008, that 83% of music buyers learn about new music through radio, television, or movies.

Anyone in the business of selling recordings is making a mistake by underestimating the reach and impact that radio has with consumers. Labels understand this, and it is why they continue to put considerable resources into influencing decisions by radio programmers and other radio gatekeepers to get their new music on the airwaves. To make the point, weekly airplay audience sizes for selected singles released to commercial radio as counted by Broadcast Data Systems (BDS) and reported in *R&R* for the week of February 6, 2009 are shown in Table 8.1.

Online and other viral promotions of recorded music are important elements of any marketing plan, but a competitive international plan with a goal of selling more than 200,000 units must account for the inclusion of promotion to commercial radio as a key element of its success.

CONTENTS

Table 8.1	Audience sizes for songs on radio	
Artist	**Song**	**National audience**
Kanye West	*Heartless*	69.5 million
Taylor Swift	*Love Story*	65.7
Blake Shelton	*She Wouldn't Be Gone*	31.9 million
Beyonce	*Single Ladies*	64.7 million

Veterans of the recording industry estimate that as many as 70% of consumer decisions to purchase recorded music can be traced directly back to exposure to the music by commercial radio. As much as consumers complain about the large number of commercials and repeated playing of the same music, radio still is the most important vehicle the recording industry has to showcase its product to the public. And despite our wireless devices, MP3 players, GPS units, dashboard video players, and other distractions, 45% of our time listening to radio occurs while we are in our vehicles (Arbitron, 2008b).

Given the role radio plays in promoting recordings to consumers, it's important to have an understanding of radio and the people who make programming decisions at those stations. They are the gatekeepers to the radio station's airwaves. When the marketing and promotion staff at the label understand what radio needs, it becomes easier for them to find a way to get their new music programmed.

THE BUSINESSES

One of the best adjectives to describe the relationship shared by the recording industry and commercial radio is "symbiotic." Though it is a term most often used in science, it means the two industries share a mutual dependence on each other for a mutual benefit. Radio depends on the recording industry to provide elements of its entertainment programming for its listeners, and the recording industry depends on radio to expose its product to consumers. No two other industries share a relationship as unique as this. However, the natures of the businesses of a record company and a radio station are very different. For a record company, it's easy to define the business: to sell recordings. Money moves from consumers to record companies when recordings are sold.

The business of radio is building an audience that it leases to advertisers. Radio uses music to attract listeners in order to attract advertising revenue. The larger the audience the station attracts, the more it can charge for its advertising. Radio, however, is not in the business of building recording careers, nor is it interested in selling recordings. The number of units that a recording is selling might be of interest to a radio programmer, but that information by itself does not necessarily affect programming decisions.

THE RADIO BROADCASTING INDUSTRY

The traditional over-the-air radio broadcast industry in the United States began consolidating when the Telecommunications Act of 1996 was signed into law. Prior to the new law, radio broadcasting companies were limited in ownership to 20 AM and 20 FM stations. The law now allows companies very broad latitude on the number of radio stations they may own, but it typically limits the control of a radio audience to less than 30% for most markets.

Radio ownership has created some of the largest media companies ever. Ranked in the order of their estimated annual earnings, here are the biggest radio companies in America.

Table 8.2	USA (2009) radio station companies
Radio broadcast company	**Number of stations**
Clear Channel Radio	1000
CBS Radio	180
Cox Radio	75
Entercom Communications	100
Citadel Broadcasting	200
ABC Radio	70

Revenue for the industry has been steadily declining, with annual earnings now eclipsed by advertising revenue on the Internet. The industry had been earning in the area of $20 billion annually, but revenue continued to erode, in part because of the global recession of 2008–2009. The largest advertiser category for commercial radio has traditionally been for automobile companies; the recession crippled the auto industry and it significantly reduced advertising spending, impacting a radio industry that was already struggling.

In the 10 years following the deregulation of radio ownership, industry consolidation caused a 39% decline in the number of owners of U.S. radio stations, from 5133 to 3126. The FCC licenses stations based upon the availability of frequencies in desired markets, but there was growth of only 7% in the number of stations from 1996–2007. In 2008 in the United States, the FCC reported there were 4786 AM stations and 6427 FM commercial stations (FCC, 2008). There were also 3040 FM educational stations (FCC, 2008).

THE RADIO STATION STAFFING

To see how decisions are made about music choices at a radio station, it is important to understand the relationships within the station. The general manager, market manager, or someone with a similar title is responsible for the business success of the station. Reporting to the general manager is a manager of administration who has responsibilities such as accounting, commercial scheduling, and keeping up with regulatory matters. The sales manager has a staff of people who sell available commercial time to advertisers. Promotions are contests and other sales-oriented activities that are often the collaborative work of the sales team as well as the programming department.

From a record marketing standpoint, the key positions at a radio station are the program director and, to a lesser degree, the music director. The **program director** (**PD**) is genuinely the gatekeeper. Without the "okay" of the programmer, there is no chance that a recording will get on the air at most large radio stations. The PD is directly responsible to the general manager for creating programming that will satisfy the target market and build the existing audience base. The programmer decides what music is played, which announcers are hired, which network services to use, how commercials are produced, and every other aspect of the image the station has within the community it serves.

Critics of radio often say program directors have too much power because they can decide whether a recording is ever exposed to listeners. Large radio chains have group programmers who play an even larger role as gatekeepers, recommending which music is appropriate for similarly programmed stations owned by the company across the country. As group owners of stations seek economies within their companies, group programmers—rather than a local program director—play a larger role than ever in the decisions regarding which music is played for the station's audience. To the radio audience, the silent partner in decisions about music programming is often a consultant who assists the PD with suggestions

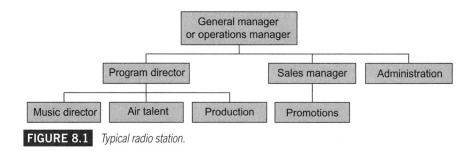

FIGURE 8.1   *Typical radio station.*

about music, often armed with research about both the music and the local audience. To the record label, the consultant becomes another important gatekeeper.

Critics of program directors and their decisions to limit the size of their music **playlists** say that radio is serving as an arbitrary filter for the massive amount of recorded music that is created every year and that it limits opportunities for newer artists. Theoretically, radio finds the most appropriate music for its audience and filters the music by choosing the best selections for the target audience. Program directors use advice, research, and their experience to find the best mix of music and information to retain and build their audiences for advertisers.

RADIO AUDIENCES

Arbitron, the audience measurement company, publishes its annual *Radio Today* in which it provides an analysis of the makeup of audiences who use commercial radio in the United States. The following chart is taken from their 2007 report and shows the percentage of persons by age and gender who listen to radio during the average week. The dark column represents males and the light column represents females. Cume is a reference to the cumulative audience for the average week. For example, this charts shows that of people who are 12 to 17 years old (P12–17), 87.2% of males listen to radio each week, whereas 93.5% of females listen to radio each week.

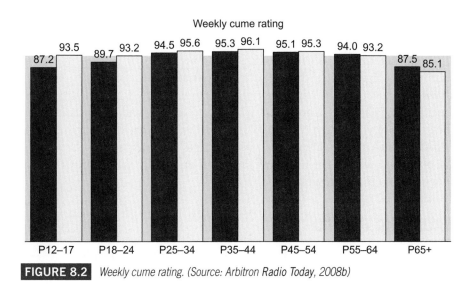

FIGURE 8.2 *Weekly cume rating. (Source: Arbitron Radio Today, 2008b)*

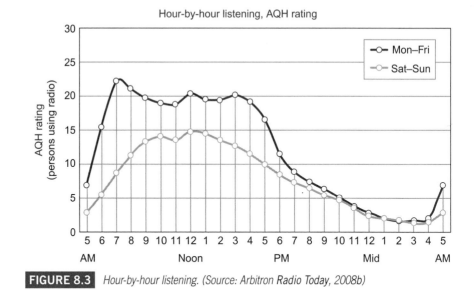

FIGURE 8.3   *Hour-by-hour listening. (Source: Arbitron Radio Today, 2008b)*

The size of the audience of a radio station is important to a label because it often determines how much time and other resources are put into promoting a song to the programming executive. Also important to the label is the time of day that the song is scheduled to be played. As you can tell by the Hour-by-Hour Listening chart, audiences are considerably larger during the week and during morning and evening rush hours. Radio refers to rush hour programming as "drive time" because audiences are largest and the most valuable to the station because of their size. AQH is a reference to the numerical size of the audience during the average quarter-hour within the timeframe measured by Arbitron.

Radio listening peaks in the morning hours, known as morning drive time. Radio listening is divided into **dayparts** of *morning drive, midday, afternoon drive, evening,* and *overnight.* From 6:00 A.M. until 9:00 A.M., listening is greatest as commuters wake up to alarms and clock radios, and they continue to listen as they drive to work. Listening picks up again around noon, declines slightly after lunch, but remains relatively strong throughout the afternoon drive time, and then tapers off drastically throughout the evening and into the overnight period.

RADIO ON THE GO

Radio listening is very popular away from home. Between 8:00 A.M. and 6:00 P.M. on weekdays, a majority of listening occurs outside the home, as much as 69%. After 6:00 P.M., the majority of listening shifts to in-home.

Distribution of AQH radio listeners
by listening location
Persons 12+

	Home	Car	Work	Other
Mon–Sun, 6 AM–Mid	38.9%	35.5%	23.0%	2.6%
Mon–Fri, 6 AM–10 AM	38.1%	37.6%	23.0%	1.3%
Mon–Fri, 10 AM–3 PM	29.0%	30.2%	38.6%	2.2%
Mon–Fri, 3 PM–7 PM	30.8%	45.1%	21.7%	2.4%
Mon–Fri, 7 PM–Mid	58.3%	28.0%	10.1%	3.6%
Weekend, 10 AM–7 PM	48.0%	37.9%	9.6%	4.4%

FIGURE 8.4 *Radio listening during the average quarter-hour. (Source: Arbitron Radio Today, 2008b)*

RADIO FORMATS

Station owners choose a radio **format**—whether it is a musical genre or a news/talk program—by finding an underserved audience that is attractive to advertisers. When the format is chosen and developed, a programmer and staff are hired, and the audience develops. The chart in Figure 8.5 shows the national radio audience sizes by format in 2007. The horizontal axis reflects the percentage of radio listeners who chose to listen to the radio formats represented in the chart.

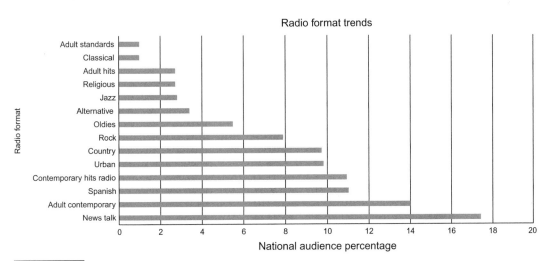

FIGURE 8.5 *Audience format trends today. (Source: Arbitron, 2008b)*

The Arbitron audience measurement service reports the national percentages of radio format shares in the chart. The top radio format is news/talk. This format has maintained its strength in recent years and continues to represent nearly 18% of all listeners.

Arbitron is a subscription service and is the only major company that measures the size and demographics of radio audiences. Although the audience share chart shows the size of the national audience, Arbitron measures the same information, radio market by radio market. The share and audience makeup of each individual commercial radio station is measured and reported to subscribing stations and advertising agencies. The size of the station's radio audience is directly related to the amount of money the station can charge for its advertising. The more listeners (or the larger its audience share), the more the station charges companies to access their audience through advertising. Arbitron charges its clients tens of thousands of dollars for its audience measurement services. Because college and other noncommercial stations do not use traditional advertising, their audience shares are not reported.

With this in mind, a programmer is very careful in choosing music for airplay because the objective is to build its target audience. The program director is not inclined to experiment with an unproven recording that will turn an audience off. More about this will be discussed in a later chapter.

Targets of Radio Formats

In order to be a commercial product, recorded music must find a target that is able and willing to buy it. Finding that target is the first step in the marketing process, followed by the development of a strategy to reach the target. This table provides some broad definitions of music formats and their targets.

The target market of a particular radio format is the logical consumer target for commercial recordings. As of this writing, there are over 26 specific radio music formats in the United States charted by *Billboard* and by MediaBase's

Table 8.3	Targets of Radio Formats*	
Format name	**Target demographic**	**Artists in the format**
Adult contemporary	Females 25–54	Coldplay, Beyonce, Nickelback
Active rock	Men 18–34	Metallica, 3 Doors Down, Seether
Alternative	Persons 18–34	Kings of Leon, Staind, Killers
CHR*/pop	Persons 18–34	Kelly Clarkson, Fray, Ne-Yo
CHR/rhythmic	Persons 12–24	Akon, T.I., Flo Rida
Country	Persons 25–54	Taylor Swift, Dierks Bentley, Tim McGraw
Hot adult contemporary	Females 18–24	Britney Spears, David Cook, Daughtry
Urban	Persons 18–34	Kanye West, Keyshia Cole, T.I.

*Artists listed in this chart are as they appear in the February 24, 2009, MediaBase America's Music Charts.

American Music Charts. Arbitron rates 57 different radio formats, which is a reflection of new music, older music, foreign language, and talk shows.

Radio station group owners often refine these gender and age targets. For example, some country radio stations owned by Clear Channel Communications specifically target females 35 to 44 years old with their country programming, while some owners target 12- to 24-year-old males with the hip-hop sounds of CHR/Rhythmic. The ability to reach the format targets of terrestrial radio stations can be a major factor in determining whether a recording will be released commercially.

One of the key components of most of these radio market targets is the 18- to 34-year-old female. Women in this age group heavily influence or actually make the purchase decisions for households, and advertisers highly value this demographic as a target for their messages.

Programmers sometimes use a clock wheel to offer a visualization of how time is allotted to the various broadcast elements. It is a wheel indicating sequence or order of programming ingredients aired during one hour (Tarver, 2005). The clock face is divided into pie pieces, and each small section of time in an hour is prescribed a very specific item to be played on the air—from a song in a specific genre to a commercial, to news and weather. For example, at the top of the format clock in Figure 8.6 is the top of the

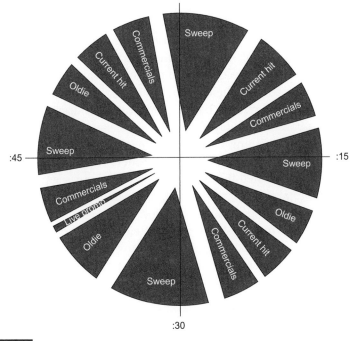

FIGURE 8.6 *Format clock.*

hour followed by a "sweep" of uninterrupted music, brief comments by the announcer, followed by a current hit with another break by the announcer, some commercials and the announcer, and then another music sweep. The clock becomes an efficient guide to the station's employees of the expected sequence and length of time for each of the timed segments within the broadcast day for each of the air shifts.

WHAT IS IMPORTANT TO PROGRAMMERS

Convincing radio to play new music is "selling" in every sense of the word. And in order to sell people anything, you must know what is important to them and what their needs are. High on that list of important things to radio is Arbitron's measurement of radio audiences because it directly impacts the earnings of the station for its owners. Understanding concepts like this and their importance to programmers will help marketers of recorded music better relate to the needs of radio and its programming gatekeepers.

RATINGS RESEARCH

The term **P-1 listeners** represents one of the prized numbers of radio programming. As Mike McVay of McVay Media puts it, "These first preference listeners . . . are referred to in radio station boardrooms, focus groups, and inside the headquarters of Arbitron doing diary reviews." McVay says the term P-1 describes "the most loyal of radio listeners," and radio programmers court this primary core of their radio audiences (www.mcvaymedia.com). The terms P-2 and P-3 refer to listeners with a lesser degree of connection and loyalty to a particular radio station.

Cume is a programming term that comes from the word cumulative, and it refers to the total of all different listeners who tune into a particular radio station, measured by Arbitron in quarter-hour segments. In other words, it is the total number of unduplicated persons included in the audience of a station over a specified time period.

AQH, or **average quarter-hour**, refers to the number of people listening to a radio station for at least five minutes during a fifteen minute period.

TSL means "time spent listening" by radio station listeners at particular times of the day. TSL is calculated by the following formula.

$$\text{Quarter-Hours in a Time Period} \times \frac{\text{AQH Persons}}{\text{Cume Audience}} = \text{TSL}$$

A radio station's **share** refers to the percentage of persons tuned to a station out of all the people using the medium at this time.

$$\frac{\text{AQH Persons Tuned to a Specific Station}}{\text{AQH Persons in Market Currently Listening to Radio}} \times 100 = \text{Share}$$

Rating refers to the percentage of persons tuned to a station out of the total market population.

$$\frac{\text{AQH Persons Tuned to a Specific Station}}{\text{Persons in Market}} \times 100 = \text{Rating}$$

A full review of terms used in audience measurement is at www. arbitron.com.

Arbitron manually measures and rates radio listener habits in 280 markets in the United States. Ratings are measured using the diary method in smaller markets. Larger markets are measured using an electronic device called the portable people meter (PPM), shown in Figure 8.8.

For the diary method of tracking listening, Arbitron selects households at random and asks members' ages 12 and older to carry the diary for one week and record their radio listening. Potential diary keepers are first contacted by telephone, and then diaries are sent to the household. Completed diaries are returned to Arbitron and the data are entered into computers and analyzed on the following characteristics:

- Geographic survey area (metro or total survey area)
- Demographic group
- Daypart
- Each station's AQH: the estimated number of persons listening
- Each station's rating: the percentage of listeners in the area of study during the daypart
- Each station's share: the percentage of one station's total daypart estimated listening audience
- Cume: the total unduplicated audience during the daypart for an average week

For those using the portable people meter (PPM), it "is a unique audience measurement system that tracks what consumers listen to on the radio, and what consumers watch on broadcast television, cable and satellite TV. The PPM is a pager-sized device that consumers wear throughout

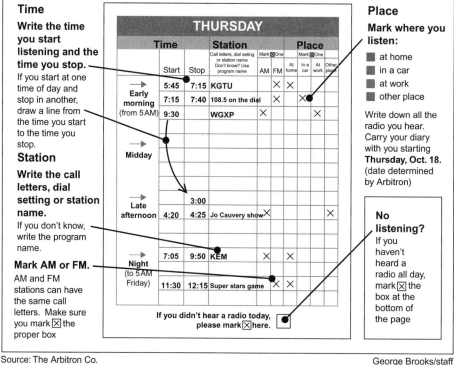

FIGURE 8.7 How to keep an Arbitron diary. (Source: Courtesy of George Brooks of The Tennessean).

the day. It works by detecting identification codes that can be embedded in the audio portion of any transmission" (Arbitron, 2009). The meter is the size of a pager and is worn at all times during the day. At night, the PPM is placed into a base unit and information is uploaded to a central database. Whereas the diary method of tracking radio listenership uses diaries targeted to a balanced yet random population sample, the PPM enlists families of people, pays them a fee, and seeks a two-year commitment for them to be part of the program.

Each Arbitron *Radio Market* report covers a 12-week period for the specified market and contains numerous pages like the example in Figure 8.9 on the following page. At the top of each page, the target demographic is

FIGURE 8.8   *Portable people meter. (Source: Arbitron, 2009)*

listed. Beneath that, the dayparts are laid out in columns. Then for each daypart, the AQH, the cume, the AQH rating, and AQH share are listed for each radio station in the area (in the left-hand column).

Ratings as defined by the audience share of a station determine advertising rates, and the example of an Arbitron report of the Atlanta radio market makes the point that ratings mean money. The chart in Appendix A of this chapter shows the call letters, the station format, the owner, and the percentage of listeners in the Atlanta market who choose each station. The table of ratings is shown beginning with the fall of 2003 as an indication of the trends of the ratings for each station in the Atlanta market. The column on the far right is the station rating for fall 2004. The station showing 9.7% of the market's radio audience charges more for advertising than the station with only 0.4% of the audience because it has more listeners and can therefore charge more for its advertising. Arbitron rating points are the targets of the audience-building efforts of a radio programmer. For example, each one-tenth of a rating point, or 0.1%, is worth $1 million in advertising rates to radio stations in the Los Angeles radio market.

Listener estimates/metro

Target listener trends

| | Persons 12+ | | | | | | | | | | | | | | |
| | Monday–Sunday 6 AM–MID | | | | Monday–Friday 6 AM–10 AM | | | | Monday–Friday 10 AM–3 PM | | | | Monday–Friday 3 PM–7 PM | | |
	AQH (00)	Cume (00)	AQH Rtg	AQH Shr	AQH (00)	Cume (00)	AQH Rtg	AQH Shr	AQH (00)	Cume (00)	AQH Rtg	AQH Shr	AQH (00)	Cume (00)	AQH Rtg
WAAA-AM															
SO '01	118	1731	1.9	9.9	167	1118	2.6	10.8	177	923	2.8	10.8	165	1172	2.6
WI '01	123	1980	1.9	10.1	155	908	2.4	9.6	186	1064	2.9	10.9	152	1115	2.4
FA '00	101	2120	1.6	9.0	110	1110	1.7	7.2	130	1207	2.0	8.8	130	1312	2.0
SU '00	115	2238	1.8	9.3	144	1233	2.2	9.3	144	1202	2.2	8.4	148	1264	2.3
4-Book	*114*	*2017*	*1.8*	*9.6*	*144*	*1092*	*2.2*	*9.2*	*159*	*1099*	*2.5*	*9.7*	*149*	*1216*	*2.3*
SP '00	126	2259	2.0	10.5	172	1272	2.7	11.1	193	1207	3.0	11.7	153	1492	2.4
					① ② ③ ④										
WBBB-AM															
SP '01	118	1731	1.9	9.9	167	1118	2.6	10.8	177	923	2.8	10.8	165	1172	2.6
WI '01	123	1980	1.9	10.1	155	908	2.4	9.6	186	1064	2.9	10.9	152	1115	2.4
FA '00	101	2120	1.6	9.0	110	1110	1.7	7.2	130	1207	2.0	8.8	130	1312	2.0
SU '00	115	2238	1.8	9.3	144	1233	2.2	9.3	144	1202	2.2	8.4	148	1264	2.3
4-Book	*114*	*2017*	*1.8*	*9.6*	*144*	*1092*	*2.2*	*9.2*	*159*	*1099*	*2.5*	*9.7*	*149*	*1216*	*2.3*
SP '00	126	2259	2.0	10.5	172	1272	2.7	11.1	193	1207	3.0	11.7	153	1492	2.4
WCCC-AM															
SP '01	118												165	1172	2.6
WI '01	123												152	1115	2.4
FA '00	101												130	1312	2.0
SU '00	115												148	1264	2.3
4-Book	*114*												*149*	*1216*	*2.3*
SP '00	126												153	1492	2.4
WDDD-AM															
SP '01	118												165	1172	2.6
WI '01	123												152	1115	2.4
FA '00	101	2120	1.6	9.0	110	1110	1.7	7.2	130	1207	2.0	8.8	130	1312	2.0
SU '00	115	2238	1.8	9.3	144	1233	2.2	9.3	144	1202	2.2	8.4	148	1264	2.3
4-Book	*114*	*2017*	*1.8*	*9.6*	*144*	*1092*	*2.2*	*9.2*	*159*	*1099*	*2.5*	*9.7*			*2.3*
SP '00	126	2259	2.0	10.5	172	1272	2.7	11.1	193	1207	3.0	11.7			2.4

1. During an average quarter hour between 6:00 AM and 10:00 AM for spring 2001, 16,700 people listened to WBBB-AM for a minimum of five minutes.
2. During the 6:00 AM to 10:00 AM daypart, 111,800 different persons listened to WBBB-AM for a minimum of five minutes in a quarter hour.
3. During the average quarter hour in this daypart, 2.6 percent of all persons in this market were listening to this station.
4. During the average quarter hour in this daypart, 10.8 percent of all persons who were listening to any radio station in this market were listening to this station.

Source: Arbitron

FIGURE 8.9 *Example Arbitron book. (Source: Arbitron).*

RADIO PROGRAMMING RESEARCH

Knowing how radio researches its audience can be helpful to marketers to understand how programmers define benchmarks for their decisions to add or remove music from their playlists, or increase or decrease the frequency of playing songs. The key research tools used by programmers are discussed in the next chapter in the section on charts.

Many stations use panels of listeners for programming research. The panel method is a research technique in which the same people are studied at different points in time. Members of the panel are selected to reflect a representative sample of a station or format's listenership, and are periodically surveyed on their opinions of music and programming. The panel members are contacted either by telephone or email and asked to respond to song hooks played either over the phone or through the Internet. The programmer then analyzes the listener response to make decisions about the continued use of the song on the air and how frequently it will be played. Another tool used by some radio programmers is auditorium research: a group—typically 100 to 300 listeners of the radio format—is asked to attend a meeting where they will react to recognizable segments of familiar songs for the benefit of the programmer's music decisions.

Another tool being used by programmers is called MScore, which uses the audience preference tracking of the portable people meter to judge whether listeners switch stations while a song is playing. For programmers, it is useful feedback about the music they program; for record labels, the information is useful to continuously monitor a song in the marketplace and modify promotional strategy (Albright, 2009; mediamonitors.com).

GETTING AIRPLAY

It is the job of the record promoter to get airplay on commercial radio stations. This has become more difficult with the consolidation of radio because there are fewer music programmers and competition for getting added to the playlist is fierce. The process of record promotion is outlined in the following chapter on charts, airplay, and promotion.

HD RADIO

Opportunities for marketers of recorded music have improved with the addition of new technology for the radio broadcast industry. Commercial stations are in the process of converting their AM and FM stations to HD radio, greatly improving the quality of the signals for both. (HD does not mean "high definition." It is a term used to brand the new service.) What this means to music marketers is that songs played on the radio will deliver near-CD quality audio and have the ability to display the artist's name and the song title on the radio receiver. Radio announcers infrequently provide artist or song information to listeners, and this new technology will help consumers of recorded music to identify artists and songs. As radio stations convert to digital broadcasting, they are also given up to two additional stations that are adjacent to their primary broadcast frequency. These new "stations" give the station owners opportunities to explore experimental programming and to broaden their listener bases. An element of a lawsuit settlement with the New York attorney general in 2005 requires several major broadcast companies to give new artists not associated with major labels the opportunity to have their music heard. It often is the HD channels that are used to allow these artists to be heard on commercial radio.

Digital radio availability is growing as major broadcast companies have begun converting stations to digital broadcast signals, ultimately making them available to listeners in every major and large radio market. As the broadcast station conversion continues, there are issues that require resolution before HD radio becomes widely accepted with consumers. Those issues include:

- HD radios are available to retrofit vehicles but they can be expensive for consumers.

- Consumers are often unaware how the cost of an HD radio benefits them.

- Arbitron shows there is a finite radio audience that is becoming smaller; radio has three times as many radio stations vying for the same size audience.

- The global recession of 2008–2009 reduced the number of new vehicles sold with the new HD technology, delaying the general introduction of the service.

- Smaller radio companies have been unable to secure funding to convert transmission equipment to provide HD to their audiences.

SATELLITE RADIO

Although satellite radio has been an opportunity for music makers to get more of their products exposed to consumers, the future of the company remains uncertain.

Sirius XM offers an array of music and other entertainment channels which are fed to proprietary radio receivers, meaning you must own a special receiver or an iPhone in order to access programming, or you must be an online subscriber. The subscriber to Sirius XM pays a monthly fee for basic services and a higher fee for additional services, much like a satellite television service. The opportunities for marketers of recordings with satellite radio are that the company has longer playlists within each genre of music; there are more opportunities for new music to be played; and every song that is played displays the artist and song title to the listener on the faceplate of the receiver. As a viable service, satellite radio has been slow to be adopted by consumers, in part because of the cost of buying special receiving equipment for a vehicle or a home audio system, and because of the monthly fees associated with the services. As part of its 2009 restructuring,

Sirius XM began offering its services to users of AT&T's iPhone, eliminating the need for special equipment with antennae. However, even with a subscription base approaching 20 million subscribers, the satellite radio company has never made a profit.

INTERNET RADIO

Internet "radio" is considerably different from radio in its traditional form. While this is promoted as "radio," it typically does not include the presence of an announcer or noncommercial information. A number of browsers and Internet service providers include bundled software, which gives users access to genre-specific music through their PCs. And, custom services like Last.fm.com and Pandora.com offer some of the features of traditional radio while giving listeners a unique listening experience closely tailored to their specific music tastes. With the expanding availability of WiFi service and the power of 3G networks, streaming radio-like music services over wireless devices brings music closer to specific consumer interests and poses a competitive challenge for traditional commercial radio. An additional advantage of Internet radio is that the song information is displayed on the computer screen as the song is played, often with a link to an online retail store to facilitate the impulse purchase.

GLOSSARY

average quarter-hour (AQH)—The number of people listening to a radio station during a 15-minute period as measured by Arbitron.

cume—The total of all different listeners who tune in to a particular radio station.

dayparts—Specific segments of the broadcast day; for example, midday, morning drive time, afternoon drive time, and late night.

format—The kind of programming used by a radio station to entertain its audience.

playlist—The list of songs currently being played by a radio station.

P-1 listeners—The primary core of listeners to a specific radio station.

program director (PD)—An employee of a radio station or a group of radio stations who has authority over everything that goes over the air.

share—The radio audience of a specific station measured as a percentage of the total available audience in the market.

TSL—Time spent listening by radio station listeners at particular times of the day.

BIBLIOGRAPHY

Arbitron and www.arbitron.com

McVay, Mike, McVay Media, www.mcvaymedia.com

MediaBase/Rate the Music (Clear Channel Entertainment)

Radio Advertising Bureau and rab.org

Radio and Records magazine and www.rronline.com

Recording Industry Association of America

Radio today: How America listens to radio. (2008). Arbitron report.

ReferenceUSA. (2009). www.libraryusa.com/referenceusa.asp Accessed 1.03.09.

Sterling, C. (June 21, 2006). Transformation: The 1996 law reshapes radio. *The Federal Communications Law Journal, 58.*

Appendix A

Owner	Station	Format	F03	W04	Sp04	Su04	F04
Cox	WSB-AM	News/talk	10.3	8.7	9.4	11.6	9.7
Infinity	WVEE-FM	Urban	9.0	7.3	7.9	7.4	7.7
ABC	WKHX-FM	Country	5.3	5.1	6.5	5.7	5.2
Cox	WALR-FM	Urban AC	4.1	4.5	4.0	4.0	4.8
Clear Channel	WLTM-FM	AC	4.8	4.0	2.6	2.8	3.7
Radio One	WPZE-FM	Gospel	5.2	5.8	4.6	5.4	5.0
Cox	WSB-FM	AC	4.1	4.0	4.5	3.1	3.5
Radio One	WHTA-FM	Urban	3.8	5.1	4.3	4.9	4.3
Jefferson-Pilot	WSTR-FM	CHR/pop	3.9	3.8	4.5	3.5	4.0
Clear Channel	WWVA-FM	Spanish cont.	1.5	1.0	0.5	0.8	4.6
Salem	WFSH-FM	Christian AC	2.8	3.1	3.3	2.6	3.0
ABC	WYAY-FM	Country	3.4	3.2	3.2	2.8	2.7
Infinity	WZGC-FM	Triple A	2.2	2.8	2.3	2.6	2.9
Clear Channel	WKLS-FM	Rock	2.8	2.8	2.6	2.9	2.3
Radio One	WJZZ-FM	Smooth jazz	2.9	3.2	2.9	3.0	2.9
Clear Channel	WGST-AM	News/talk	2.3	2.7	2.7	2.8	2.7
Cox	WBTS-FM	CHR/rhythmic	2.6	2.8	3.2	3.1	2.4
Susquehanna	WNNX-FM	Alternative	3.4	2.8	3.3	2.9	2.6
Susquehanna	WWWQ-FM	CHR/pop	2.4	2.2	2.5	2.3	1.6
Clear Channel	WLCL-FM	Oldies	2.0	2.9	2.5	2.2	2.4
Radio One	WAMJ-FM	Urban oldies	1.7	1.5	1.6	1.3	1.8
Cox	WFOX-FM	Urban	2.0	1.7	2.4	1.5	1.5
Jefferson-Pilot	WQXI-AM	Sports	1.2	1.2	0.8	1.1	1.1
Dickey	WCNN-AM	Sports	0.8	0.8	0.8	1.0	1.1
Davis	WLKQ-FM	Reg. Mex.			0.5	1.2	1.0
Clear Channel	WBZY-FM	Alternative	0.6	0.7	1.1	0.8	0.5
GA-Mex	WAZX-AM	Reg. Mex.	0.8	0.4	0.7	0.4	0.4
La Favorita Inc.	WAOS/WXEM	Reg. Mex.	1.0	0.9	1.1	0.6	0.4

This chart shows the owner of the Atlanta radio station, its call letters, its format, and the percentage of the Atlanta radio audience the station had during the average quarter-hour of the radio broadcast day. Ratings are reported in three month periods. For example, F04 is the overall audience "share" for the fall of 2004.

Williams, G. (2007). Review of the radio industry, 2007. *Federal Communications Commission*.

I wish to express a special thanks to Bob Michaels of Arbitron for his permission for extensive use of Arbitron data and images for this chapter.

Thanks also to Tom Baldrica, Sony Nashville; Bill Mayne; Erica Farber; Lee Logan, Shane Media; Joe Redmon; Larry Pareigis, Nine North Records; and Jaye Albright, Albright & O'Malley.

BIBLIOGRAPHY

Albright, J. (2009). "The PPM Music Test 'MScore Switch' Is On." *Presentation at A&O Pre-CRS Seminar*, March 3, 2009. Nashville, TN.

Arbitron. (2008a). "How America Listens to Radio," Radio Today 2008 Report. http://www.arbitron.com/home/content.stm.

Arbitron. (2008b). Radio Today, 2008 Edition. www.arbitron.com.

Arbitron. (2009). The Portable People Meter System. www.arbitron.com/portable_people_meters/home.htm.

Brooks, G., with special permission of *The Tennessean*. Nashville, TN: Gannett Company, Inc.

Federal Communications Commission (FCC). (2008). Broadcast stations totals as of December 31, 2008, News release issued February 27, 2009, at www.fcc.gov.

Tarver, C. (2005). www.udel.edu/nero/Radio/glossary.html.

Webster, T. (2004). *What's Wrong with Focus Groups?* Edison Media Research www.edisonresearch.com/home/archives/000426print.html.

CHAPTER 9

Charts, Airplay, and Promotion

Paul Allen

CONTENTS

From the earliest days of product marketing, sales people have constantly sought as many ways as are possible to say their wares are the best available and that their products are number 1 in the eyes of consumers. In music especially, bragging rights of having a "number 1" provide leverage for promoters at the label to ask the chart makers to "join the crowd" and move their single or album higher on the charts, ultimately impacting sales. In this chapter, we will explore some of the important charts that are helpful in getting exposure on radio, television, and the web.

Airplay in the traditional sense of music being played on **terrestrial radio** is viewed by some as being a dying medium and a waste of promotion money. As we saw in the last chapter, however, radio continues to draw huge audiences, exposing recorded music to tens of millions every day. During one week in October 2008, the top single on *R&R*'s rap airplay chart was heard by a U.S. audience of over 100 million. Any medium that connects our product with its target market with such impact is far from dead and must be an important component to our marketing mix.

Also in this chapter, we will look at the importance of promotion at a record label. We have all been to the doctor's office and seen drug company marketers drag their wheeled bag of samples and a box of donuts into the office area "behind the door." They are promoting the benefits of a company's medicines and building relationships with doctors so they can earn bonuses based on how many prescriptions are written in their territories. Doctors do not need medicines but their patients do. In the music business, promoters seek to convince program directors to "prescribe" the label's music to the station's listeners. And like the promoter of medicine, the music promoter is paid bonuses based on how many programmers are convinced to choose the label's music to play on the air.

THE CHARTS

This section of the chapter looks at charts of all types from the viewpoint of their value to marketers of music in promoting music to the gatekeepers. From a historical perspective, music charts have been among the most valued information provided by trade magazines and newsletters, which is why our discussion of charts begins with the **trades**.

THE HISTORY OF TRADE MAGAZINES

Billboard magazine has, for more than a century, been the leading music business publication providing comprehensive weekly views of the recording industry, the music business, and commercial radio.

Billboard was first introduced in 1894 as a publication supplying its readership with information about advertising and, a few years later, about the carnival industry. Its reporters then began writing stories about sales of sheet music, and early in the last century it added regular features about silent films and commercial radio. In 1936, it published its first "hit parade," a term used at the time to rank popular songs that then became a term used by radio to denote its most popular music. In 1940, *Billboard* compiled and published its first music popularity chart called "Best-Selling Records Chart." The first national number 1 recording reported by *Billboard* was Tommy Dorsey's *I'll Never Smile Again* with vocals by Frank Sinatra. In the 1940s, *Billboard* became a leader and ultimately the icon for music charts by publishing numerous charts for various genres of music. The *Billboard* Hot 100 was launched in 1958, and it ranked single releases based on both sales and airplay. That chart remains one of the continuing staples of the magazine (Billboard.com, 2009).

Other trade magazines have competed with *Billboard* over the years, including the *Gavin Report*, which closed in 2002; *Cashbox*, which closed in 1995; and *Radio & Records* until it was acquired by VNU—the parent company of *Billboard*—in 2006. *Gavin* and *Cashbox* magazines were key industry trade publications for many years until they were retired for economic reasons. Both of these trades relied upon "reported" airplay by radio stations, which gave the publications their airplay charts for the coming week, whereas *Billboard* relies on electronically monitored airplay to generate some of their charts. *Radio & Records* was launched in October 1973 and remained a genuine competitor for *Billboard* until it was purchased by VNU, but the fate of *R&R* was signaled with the resignation in January 2009 of its veteran publisher and industry icon, Erica Farber. After 36 years of publication, *Radio & Records* was shut down by VNU in June, 2009.

THE IMPORTANCE OF CHARTS

Perhaps the most important piece of real estate a record label can own is a high position on a chart which influences gatekeepers. For example, a top-selling digital single on the SoundScan digital sales chart may influence a radio programmer's decision to add music to its airplay list or a decision by a respected blogger to mention it online. Credible, influential charts of all types are key tools used by record labels to promote music to people who can provide access to their audiences. Airplay, sales, and download charts can have the effect of demonstrating "word-of-mouth" interest between and among audiences, and give a label promotion department the information they need to encourage gatekeepers to engage the music. Airplay charts are especially important for use by radio programmers to give them a basis to compare their audience offerings with those audiences of similar cities, and adjust their programming if necessary.

Creating the Airplay Charts

Charts in trade magazines are defined in numerous ways, and it is important to be sure the distinction is made between sales charts and airplay charts. If a label has a number 1 album on the *Billboard* 200, it earns that designation based solely on its position on that sales chart. If a label has a number 1 single on the radio, that position is based upon any or all of the following criteria: the number of times a single is played on the radio during a specific week, how big the cities are in which the song is played, and how many copies it sold.

Nearly every major music genre has an airplay chart in *Billboard*. The airplay charts compile the national airplay of singles on radio stations as detected by a proprietary system owned by VNU, called BDS. Most genres are also represented in MediaBase airplay charts, but first let's look at BDS.

BROADCAST DATA SYSTEMS (BDS) AND *BILLBOARD* CHARTS

Broadcast Data Systems (BDS) is the technology used by *Billboard* and *Canadian Music Network* magazines to detect each **spin** of a recording on radio in cities in which BDS has installed a computer to monitor airplay. As the spins are detected, the computers upload the number of detections to a main database that is then used to create the weekly airplay charts. Geoffrey Hull cites *Billboard*, describing the system as:

> A proprietary, passive, pattern-recognition technology that monitors broadcast waves and recognizes songs and/or commercials aired by radio and TV stations. Records and commercials must first be played into the system's computer, which in turn creates a digital fingerprint of that material. The fingerprint is downloaded to BDS monitors in each market. Those monitors can recognize that fingerprint or "pattern," when the song or commercial is broadcast on one of the monitored stations. (Hull, 2004)

As the computerized airplay monitor "listens" to a song being played on the radio, it compares its digital fingerprint to that on file and then logs it as a detected play of the song.

BDS has monitors for airplay in 140 radio markets and claims to listen to over 1600 stations and detect over 100 million songs each year (BDS, 2009). These detections are used to compile 25 airplay charts for *Billboard*. Additionally, the service compiles detections of airplay on satellite and cable music channels. Label marketers must be sure they register their recorded music with BDS or there will be no detections of airplay. BDS provides information on its web site on how to register a song and get a digital fingerprint created for the airplay monitoring system. Without the airplay statistics, the radio promotion department will be without some of the bragging rights they need to continue to promote the single, and the marketing department will be without one of its key tracking tools (BDS, 2009).

BDS gives the label's marketing department considerable information about which radio stations are spinning a single and how frequently. Combining this information with SoundScan data on sales of singles and albums, label marketers have continuing feedback on the performance of their recorded music projects. And most importantly, this feedback gives marketers the information needed to modify marketing plans in order to draw as much commercial activity out of the marketplace as possible.

UNDERSTANDING THE *BILLBOARD* CHARTS

Weekly charts representing airplay, sales, and a combination of both appear in *Billboard*. Two charts which use only SoundScan sales data are the *Billboard* 200—which is a mirror of the Top 200 current albums sold weekly as tracked by SoundScan—and the R&B/Hip-Hop Albums chart. Charts that use mixed data, meaning both sales and airplay information, are the *Billboard* 100, the Pop 100, and Hot R&B/Hip-Hop Songs (Billboard.com, 2009).

Some of *Billboard*'s weekly airplay charts rank singles based upon the number of times a song is played on monitored radio stations and the resulting size of the combined audience that heard the song each time it was played. For example, a song played once by Arbitron's major R&B station in Los Angeles represents a larger audience than the song played once in Chillicothe, Ohio. But the combination of all plays multiplied by all audiences as measured by Arbitron dayparts determines the airplay ranking for a single. Charts using this method are Hot Country, Hot 100 Airplay, Hot Rap Songs, and Hot Latin Songs. The remaining airplay charts base the ranking of singles only on the number of times the single is played by all monitored stations in one week, with weighting ranking by audience size (*Billboard*, 2008).

Within the charts, *Billboard* indicates recognition for accomplishments each week with the use of special awards. Designations are as follows:

Table 9.1	
Designation	**Description**
Greatest gainer	Indicates the album with the greatest increase in sales
Pace setter	Notes the album with the largest percentage sales gain
Heatseeker graduate	Is an album by a new artist that was removed from the Heatseekers chart into the top half of the *Billboard* 200 chart
Highlighted position number	This identifies a single or an album that has shown growth in either audience or sales over the previous week

(Source: Billboard, 2009)

A heatseeker is a special designation by *Billboard* for developing artists and is described as:

The Top Heatseekers chart lists the bestselling titles by new and developing artists, defined as those who have never appeared in the Top 100 of The Billboard *200 chart. When an album reaches this level, the album and the artist's subsequent albums are immediately ineligible to appear on the Top Heatseekers chart*

—Billboard, 2008

The Hot 100 has been a part of *Billboard* magazine since 1955, first as the Top 100 and then in its current form since 1958. Since then, it has spawned numerous other genre-specific charts. The Hot 100 is a chart that is developed each week using a formula that combines the physical sales of singles and sales of digital downloaded singles, as well as the number of spins a song receives on radio, regardless of the genre of music or the radio format in which the song is programmed.

Other charts published weekly by the magazine include rankings by genre, created by using data gathered for airplay through its BDS reporting system. Also, each week *Billboard* publishes its Top 200 chart, which is a compilation of the best-selling albums ranked by the number of unit sales for the previous week. Both of these charts are especially important to the marketing effort by a label on behalf of its active new music projects. Tracking the impact the music is having at radio and at retail gives label marketers information that is helpful to control the success of singles and albums.

Among the newer charts is *Billboard*'s Pop 100. This chart lists only the top songs being played at top 40 radio stations and includes both airplay and sales to determine chart positions. Songs are ranked, in part, based on the number of gross impressions. This is compiled by cross-referencing the exact times the songs are played against Arbitron's data on listenership at that time.

MEDIABASE 24/7

MediaBase is a service owned by Clear Channel Communications that monitors the airplay of recordings on over 1000 U.S. radio stations, as well as 125 in Canada (Hoovers, 2009). The airplay detections by the service are then published as "America's Music Charts" in a full-page advertisement in *USA Today* each Tuesday (Leeds, 2006) as well as on the MediaBase web site.

While the information is very similar to that provided by BDS and is used by record labels in the same ways, there are some differences between the services. Where BDS uses computers to detect airplay, MediaBase employs people to actually listen to radio stations and log the songs played. Employees of the company who are paid to detect airplay are experts in their genres of music, often work from their homes, and are provided the necessary hardware and software by the company. Employees who work in airplay detection are often responsible for logging songs for the 24-hour broadcast days of eight radio stations.

Another difference in the services lies in the particular stations whose airplay is monitored. MediaBase monitors an estimated 80% of the same stations as BDS. Some record labels, artist managers, and artists see the need to subscribe to both services to be sure they are getting accurate feedback on the performance of single releases at radio (Radio-media, 2009; Rhodes, 2009).

Other Charts That Measure Popularity
CMJ *Charts*

The *College Music Journal*, now simply called *CMJ*, includes music charts created through the *CMJ* Network, which refers to itself "as connecting music fans and music industry professionals with the best in new music through interactive media, live events and print" (*CMJ*, 2009). Charts for *CMJ* are created by reports of airplay from their "panel of college, commercial and non-commercial radio stations," and tend to represent music that is not a part of the commercial mainstream. Its charts are reported in publications by the company, as well on its web site and chart titles. Those charts include:

> *CMJ* Top 20
> *CMJ* Radio 200
> Hip-Hop Top 20
> Triple A Top 20
> Loud Rock Top 20
> Jazz Top 20
> New World Top 20
> RPM Top 20
> *CMJ* Loud Rock Select
> *CMJ* Radio Select
> *CMJ* Independent Only Radio Select
> (*CMJ*, 2009)

SoundScan Current Hot Digital Tracks

This chart tracks the weekly sales of digital singles from all participating sources that report their sales to SoundScan using a registered UPC symbol. The chart's place on overall marketing strategy was discussed in Chapter 7 about SoundScan, but its importance in promotion also becomes obvious. Label radio and video promoters point to the number of sales as a strong indicator of the connection of the music with consumers' pocketbooks. It represents the conversion of passion towards music into commerce, and is thereby worthy of consideration by radio programmers to include the music at their radio station or increase the number of times they present it to listeners on the air.

iTunes Charts

iTunes issues its continuing charts of music sales, which include 20 charts of the best-selling albums based on genre. It also publishes its top-selling singles within the same 20 categories. With iTunes being the largest retailer of recorded music, its charts carry considerable weight with those who would use their information to create buzz among gatekeepers about the popularity of any album or single. The iTunes Top Songs chart provides a listing of 100 most-purchased songs on their web site.

iTunes music sales charts also expand the vision of sales to international music downloads, underscoring the meaning of the World Wide Web as well as the global impact of music. Traditional charts are often updated weekly, whereas iTunes presents a real-time ranking on the sales of music through its downloading service.

BigChampagne

BigChampagne is a subscription service that tracks the movement of music Internet downloads from retailers as well as within peer-to-peer (P2P) networks. The company tracks an estimated 1 billion P2P shared music files each month. Genre-specific charts of swapped music files from BigChampagne are available at www.allaccess.com, and these tracking data are used by both record labels and radio programmers to maintain a current view of the interest file sharers have in new music. For radio programmers, it becomes one of the tools available to track consumer interest in new music as they consider airplay decisions for stations they program (Taylor, 2006). For some record labels, data from BigChampagne are used for their direct promotion of the popularity of shared music to radio.

RateTheMusic.com

An extension of MediaBase is RateTheMusic.com (2009), which is a music rating service used by radio programmers and record labels to capture target market preferences of current music by using their Internet site to capture data. Fans of music register with the site for free and are asked to provide their opinions on current music, which then creates local and national charts for the most popular songs being played on radio. Those who are participants in RateTheMusic are asked to give the service some basic demographic information that is then used to direct the appropriate music to respondents for their input in the weekly chart. From a label perspective, a promotion department can use ratings from RateTheMusic to help make the case to a local or national programmer that a particular song should be ranked higher on a station's **playlist**. Pinnacle Media Worldwide has a similar service being used by BDS for their clients. The Pinnacle calls its product RadioSurvey.com and offers prizes to music fans who sign up to be participants in surveys of music, which then generate popularity charts (Allbusiness.com, 2005).

Amazon Charts

One of the oldest online retailers, Amazon.com (2009), creates hourly charts that rank music being purchased through its web site in 25 ratings categories. Their overall chart, Bestsellers in Music, is not genre specific like their other 24 charts. Unlike iTunes, Amazon sells digital versions of recorded music as well as physical CDs, vinyl albums, and used CDs. To the record label, the near real-time charts provided by Amazon give the promotion and publicity departments another tool to use in acquiring more exposure for the music.

YouTube

YouTube's daily charts of the most popular music videos create yet another opportunity for the record label to promote its music to gatekeepers. As most users of the web site have discovered, it provides an efficient way to find the most popular music videos in a variety of genres as well as videos of those who are rising stars, whether they are on independent labels or the majors. The real-time tracking of video views creates an ongoing "chart" of sorts and another talking position for those promoting a label's music.

Other Sources to Track Music Popularity

The number of sources available to the label to help promote music past the gatekeepers continues to grow; some require a fee to subscribe to their services.

Social networking sites track views of videos and streams of songs with data that are available to the label at no cost. AOL provides charts of audio and video streams of the most popular music at Billboard.com, and Yahoo! creates charts of the most popular audio and video streams from its web site. Troy Research is a subscription research service offered to radio stations and record labels to track the popularity of music, with results creating custom charts for specific markets.

Pollstar magazine tracks the popularity of recording artists based on their ticket sales at performance venues. The magazine's online web site exists to connect music fans with concert tickets; however, the magazine is a weekly subscription-based publication that lists artists, venues, ticket sales percentages, and gross revenue from single appearances. Continued sell-outs by an artist can make an important statement and talking point about an artist's growing fan base and popularity. For example, a label promotion person might use the information from the Box Office Summary chart for one of 2009's new artists like the Eli Young Band to tell a programmer that the group sold out its performance at Whiskey Dick's in El Paso, Texas, on January 16, 2009. *Billboard*'s Box Score chart is published weekly but only reports on the top 35 grossing shows, making it useful to promote major hit makers but less useful for artists that are in the developing stages of their recording careers (*Pollstar*, 2009).

While BigChampagne is the veteran tracking service for online music activity, other related services such as torrent trackers are continuing to develop; some labels find it helpful to also track blogs as sources for positive information to draw from in an attempt to influence charts.

RADIO PROMOTION

Lobbying and lobbyists have been around as long as any one person has been responsible for a decision or a vote, and people have always wanted to influence that decision or vote in their favor. Every day, lawmakers at the national and state levels meet with representatives of special interest groups who ask them to vote on matters that are in the best interests of their groups or their clients.

The same thing happens between a radio programmer and a record promoter. Record promoters are lobbyists in the purest form. They are either on the staff of a record company or they are part of a company specifically hired by the label to promote new music.

The people lobbied are usually the program directors of radio stations that report their charts to the major trade magazines. Program directors, also referred to as PDs, have the ultimate responsibility for everything that a radio station broadcasts—banter by personalities, advertising, information such as news and traffic reports, and all music played by the station. Simply said, radio programmers can decide whether a recording ever gets on the air at their station.

Decisions by radio programmers are one of the keys to the life of a recording and have become the basis for savvy, smart, and creative record promotion. Programming decisions about music determine:

- Whether a new recording is added to the playlist of a radio station that reports its chart and airplay to trade magazines

- Whether the recording receives at least **light rotation** on the playlist

- Whether it eventually receives **heavy rotation** on the station's playlist

PROMOTION

Traditional marketing texts teach the four P's, one of which is promotion. Those same texts tell us that the promotional mix consists of public relations, advertising, sales promotion, direct marketing, and personal selling.

The record promotion by a label comes closest to being personal selling than any other aspect of the traditional promotional mix at a label, except perhaps the label sales department. Personal selling is one on one, and the work of a record promoter is exactly that. The radio promotion department at a record label has the responsibility of securing radio airplay for the company's artists. Quite simply, they lobby the station decision makers—gatekeepers—to add recordings by the label's artists to the station's playlist. With commercial radio creating hundreds of millions of music impressions every week, some label executives claim that radio is responsible for 70% of the sales of recorded music. Label record promotion people can be critical to the success of an album project. It is this close connection between airplay and record sales that creates the need for record promotion to radio as a key element of the marketing plan.

The effectiveness of a record promotion person hinges on the strength of the relationships she has with radio programmers. These relationships are built much like those of anyone in business who has a client base requiring regular service. The promoter makes frequent calls in person and by telephone to the programmer, arranges lunch or dinner meetings, provides the station access to the label's artists, and helps the station promote and market itself with contests, performances by artists, and giveaways. The promoter who has effectively developed a good relationship with a programmer can make the telephone call and ask the programmer to treat her current record project favorably—which is difficult for smaller labels to do because of their limited promotional resources.

If the promoter has no relationship with the radio programmer from a reporting station, it is highly unlikely that telephone calls will ever be returned. Programmers today have too many things to do and little time to listen to promotional pitches from people and companies they don't know. And because of centralized programming by some radio groups, there is more pressure on fewer programming personnel to find time to take a call from a promoter.

PROMOTING

It is important to understand that promotion by a label is focused on radio stations that program current, new music. Stations airing older music are using record company catalogs as the basis for their entertainment programming, and many fans of the artists they program already have the music in their libraries or on their MP3 players. Older albums and their related singles have been around for years, sometimes decades, and there is limited interest on the part of a label to promote them to radio. Most energy and money is invested by record companies in promoting the newest singles and album projects.

Record promotion to radio by labels typically must answer four questions:

1. How does it help us sell recordings?
2. What does it do for our artist?
3. What does it do to further our agenda and help us market the artist?
4. Does it make sense for the radio station?

As you create a promotional tour for an artist with a new CD, emphasis should be on the most popular radio stations in targeted cities, especially those "reporting" stations that report their airplay charts to major trade magazines. To determine which stations are rated the highest in audience shares, you will find Arbitron's current reported findings at www.allaccess.com, and there are other online sources. Specific information about the station programmer's name, address, and telephone number is available through Bacon's MediaSource.

After a single has been released to radio, nearly everyone at the label tracks its progress. For the promotion staff, the monitoring tools and charts discussed in this chapter help guide their work. Some labels also have staff members who monitor the "buzz" on their new music in chat rooms and in blogs. All of this information helps the label determine whether they have a hit on their hands and should step up the promotional effort, or whether it's time to cut their losses and pull the project from the market.

The typical business week for record promoters begins early in the week as the trade chart closings show the successes and failures of singles from the previous week and give the promotion team the information they need to allocate their time for the next 7 days. With the plan in place, the promotion staff then begins its weekly cycle of contacting radio programmers to build airplay for the label's products.

One label executive says his regional label promoters are always on the job. The only time he allows a break is when his promotion team members are on vacation. Regional promoters with some experience can go to work at a starting salary in the range of $70,000 to $80,000 US, plus expenses. Vice presidents of label promotion for larger labels can easily earn well over $200,000, plus bonuses and other incentives.

HISTORY

Record promotion and its regulation by the federal government began not long after the advent of commercial radio broadcasting. In 1934, Congress created and passed the Communications Act, which restricted radio licensees—the stations themselves—from taking money in exchange for airing certain content unless the broadcast was commercial in nature. However, this early act contained nothing that prohibited disc jockeys (DJs) from taking payments in exchange for airplay. During the big band era of the 1940s and the rock 'n' roll days of the '50s, DJs were routinely taking money from record promoters in exchange for the promise to play a record on the air. Disc jockeys during this time often made their own decisions about which records would be included on their programs, and promoters would approach them directly to influence their record choices.

Lawmakers railed against the rampant bribes being given to DJs to play records. In 1960, Congress amended the 1934 act to include a provision that was intended to eliminate illegal bribes to play music, so-called **payola**. Under the revised law, disc jockeys and radio stations were permitted to receive money and gifts to play certain songs, but the amendment placed a requirement that these inducements must be disclosed to the public on the air. If this disclosure was not made, it exposed the DJ and management to possible fines and imprisonment (Freed, 2005). The change in the law also created the requirement that record labels must report their cash payments and major gifts for airplay to the station. This 1960 amendment continues to guide the radio and records industries today.

Despite the stronger laws against payola, federal investigators were called upon to investigate scandals within the record promotion business in the 1970s and 1980s. There were no major convictions despite the appearance that money, drugs, and prostitution were being used as leverage by promoters to get radio airplay for recordings (Katunich, 2002).

Entanglement by the payola laws began when radio stations asked the label promoters for something of value, whatever it was. The first question to the record company then becomes "How does this help me sell copies of the artist's record?" One label says that everything they do today with radio stations to promote an artist "has to really pass the smell test." They require proof that promotional prizes and free concerts by artists are acknowledged on the air as being given by the label. They expect to receive affidavits from the station showing when the announcements were aired and how much money those announcements were worth, underscoring the disclosure requirement for radio stations taking payments as well as the record labels providing payment. In the late 1990s and early 2000, stations

From:
Sent: Tuesday, January 14. 2003 8:53 AM
To:
Subject: RE: WLIR/Long Island in on Everclear

unfortunately we had to pay his indie 1000 dollars, so there is no money for additional promo, let him know.

-----Original Message-----
From:
Sent: Tuesday, January 14. 2003 6:59 AM
To:
Subject: WLIR/Long Island in on Everclear
Importance: High

Gary Cee is asking for some promotional support for this add and moving Coldplay to 5x/day....he asked for $1,500 in tee shirts, I said there was no way, he then asked for $1,000-I said I would check, lemme know what I can do.........
please advise....

cheers,

FIGURE 9.1   *One of hundreds of emails acquired by and used by the New York Attorney General's office in his investivation of alleged pay-for-play violations of Payola Laws. (Source: New York Attorney General)*

and labels began to break the payola law and in 2005 came under investigation by the New York Attorney General's office. The examination into these practices resulted in Sony BMG, EMI, Warner Bros., and Universal Music paying over $30 million in penalties for paying radio stations to play their music. The Federal Communications Commission followed the New York inquiry, which resulted in $12.5 million in fines levied against Citadel Broadcasting, CBS, Entercom, and Clear Channel Communications (Barbington, 2007). Following the investigation, labels of all sizes spent considerable time and fees on lawyers to ensure that promotion stayed within the guidelines of the law.

GETTING A RECORDING ON THE RADIO

The consolidation of terrestrial radio has concentrated some of the music programming decisions into the hands of a few programmers who provide consulting and guidance from the corporate level to programmers at their local stations. In many cases, local programmers have the ability to add songs to their playlists based upon the preferences of the local audiences. Here is how songs are typically added to stations' playlists for those stations that program new music:

1. A record label promoter or an independent promoter hired by the label calls the corporate programmer, the station music director (MD), or program director (PD), announcing an upcoming release. Radio music directors have "call times." These are designated times

of the week that they will take calls from record promoters. The call times vary by station and broadcasting company and are subject to change. For example, an MD may take calls only on Tuesdays and Thursdays, 2:00–4:00 P.M.

2. Leading up to the **add date**, meaning the day the label is asking that the record be added to the station's playlist, the promoter will call again, touting the positives of the recording and asking that the recording be added to the playlist.

3. The music director or program director will consider the selling points by the promoter, review the preferred charts for performance of the recording in other cities, consider current research on the local audience and its preferences, look at any guidance provided by the corporate programmers/consultants, and then decide whether to add the song.

4. The PD will look for reaction or response to adding the song.

An important component of promoting a recording to radio is the effectiveness of the record company promotion department or the independent promoters hired to get radio airplay. This would appear to be a simple process, but the competition for space on playlists is fierce. Thousands of recordings are sent to radio stations every year, and the rejection rate is high because of the limited number of songs a station can program for its audience. Some of the recordings are rejected from being included on playlists because they are inferior in production quality; some are inappropriate for the station's format; and some may never connect with a commercial radio audience and ultimately will lose their label support if they fail to quickly to become favorites with the radio audience.

Record promotion has been a big-dollar investment, which made it a key marketing element necessary to stimulate consumers' interest in buying new music. Costs for a label to market and promote a single easily reach $1 million, not including the production of the recording or any advances to the artist or producer. Even in the world of country music, where annual sales are often less than 10% of all recorded music, labels invest as much as $300,000 just to get the single of a new artist into the Top 20 of an airplay chart.

However, the marketer's litmus test for the viability of a recording is to honestly compare it with other songs achieving success on the various charts. If it is not at least as good as those listed on those charts, then it doesn't have any chance at all, even if it has a competitive marketing budget.

LABEL RECORD PROMOTION AND INDEPENDENT PROMOTERS

Most large labels have a promotion department whose sole purpose is to achieve the highest airplay chart position possible. While most consumers assume a number 1 song is the biggest seller at retail, the number 1 song on most *Billboard* charts actually is the song that has the most airplay on radio. The connection between airplay and sales is well documented, so a high chart position is critical to the success of a recording and becomes the heart of the work of a record promotion department.

Labels often have a senior vice president of promotion who usually reports directly to the label head. The senior vice president of promotion at a pop music label typically has several vice presidents of various music types based on radio formats. These vice presidents then have regional promotion people who are viewed by the label as field representatives of the promotion department. They are liaisons to key radio stations in their region. The vice president and the so-called regionals are the front line for the label attaining airplay. It is their responsibility to create and nurture relationships with programmers for the purpose of convincing them to add the company's recordings to their playlists. At a country label, the senior vice president of promotion typically has a national director of promotion and several regionals (see Figure 9.3).

Labels sometimes hire independent promoters (**indies**) to augment their own promotional efforts. Because promoting songs for airplay relies on well-developed relationships, indies may have developed stronger relationships with some key stations than the label has, and the record company is willing to pay indies (half of which is often recoupable from the artist) for the value of those relationships. In the 1980s, a practice of labels hiring independent promoters, or "indies," to attain airplay at radio came under very tight scrutiny by federal law enforcement. Fred Dannen (1991), in his book *Hit Men*, found that CBS Records was paying $8–10 million per year to indies to secure airplay for their acts. By the mid-'80s, he says that amount was $60–80 million for all labels combined. The resurgence of independent promotion in the 1990s caused the New York Attorney General in 2004 to look closely at these lucrative agreements, with the result that most indies shuttered their doors or went into other businesses.

Independent promoters have typically made their money this way: they would sign an agreement with a radio station to be the station's exclusive consultant on new music for a year and then pay the radio station for that right. The indie promoter did not require the station to play specific songs, according to the typical agreement, but the station did promise to give its playlist for the following week to the indie promoter before anyone else. Then, the promoter would send an invoice back to the record company for $3000 to $4000 for each single that was charted on stations that they represented. Because of this "exclusive" arrangement with the radio station, record label promotion people had no dealings with the radio stations represented by indies. One label promotion vice president said the $4000 could easily turn into $30,000 per single if the indie had to provide contest prizes to help promote the single at radio and to pay for other promotional expenses. Many arrangements with indie promoters included provisions for bonuses based on their success at charting records with individual stations (Phillips, 2001, 2002).

The use of independent promoters provided the record company with a layer of insulation between themselves and radio. The record companies did not deal directly with certain radio stations in matters of adds and spins; rather, they dealt only with independent promoters who promoted to these stations. As one executive said, "The use of independent promoters creates a clearinghouse by removing the label one step away from . . . making any compromises that some might make to get a song on the air." Another says, "This way the money doesn't go directly from the label to the radio station" (personal interviews, fall 2004).

Critics of the independent promotion system claimed it tended to shut off access to radio airplay by independent labels and artists. Alfred Liggins is CEO of Radio One, a company that owns nearly 70 radio stations targeting African-American audiences. He acknowledged that their exclusive relationships with independent promoters means that labels without an indie promoter are less likely to get a record played on his stations (*20/20*, 2002). Nearly every record label has discontinued its relationship with independent promoters who pay radio stations for access. While the current trend is away from independent promotion, history has shown that this might just be temporary.

SATELLITE RADIO PROMOTION

Satellite radio has expanded the horizon for promoting recordings to radio. The services provided by Sirius XM have a relatively low monthly subscription fee and offer scores of music channels. Though the service requires that a vehicle must be equipped with a special receiver, many newer cars and trucks have satellite receivers integrated into their standard radios. Home music systems can be fitted with receivers for satellite radio, and iPhones also offer the service on a subscription basis without the need for additional equipment. Satellite radio delivers an audience near 20 million, and offers more subgenres and niche formats, which gives labels great opportunity for airplay. Though record company promoters actively work with satellite music channel programmers seeking adds to their playlists of current music, the future of satellite radio is a question mark. As a company, it has never made a profit, and the worldwide economic downturn in 2009 virtually halted growth because sales of new vehicles equipped with satellite receivers had a major decline. *Record Label Marketing* maintains a web site for this book at www.recordlabelmarketing.com, and the authors will update this section regarding satellite radio with links on the site.

APPENDIX

Figure 9.2 provides an example of an organizational chart for a typical pop label record promotion department. Note that the VPs of the various formats are specific to the music. However, the individual regional promoters typically promote all current music types marketed by the label.

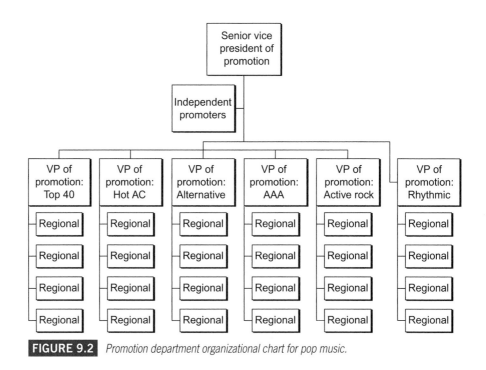

FIGURE 9.2   *Promotion department organizational chart for pop music.*

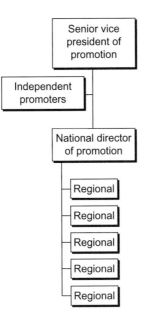

Organizational chart for a typical country label record promotion department.

GLOSSARY

add date—The day the label is asking that the record be added to the station's playlist.

heavy rotation—These recordings are among the most popular songs played on a radio station.

indie—A shorthand term meaning an independent record promoter who works for radio stations and record labels under contract.

light rotation—These are recordings that are played fewer times on a radio station than songs in heavy rotation.

payola—The illegal practice of giving and receiving money in exchange for the promise to play certain recordings on the radio without disclosing the arrangement on the air.

playlist—The weekly listing of songs that are currently being played by a radio station.

recurrents—Songs that used to be in high rotation at a station, but are now on the way down, reduced to limited spins.

rotation—Mix or order of music played on a radio station.

spin—This is a reference to the airing of a recording on a radio station one time. "Spins" refers to the multiple airing of a recording.

terrestrial radio—This is traditional commercial radio broadcasting using a transmitter and tower consisting of AM, FM, and HD radio.

trades—This is a reference to the major music business trade magazines.

BIBLIOGRAPHY

20/20 (November, 2002). ABC Television. http://www.abcnewsstore.com/store/index.cfm?fuseaction¼customer.product&product_code¼T020524%2002.

Allbusiness.com. (2005). www.allbusiness.com/services/motion-pictures/4466370-1.html.

Amazon.com. (2009). www.amazon.com/gp/bestsellers/music/all.

Barbington, C. (March 6, 2007). Big Radio Settles Payola Charges. *Washington Post*, D1.

BDS. (2009). http://www.bdsonline.com/about.html.

Billboard. (October 4, 2008). Charts Legend. p. 59.

Billboard.com. (2009). Chart Methodologies. www.billboard.com/bbcom/about_us/bbmethodology.jsp.

Box Office Summary. (January 16, 2009). *Pollstar*. p. 28.

CMJ. (2009). http://www.cmj.com/company.

Dannen, Fred. (1991). *Hit Men*. New York: Vintage Press.

Freed, A. (2005). http://www.historychannel.com/speeches/archive/speech_106.html.

Hoovers, Inc. (February 17, 2009). *Company Reports*. CC Media Holdings, Inc.

Hull, G. (2004). *The Recording Industry*. London: Routledge. pp. 201–202.

Katunich, L. J. (April 29, 2002). *Time to Quit Paying the Payola Piper*. Loyola of Los Angeles Entertainment Law Review.

Leeds, J. (October 2, 2006). Song Tracker Finds a New Way to "Publish" Its Charts. *New York Times*, Section C, 6.

New York Attorney General. http://www.oag.state.ny.us/media_center/2006/jun/EMI-Exhibits.pdf.

Phillips, C. (May 29, 2001). *Logs Link Payments with Radio Airplay. Los Angeles Times*.

Phillips, C. (May 24, 2002). *Congress Members Urge Investigation of Radio Payola. Los Angeles Times*.

RateTheMusic.com. (2009). http://ratethemusic.com/faq.

Rhodes, R. (February 2009). Account Executive with MediaBase.

Radio-media. (2009). http://www.radio-media.com/song-album/articles/airplay26.html.

Taylor, C. (February 17, 2006). *Tapping into Digital Download Data*. Billboard Radio Monitor.

CHAPTER 10

Publicity of Recorded Music

Tom Hutchison and Paul Allen

An important element of marketing planning and execution by a label is promotion through the use of publicity. Labels often handle publicity for the artist's recording career as well as for news and press releases about the label itself. Sometimes artists will hire a personal publicist to handle other areas of their life and career.

This chapter is designed to give an overview of the publicity department at a record label, including its responsibilities and how publicity contributes to the success of a recorded music project.

LABEL PUBLICITY

The objective of label publicity is to place nonpaid promotional messages into the media on behalf of the artist's recorded music project. That can range from a small bullet point in *Rolling Stone* to a mention in a music blog, to an appearance on the *Late Show with David Letterman*. Appearances in the new and traditional media contribute to the success of the label's promotional plan to put the artist in print and on the air to support the marketing of the label's music.

The theory is that the more positive impressions consumers receive about a recording, the more likely they are to seek additional information about the recording and to purchase it. Advertising planners use the term "reach and frequency" as they compile a strategy and its related budget. This means they plan an affordable ad campaign that can "reach" sufficient numbers of their target market with the "frequency" necessary for people to remember the message and act by purchasing. Publicity becomes a nice complement to that strategy without the direct costs of paid advertising.

Label publicity on behalf of a recording artist, like any positive publicity, has a certain credibility that paid advertising does not. After all, a journalist thought the artist was interesting enough to write an article about or to review the music, and a publisher thought it was interesting enough to make editorial space online or in a magazine or a newspaper, to present the story or run the review.

Advertising is the handcrafted, paid message of the record company's marketing department, designed to sell recordings. It is a message created with the intent of getting into consumers' wallets. However, an effective publicity campaign can create an interest by journalists in writing articles about artists and their recordings in the not-so-commercial setting of a feature article. An online article or one in a newspaper or magazine can suggest to the consumer that there is something more to the label's artist than just selling commercial music. Published articles and TV magazine–style stories (for example, *60 Minutes*) tend to add credibility to the artist as an "artist" in a way that paid advertising cannot.

There are key differences between publicity from the label and the advertising placed by the label. Label publicists generally create and promote messages to the media that are informative in nature and do not have a hard "sell" to them. On the other hand, advertising is designed to influence and persuade the consumer to purchase CDs. Chapter 11 discusses the message style and the targeting aspects of record label advertising.

PUBLICITY IN THE MUSIC BUSINESS—A HISTORICAL PERSPECTIVE

The earliest music promoters were in the publicity business at the beginning of the last century, primarily helping to sell sheet music that was heard on recording playback devices or at public performances. Those who worked in the publicity profession in the early 1900s relied primarily on newspapers and magazines to promote the sale of music.

In 1922, the federal government authorized the licensing of several hundred commercial radio stations, and those in the recorded music business found their companies struggling as a result. People stopped buying as much music because radio was now providing it, and newspapers and magazines were no longer the only way the public got its news. Radio became the entertainer and the informer. But publicists found themselves with a new medium and a new way to promote, and quickly adapted to it, much in the way they did in 1948 with the advent of television as a news and entertainment form.

The work of the label publicist today involves servicing online music information blogs and web sites, many of which are extensions of magazines, newspapers, and video channels. Label publicists also work with traditional print media for feature articles, and they work with television program **talent bookers** to arrange live performances.

THE LABEL PUBLICITY DEPARTMENT

The work of a label publicity department (sometimes called the media relations department) is very much like a sales department. Rather than selling "things," they are selling ideas and an image. In particular, they are "pitching," or selling story ideas to the media, attempting to stimulate an interest by the correspondent in writing or producing a story about the artist's recording. Labels which use multiple rights recording contracts—sometimes referred to as "360 deals"—use their publicity departments to promote all aspects of the artist's career including touring, merchandise, web sites, songwriting, and their recordings. Among high-profile artists with such arrangements with their label are Madonna, U2, Nickelback, and Shakira.

Staffing the Publicity Department

Large labels often employ a director or manager of publicity. The full responsibility of media relations and publicity rests with the directors. They are accountable to the media outlets they serve, and they are also accountable to the various departments at the label such as A&R, marketing, and sales. Sometimes that accountability stretches to the artist and the artist management team.

Structurally, some managers of publicity report directly to the president/chief operating officer of the label, and they are often called directors or vice presidents of publicity, or some similar title. Some labels align the publicity department's accountability directly to the marketing department, while others make publicity a component of the **creative services department**. (Creative services typically handles design and graphics work for albums and point-of-purchase [POP] material, as well as imaging for the artist and the recorded music project.) Those who are hired as independent publicity companies for label projects typically report to the vice president of marketing.

Staff publicists work for the director and handle the day-to-day planning, coordination, and execution of the department's work. Labels that have a large roster can stretch the time and energy of staff publicists, and sometimes the limitations of staff time require the employment of **independent publicists**. It is not uncommon to find that staff people are handling ongoing publicity efforts for several active artists at the same time.

Large label publicity departments sometimes hire independent or freelance publicists because these publicists have relationships with key media gatekeepers. They are hired because they have important contacts that the label does not have, and they can be effective in reaching these media outlets on behalf of the artists of the label. In some ways, this is similar to the use of independent radio promoters by record companies to reach programmers that are key to securing airplay for a recorded music project.

Independent and smaller record labels sometimes handle publicity in-house, but they often hire a publicity firm on a project-by-project basis or pay a retainer fee in order to have access to their services as they are needed. The cost of paying an employee to handle publicity at a small label becomes a financial burden at times when business is slow. During down cycles in the record business, labels have laid off entire publicity departments and hired independent firms to handle the work for them.

TOOLS OF THE LABEL PUBLICITY DEPARTMENT

The Database

The publicity department creates and distributes communications on a regular basis, so the maintenance of a quality, up-to-date contact list is critical to the success of that communication effort. Some publicity departments maintain their own contact lists; some use their own lists, plus lists through a subscription service; others rely entirely on subscription database providers.

An example of a subscription, or "pay" service, for media database management is Bacon's MediaSource. Bacon's updates its online database daily with full contact information on media outlets and subjects on which they report. The value of maintaining a quality database by the publicity department is that the information enables them to accurately target the appropriate media outlet, writer, or producer. Services like Bacon's can literally keep a publicity department on target. Most labels and their independent publicist partners maintain several lists within their databases to ensure they are not sending news releases to people who are not interested in the subject matter. These lists are used to distribute press releases, press kits, promotional copies of the CD for reviews, and complimentary press passes to live performances.

Another service used by labels and publicity companies is Cision. This company offers an array of services including the Bacon's database as well as in-depth information on media writers regarding their personal preferences and peeves as journalists. Journalists frequently change jobs, so having a service like Cision can make database maintenance easier for publicity departments. Publicity departments also use paid wire services, such as PR Wire and Business Wire, for news release distribution for major stories.

The most effective way to reach media outlets is through email, because "it's inexpensive, efficient, and a great way to get information out very quickly" (Stark, 2004). A few media outlets still prefer regular mail or expedited delivery services, but the immediacy of the information is lost. An effective publicist learns the preferred form of communication for each media contact. Bacon's MediaSource and Cision provide some of that information, but it is always preferable to check with the journalist.

Internet distribution of press information from a label requires the latest software that will be friendly to spam filters at companies that are serviced with news releases. A spam filter is an electronic filter used by many large companies to prevent the sending of unsolicited email to company employees. The most reliable way to ensure news releases and other mass-distributed information are received by a media contact is to ask about any spam filters and the ways to bypass them. Some companies use Emma (www.myemma.com) to track whether a news release was received and whether the receiver opened the email containing the news release. It becomes an effective way to be sure news releases are accurately targeted to interested journalists and that they were able to get through spam filters.

The Press Release

The **press release** is a standard tool in public relations and is a brief, written communication created for and targeted to journalists. It frames an event or a story into several paragraphs, with the hope that media outlets will find it interesting enough to use as a basis for a story they will create. The press release is used to publicize news and events. Following are some examples of when a press release should be used:

1. To announce the release of an album
2. To announce a concert or tour
3. To publicize an event involving the artist or label
4. To announce the nomination or winning of an award or contest
5. To offer other newsworthy items that would be appealing to the media

The press release should be written with the important information at the beginning. Today's busy journalists don't have time to dig through a press release to determine what it is about. They want to be able to quickly scan the document to determine whether this is something that will appeal to their target audience.

The Anatomy of a Press Release

The press release needs to have a **slug** line (headline) that is short, attention-grabbing, and precise. The purpose or topic should be presented in the slug line. The release should be dated with contact information including phone and fax numbers, address, and email. The body of text should be double-spaced.

The lead paragraph should answer the five W's and the H (who, what, when, where, why, and how). Begin with the most important information; no unnecessary information should be included in the lead paragraph (Knab, 2003a). In the body, information should be written in the inverse pyramid form: in descending order of importance.

An electronic news release sent via email should include embedded links where appropriate. Some links that should be considered to be included in the text of the news release should take the reader to the artist's web site, the label's web site, or to any other site that contributes to the journalist's understanding of the importance of the story in the news release. The added benefit of embedding links in electronic news releases is that they often become featured in blogs and music web sites, which can improve page rankings by search engines on subjects relating to the artist, as well as help fans find their ways to sites maintained for the artist's benefit.

The Bio

The artist biography provides a window into the artist's persona and includes information that sets the artist apart from others. Before writing the bio, an examination is done on the artist's background, accomplishments, goals, and interests to find interesting and unique features that will set the artist apart from others (Knab, 2003b). Keep in mind the target readership of the bio. The **bio** should be succinct and interesting to read (Hyatt, 2004), and should create an introduction that clearly defines the artist and the genre or style of music.

The Press Kit

Another primary tool of the department is the press kit they develop on behalf of the artist and the recorded music project. Although there are occasions when a physical press kit may be useful, perhaps at showcases or for talent buyers and sponsors, they are generally files that are sent via email to journalists and others for whom the information is targeted. Magazines often require a more extensive pitch than a standard news release accompanied by an electronic press kit, but the basic components of the kit continue to include the following (Walker, 2009):

- A press release announcing the release of the album and a tour to support the music

- The artist's bio, which is often created by someone hired by the label for this specific purpose

- A music file that contains the single or the album

- High-resolution color jpegs of the artist on disc

- A "**cut by cut**," which is a document listing the tracks included in the music file with the artist's personal comments about each track

- A **discography**, or bibliography of music recordings

- Sometimes, a video file of the artist's music video, showing him discussing the music and the project, which helps journalists get a sense of what the artist looks like and how he relates other than through his music

- Scans or copies of **clippings** (**tear sheets**): articles printed or video stories created by other media

Journalists say that the most useful parts of the press kit, sometimes called a pitch kit, are the audio files, the cut-by-cut, and the photos.

The email with the attached press kit and news release should be personal to the journalist. The purpose of the email must be very clear in the subject line, and the purpose outlined in the body of the email should be specific, not general. From the journalist's perspective, it must answer the questions: "What do you want?" and "Why is this important to me?" Otherwise, the label publicist risks having the journalist set the news release and press kit aside to figure out, perhaps later, what it is all about. And later may never happen.

The online service Sonicbids.com creates and stores electronic press kits for artists and can be a good online resource for smaller labels with limited budgets for publicity.

Photos and Video

A good photograph can generate a lot of publicity. It can be the most striking and effective part of a press kit. A good-quality photo has a much better chance of being run in print media or included on a web site and is worth the extra effort and expense. An experienced professional photographer can bring out the true personality of the artist in a photograph. Sometimes, location is used to help portray the artist's identity, but studio shots are easier to control (Knab, 2001). Publicity shots are not the same as publicity photos. A publicity shot is one taken backstage with other celebrities, or at events. The publicity photo is the official photographic representation of the artist. Publicity photos should be periodically updated to keep current with styles and image. But, once a photo is released to the public, it is fair game for making a reappearance at any time in the artist's career, even if she has moved on and revamped her image. Some photographers prefer the use of film over digital photography for artistic reasons, like some music producers prefer analog over digital tracks and some audiophiles prefer vinyl over digital downloads. However, the elimination of the cost and time of film processing coupled with the immediacy of digital imaging have considerably reduced the cost of photo shoots by record labels.

The music video also becomes an important part of the continuing effort of the publicity department to promote the album as consecutive singles are released. Though many labels have video promotion departments, the artist's music video is also a valuable tool used by publicists in securing live and taped appearances in television programs. (Other uses of record label videos are discussed in depth in Chapter 15.)

Working with the Artist's Image

Any aspect of the entertainment business relies on the created perception of the artist or event. After all, it is show business. It is what consumers and fans *think* about the artist and how they *feel* about him, rather than what the reality may be. It is important to the label to know the perception of the recording artist in the minds of music buyers. One of the most important contributions a label publicity department can make toward defining the public's perception of the artist is carefully helping the artist develop his image.

Recording contracts that include comprehensive and multiple rights over career management of their newest artists create a number of impacts on the image of the artist around which the publicist must work. They include:

- The name chosen by the artist
- Physical appearance of the artist
- The artist's recording style and sound
- Choices of material and songwriting style
- The artist's style of dress

- The physical appearance of others who share the stage
- The kind of interviews done on radio and TV
- Appearance and behavior when not on stage

(Frascogna and Hetherington, 2004)

Some labels hire hair stylists and clothing and costume consultants, some will pay for dental work, and some are rumored to pay for cosmetic surgery in order to polish the artist's image to prepare her for her expanded public career (Levy, 2004).

Hiring a **media consultant** is a judgment call for the label, and they may or may not feel that an artist should receive training. A media consultant trains artists to handle themselves in public interviews and other nonmusical performance occasions. Working with the artist, the consultant prepares her for interviews by taking the unfamiliar and making it familiar to her, teaching her to know what to expect, and giving her the basic tools to conduct herself well in a media interview. The debate comes from critics who say media consultants go too far by preparing artists with suggested answers and thereby create cookie-cutter interviewees with nothing new or interesting to say. Former Equity Records president Mike Kraski acknowledges that some record labels attempt to "sand off rough edges" of new artists they sign. In these cases, they may go too far by guiding an artist in a direction creating "something that is not true to the artist" for the sake of making something more commercial (Havighurst, 2007; Allen, 2007).

Publicists and others at the label should take care not to compromise the unique qualities an artist has by developing an image that isn't consistent with who the artist is. The perceived personal values of the artist should be apparent in his music and his public image in order to tighten the connection with record consumers who share those values. Ideally, the label and the artist manager will work with the artist to help him define who he is personally and creatively, and then coach him to an image which is consistent with who he is and which is commercial, but not artificial.

For an established artist, care must be taken not to radically change her image. Fans are quick to pick up on efforts to radically retool an image, and the result can ultimately turn fans away from an artist.

EVALUATION OF PUBLICITY CAMPAIGNS

Top management and artists themselves will seek feedback about the effectiveness of a media campaign on behalf of an album project. Clipping services are available that will search publications and television shows for copies of articles or video clips that will demonstrate the news item connected with the media. AristoMedia, an independent publicity company in Nashville, often uses Google Alerts as an efficient way to check the impact of its news releases a few days following dissemination. It is good for record label–artist relations to present the artist and manager with a box of newspaper and magazine clippings and a list of links to indicate the successful efforts of the publicity department in securing media coverage.

TARGET MARKETING THROUGH THE PUBLICITY PLAN

Label publicity, as in any part of the marketing process, must keep its focus on the target market for the recording. The starting point is to get a clear understanding of what that target market is and how publicity fits within the project's marketing plan, then to develop a plan to reach the market through publicity. The plan created by the publicity department is coordinated with the other departments within the label. For example, A&R will define the artist, the music, and the genre. Marketing will provide the overall plan that will include sales objectives, the target market, the marketing communications plan, and an estimated release date for the single and the album. Radio promotion will provide its timeline to work the record to radio, and let publicity know whether the music will be promoted to multiple formats. The New Media Department will create its timeline and contact schedule for promotion to online radio, iTunes, social networking sites, ezines, and other online and wireless points of contact. That helps the publicist know where to place effort. Sales will give its timetable so that publicity efforts will be timed to maximize sales at retail.

New artists especially require a major push by publicity to create that invaluable buzz among the electronic gatekeepers of web sites, radio, and television. The trades, online sources, and trade newsletters must be worked before the release of the album, and radio should be worked prior to the release of the first single. Press releases are sent to trade publications and online journalists in the hope of getting an article or at least a brief mention about the upcoming release.

The most effective publicists research the magazines, newspapers, ezines, web sites, blogs, and podcasts before submitting materials for publication or inclusion in their programs. The types of stories, reviews, and features that each publication prints should be noted. Then, only those that match the target market and that regularly feature the types of stories being pitched should be approached. It annoys writers and editors when a publicist pitches inappropriate material to their publication, primarily because of the time it wastes for the journalist. It is generally easier to get placement in a niche magazine or ezine than in a more general publication. Genre-based music magazines are more receptive to publishing suitable material, and there is less competition than one would find with a general-reading publication such as *People*.

Reaching the target audience with press releases requires an understanding of the actual audience of the media. For example, Figure 10.1 outlines the traditional print media type and its intended audience.

Trade publications, like those noted above, offer the benefit of setting up a recorded music project to garner the interest of radio and video channels. Web site versions of music publications have considerably expanded the music information universe and have helped publicists highly target consumers by the niches they serve. A list of some of those sites is provided as an appendix to this book.

EZINES

Electronic, or online, magazines offer a good way to introduce a label's artist to the target market. MarketingTerms.com defines an *ezine* as an electronic magazine, whether posted via a web site or sent as an email newsletter. A more complete definition describes them as online versions of magazines, complete with magazine-style formatting. Some are electronic

Media	Targets and examples
Daily newspapers	General readers from a national, regional, local base
Lifestyle, entertainment magazines	Broad national readership with a certain editorial focus *People, Entertainment Weekly, Cosmopolitan*
News weeklies	Young, urban, sophisticated, culturally aware readers *Village Voice, SF Weekly, Boston Phoenix*
Music and pop culture magazines	Young, affluent adults with strong interest in music *Rolling Stone, Spin, Vibe*
Genre-based music magazines	Customers of chain record stores *Country Weekly, The Source, Down Beat*
Magazines for music hobbyists and pros	Players of instruments, recording engineers *Guitar World, Mix, Musician*
Magazines for record collectors	Collectors and aficionados of oldies and rare discs *Goldmine*
Trade publications	Music industry professionals *Billboard, Variety, CMJ*
Fanzines	Pre-teen and teenage fans of new artists *Tiger Beat*

FIGURE 10.1

versions of existing print magazines like those noted above, whereas others exist only in the digital format. The web-posted versions usually contain a stylized mixture of content including photos, articles, ads, links, and headlines, formatted much like a print equivalent. Most ezines are advertiser supported, but a few charge a subscription fee.

Many established music ezines are genre specific or have particular subject areas dedicated to genres. They may feature music news, concert and album reviews, interviews, blogs, photos, tour information, and release dates. As a result, their readers are predisposed to be receptive to new and unfamiliar artists and their music, provided that the artist is within the genre that the e-zine represents. A study on readers of the Americana music magazine *No Depression* found that 90% of their readers learned about new music from an article, one published either in a print magazine or in an online version.

What to Send

Ezines are mostly interested in feature articles and press releases pertaining to some newsworthy item (such as an album release or a tour schedule announcement). A feature story for an ezine includes biographical information as well as the newsworthy information—in other words, it is part bio and part press release. Label publicists write the article with an assumption that it will appear unedited in the online publication, in the inverted pyramid style. Articles like this also include an attached publicity photo or two along with the article for submission. The most effective articles and press releases include plenty of links to specific pages on the artist's and the label's web site that pertain to the featured topics (tour page for touring news, product page for record release news, etc.).

Label publicists take care not to send a news release to an ezine or other publication if there is nothing that is considered newsworthy. Like all journalists, those at ezines will look at the news release to determine its relevance and timeliness for their readership, and an irrelevant news release strains the relationship the publicist has with his contacts.

Where to Send It

The Ezine Directory has a listing of many of the better-known music ezines, along with descriptions and ratings of each (www.ezine-dir.com/Music). The goal, like any target of publicity, is to find those with the correct target market and submit articles, music, and photographs to the editor, encouraging her to include a link to the artist's web site. Some ezines have submission forms available on their web site, whereas others are not as specific about their submission policy. When an article does appear online, the publicist will link the artist's and sometimes the label's web sites to it.

RESOURCES FOR EZINES AND DISTRIBUTION OF PRESS RELEASES

The Ultimate Band List: www.ubl.com
CD Baby: www.cdbaby.org
BeatWire: www.beatwire.com
The Ezine Directory: www.ezine-dir.com
Music Industry News Network: www.mi2n.com
MusicDish: www.musicdish.com (try the Open Review and Submit Article)
John Labovitz's E-zine-List: www.e-zine-list.com/titles_by_keyword/
    music/page1.shtml
Zinester Ezine Directory: www.zinester.com
DMusic: http://news.dmusic.com/submit
PRWeb: www.prweb.com

THE PUBLICITY PLAN

Targeting a publicity plan to the specific music culture(s) means the publicist understands the audience of the music and understands how they consume entertainment and information (Spellman, 2008). The publicity plan is designed to coordinate all aspects of getting nonpaid press coverage and is timed to maximize artist exposure and record sales. The plan is usually put into play several weeks before the release of an album. In the case of music magazines, the plan begins months in advance due to the long lead time necessary to meet their deadlines for publication.

Planning for a publicity or media plan begins with coordination with the label's marketing plan and linking the plan timetable with the marketing calendar. The media marketplace is then researched, and media vehicles targeted. The materials are developed and pitching to journalists and talent bookers begins. **Lead time** is the amount of time in advance of the publication that a journalist or editor needs to prepare materials for inclusion in the publication. A schedule is created to ensure that materials are created on schedule and then are provided in a timely manner to make publication deadlines. Long-lead publications are particularly problematic for the publicist as they must have materials prepared months in advance of the release date, and

The publicity plan
• Set publicity goals
• Identify target market
• Identify target media
• Create materials
• Set up timetable w/deadlines
• Pitch to media
• Provide materials to media
• Evaluate

FIGURE 10.2 *The publicity plan.*

sometimes those materials are not yet available. If an artist suddenly breaks in the marketplace, it is too late to secure a last-minute cover photo on most monthly publications.

Pitch letters are then sent out to media requesting publication or other media exposure. The pitch letter is a carefully thought out and crafted document specifically designed to grab the interest of a busy, often distracted journalist, TV producer, blogger, or online web site editor. It is never emailed to a bulk list but is specifically tailored to each media outlet being contacted (D'Vari, 2003). The pitch letter should begin with a few words presenting the publicist's request and then quickly point out why the media vehicle being contacted should be interested in the artist or press material. Prep sheets are also developed and sent to radio programmers and their consultants so that DJs can discuss the artist as they prepare to play the music on the air. As the publicity plan unfolds, it is necessary to evaluate the efforts through clipping services and search engines (see the section on evaluation of publicity campaigns).

Time frame	Publicity task
Upon signing the artist	Schedule meetings with artist Press release announcing signing
During the recording	In-studio photos Invite key media people to studio
Also during this time period	Schedule media training if needed Schedule media photos Determine media message
When masters are ready	Hire bio writer Create advance copies for reviews Create visual promo items
When advance music is ready–ideally four months out	Send advances to long-lead publications Send advances w/bio and photo VIPs–magazines, TV bookers, and syndicators Begin pitch calls to secure month-of-release reviews Start servicing newsworthy bits on the artist on a weekly basis to all media
Advance music–one month out	Send advances/press kits to key newspapers, key blogs, and TV outlets Begin pitch calls to guarantee week-of-release reviews
One week out	Service final packaged CDs to all media outlets Continue follow up calls and creative pitching
After release	Continue securing coverage and providing materials to all media outlets

FIGURE 10.3 *Example of a publicity timeline for a major label. (Source: Amy Willis, Media Coordinator, Sony Music Nashville)*

Preliminary publicity timeline for indie artist prototype		
Day	Date	Event
Fri	12/6	Advance promo copies at *The Rocket, UW Daily, Pandemonium, Seattle Times*, and the *Seatttle Weekly*
Mon	12/9	Arrange interviews on KCMU, Seattle, and KUGS Bellingham
Fri	12/13	Deadline for completing database of print/broadcast mailing list
Fri	1/10	Single sent to college radio
Fri	1/17	*Calling It A Day CD release day*
Mon	1/20	Album mention in *UW Daily/Seattle Times*
Tues	1/21	Album mention in *Seattle Weekly*
Thurs	1/24	Tour begins in Seattle
Thurs	1/24	Album mention in CMJ/*Hits*
Tues	1/28	Album mention in *The Rocket* and *Pandemonium*
Tues	1/27	Album mention in Spokane and Tri-City daily papers
Mon	2/2	Album mention in *Cake* and *Fizz*
Fri	2/6	Album mention in *Flipside* and *Village Noize*
Wed	2/15	Album mention in *Spin*
Wed	2/15	Album mention in *Virtually Alternative*
Fri	2/17	Album mention in *Next*
Fri	2/24	Album mention in *Magnet*
Mon	2/27	Interview on KUGS Bellingham
Mon	2/27	Interview on KCMU Seattle
Tues	2/28	Feature story in *The Rocket*

FIGURE 10.4 *Preliminary publicity timeline for an indie artist prototype. (Source: Christopher Knab)*

BUDGETS FOR MONEY AND TIME

A budget for the publicity campaign is developed based on the objectives of the project, the expectations of the label for the part publicity will play in stimulating interest in the artist's music, and the degree to which the label is managing the artist's career. If this is the first album for the artist, the development of new support materials may be necessary, such as current photos and a bio. If the new artist is working under a multiple rights contract, tour press support will be necessary. If it is an established artist, budgets could be considerably higher, in part because of the expectations of the artist to receive priority attention from the director of publicity.

Publicity costs include the expense of developing and reproducing materials such as press kits, photos, bios, video, and so forth; communication costs (postage and telephone bills, maintenance of contact lists); and staffing costs. The minimum cost for an indie label would run about $8000, with $3000 of that for developing press kits and $2800 for postage. Adding an outside consultant to the project would add another $1500 or more per month. For major label projects, an outside publicist can be hired for

6 months for $25,000 to provide full support to a single, an album, and tour publicity for $25,000; this includes out-of-pocket costs such as postage, press kits, web site maintenance, and anything else the label requires to support the publicity effort.

An equally important part of the plan is to budget adequate time to support the album, based on when it will be released during the annual business cycle of the label. If the in-house staffing is adequate—given the timing of the project—the plan can be created. If, however, the publicity department is overloaded, the director may consider hiring an independent company to handle publicity for the project. This seemingly removes the burden from the director, but it adds oversight duties because the director must be sure the outside company is working the plan according to expectations. The ultimate success (and failure) is still the responsibility of the director of publicity.

TELEVISION APPEARANCES AND NEWS SHOWS

Major entertainment television news shows, including syndicated news shows on major network affiliates and cable channels, are most often interested in major acts. Their viewers want to know the latest information about their favorite recording artists. Stars that are easily recognizable—those with the highest **Q factor**—are those most often sought for their entertainment news stories because they draw a lot of interest and big audience ratings. With major artists filling prime interview opportunities, it becomes a genuine challenge for the record label that is trying to publicize a new artist with consumers. A new act with a new single or album must have an interesting connection with consumers that goes beyond the music in order to compete with the superstars who will always get airtime. There are many more new artists looking for publicity, forcing the best label publicists to be as creative as they can be on behalf of the few new acts that can break through.

Television interviews with new artists require a back story that sets them apart from every other "new artist with a great voice" who is seeking the media spotlight. Shows look for that added dimension to a new artist that makes her interesting to the viewers, and they often look for the nontraditional setting in which to present the story. Though at times it is overdone, connecting an artist with his charity work becomes an interesting angle for television.

The challenge to the label publicity department is to find those key personal differences that make their recording artists interesting beyond their music. Label publicists are sometimes criticized for citing regional radio airplay, chart position, YouTube views, or label financial support as the only positives that make their newest artists stand out. Those in the media say they look for that something special, different, and newsworthy that gives an angle for them to write about. In that light, it puts the responsibility on the label publicist to find several different angles to offer to different media outlets to generate the interest needed to get a story placed. Writers for major media want their own angle on an artist when possible because it demonstrates to media management that an independent standout story has been developed, making them different from their competition. Sometimes, though, the story angle about an artist is different enough that it stands on its own, and most media will see the value it has for their audiences. Entertainment writers and producers are often self-described storytellers, and delivering that unique story to them is a continuing challenge to the successful label publicist (Lathrop and Pettigrew, 2003).

Talk-Entertainment Shows

Label publicists are often the facilitator of an artist's appearance on popular talk and entertainment shows, frequently with the result of introducing an artist to an audience that does not use other media to find music. Among the most popular shows include those with hosts Jay Leno, David Letterman, Conan O'Brien, Jimmy Kimmel, and Oprah. Other television shows that publicists seek as targets for the label's artists include the early morning shows like *Today*, and *Saturday Night Live*. Great performances on shows like these are expected of the artist, but Ashley Simpson's ill-fated 2004 appearance on *Saturday Night Live* will probably follow her forever. (Simpson had an embarrassing performance when tracks for the wrong song were cued up and played, revealing prerecorded vocal tracks.)

The payoff to the label for an artist's appearances on television shows comes in the form of sales. For example, after spending a week appearing on an array of television shows like *The View* and *Today* in the week of February 24, 2007, Josh Groban's album *Awake* rose in the *Billboard* 200 sales chart from number 33 to number 17 (Paoletta, 2007).

Bookings to programs like these are handled by the publicist and are based upon their relationships with talent bookers on these shows. It is not uncommon for a publicist to precede a pitch for an artist to appear on one of these shows by sending a big fruit basket. However, the success of placing the label's new artist on one of these shows is also based on the ability of the publicist to build a compelling story for the artist that will interest the booker. Often the publicist will offer another major artist for a later appearance in exchange for accepting the new artist now.

Major labels have the benefit of their high-profile roster and artists under development, and the financial resources to promote live performances to major network and cable shows. Independent labels, with their much smaller promotion budgets, must by necessity approach a pitch for a live performance keeping those limited dollars in mind. Cole Wilson is the music booker for *The Late Late Show With Craig Ferguson* and offers these points to the indie publicist seeking a performance on the show.

- The artist needs an online presence on MySpace or YouTube where she can see performances and read the comments left by fans.

- Talent bookers for late night shows in New York and Los Angeles frequently spend time visiting live entertainment venues in those cities, providing the bookers an opportunity to see a prospective artist.

- If an artist is "different from the norm" it gives the booker an opportunity to present something fresh to the audience. She says an artist who sits on a stool and sings for three minutes lacks visual appeal.

- The publicist should remember that it isn't just the talent/music booker who must be convinced the artist should be invited to the show. Often it is a committee who will want to view the artist from as many perspectives as the publicist can present.

- The artist should remember that an appearance on television does not mean that he can make "outrageous demands" from the show. (Donahue, 2008)

The label publicist who is responsible for pitching an artist to news and entertainment shows gets some advice from *Today* show senior producer Melissa Lonner. Here are her quoted tips about the business etiquette of pitching an artist for an appearance on the NBC morning show.

- Keep your pitches short over the phone and/or email.
- Don't pitch on voicemail.
- Send a CD of music with selective press clippings.
- Send an email to follow up and recommend a track.
- Don't send the deluxe press clippings collection.
- Don't say why the artist should be on *Today*.
- Provide the music, stats, and the facts—not the hype.
- Don't say that the artist is the next "_____."
- Follow up on pitches via email or phone.
- Be kind, calm, and honest.
- Don't stalk, threaten, or demand.

(Paoletta, 2007)

Award Shows

The value of having an artist perform on an award show is obvious—it sells recordings. These slots are coveted by all the record labels, and lobbying efforts may pay off in a big way. Many awards shows are showing modest declines in viewership, but artists who are nominated or who perform on music award shows can see spikes as high as 700% in the sales of their music. For example, the week after the announcement of the 2009 Grammy nominations, Alison Krauss and Robert Plant's album, *Raising Sand*, had an 88% increase in sales from the resulting publicity (Ben-Yehuda, 2009).

COMPARING PUBLICITY AND RECORD PROMOTION

> The savvy labels recognize how important publicity is to the mix—it's almost as important as record promotion.
> —Phyllis Stark, Former Nashville Editor, *Billboard* magazine

The preceding chapter in this book looked at record promotion and its impact on the marketing of a recorded music project. The chart in Figure 10.5 is a look at the relationship that publicity has with its counterpart in the record promotion effort for an album.

Record promotion	Publicity department
Develops and maintains relationships with key radio programmers (gatekeepers).	Develops and maintains relationships with key writers, news program producers, and key talent bookers for network and cable channel TV shows.
Tells radio programmers that a new single or album is about to be released and to prepare for "add" date; sends promo singles and albums.	Prepares and sends a press kit to journalists announcing the new single or album project.
Schedules the new artist for tours of key radio stations for interviews and meet 'n' greets with station personnel.	Schedules the artist and sometimes the album producer for interviews with both the trade press and consumer press.
Employs independent radio promotion people who have key relationship with important radio programmers.	Employs independent or freelance publicists who have key relationships with important media outlets.
Effectiveness of their work is measured by the number of "adds" they receive on the airplay charts of major trades.	Effectiveness is measured by the number of "gets" they receive, meaning the number of articles placed, number of TV news shows in which stories run, the number of talk/entertainment shows on which the artist performs (Phyllis Stark).
Gets local radio publicity and airplay for new artist based on the promise of an established artist making a local appearance sometime in the future.	Gets new artists booked on major talk/entertainment shows based on the promise of making an established artist available to the show sometime in the future. Supports local press during touring.

FIGURE 10.5 *Radio promotion compared to publicity.*

The author extends a special thanks to Jeff Walker of AristoMedia for his assistance with information presented in this chapter.

GLOSSARY

bio—Short for biography. A brief description of an artist's life and/or music history that appears in a press kit is a bio.

creative services department—A work unit at a record label that handles design, graphics, and imaging for a recorded music project.

clippings—Stories cut from newspapers or magazine. See *tear sheets*.

cut by cut—A listing of comments made by an artist about each of the songs chosen to be included on an album project.

discography—A bibliography of music recordings.

independent publicist—A person or a company that performs the work of a label publicist on a contract or retainer basis.

lead time—Elapsed time between acquisition of a manuscript by an editor and its publication.

media consultant—Trains artists to handle themselves in public interviews.

press release—A formal printed announcement by a company about its activities that is written in the form of a news article and given to the media to generate or encourage publicity.

Q factor—A general term used to determine the overall public appeal of an artist in the media. A high Q factor means an artist is able to draw large television audiences.

slug—A short phrase or word that identifies an article as it goes through the production process; usually placed at the top corner of submitted copy.

talent bookers—People who work for producers of television shows whose job it is to seek appropriate artists to perform on the program.

tear sheets—A page of a publication featuring a particular advertisement or feature, which is sent to the advertiser or PR firm for substantiation purposes. See *clippings*.

ACKNOWLEDGMENTS

In addition to Jeff Walker, the author also extends thanks to Tom Baldrica, Bill Mayne, Phyllis Stark, Ed Benson, the folks at Starpolish.com, and the CMA for their assistance in providing added information and insight into this chapter.

BIBLIOGRAPHY

Allen, P. (2007). *Artist Management for the Music Business*. Boston, MA: Focal Press.

Ben-Yehuda, A. (2009). The real prize Billboard.com. *Billboard*.

D'Vari, M. (2003). How to create a pitch letter. www.publishingcentral.com/articles/20030301-17-6b33.html.

Donahue, A. (June 28, 2008). The indies issue: How to get on a late night show. *Billboard*, *120*(20), 27.

Frascogna, X., & Hetherington, L. (2004). *This business of artist management*. New York, NY: Billboard Books.

Havighurst, C. (2007). All things considered: Nashville band leaves label and thrives. *National Public Radio, Feb. 7*.

Hyatt, A. (2004). How to be your own publicist. http://arielpublicity.com.

Knab, C. (2001). How to write a music-related press release www.musicbizacademy.com/knab/articles/. *Promo Kit Photos*.

Knab, C. (2003a). www.musicbizacademy.com/knab/articles/pressrelease.htm. (November).

Knab, C. (2003b). www.musicbizacademy.com/knab/articles.

Lathrop, T., & Pettigrew, J. (2003). *This business of music marketing and promotion*. New York, NY: BPI Publications.

Levy, S. (2004). CMA's music business 101. Unpublished.

Paoletta, M. (April 21, 2007). As seen on TV. *Billboard*, *119*(16), 27.

Spellman, P. (2008). Creative marketing: Making media waves: Creating a scheduled publicity plan. Music business insights (Issue I:3). www.musicianassist.com/archive/newsletter/MBSOLUT/files/mbiz-3.htm.

Stark, P. (2004). Personal interview.

Walker, J. (2009). Personal interview.

CHAPTER 11

Advertising in the Recording Industry

Amy Macy and Tom Hutchison

BASICS OF ADVERTISING

Advertising is a form of marketing communication. It is described by Bovée and Arens (1986) as "the nonpersonal communication of information, usually paid for and usually persuasive in nature about products, services, or ideas by identified sponsors through the various media." The fact that advertising is paid for and persuasive (i.e., the seller controls the content and message) separates it from other forms of mass-media communication such as publicity, news, and features. Advertisements are usually directed toward a particular market segment, and that dictates which media and which vehicles are chosen to present the message. A **medium** refers to a class of communication carriers such as television, newspapers, magazines, outdoor, and so forth. A **vehicle** is a particular carrier within the group, such as *Rolling Stone* magazine or MTV network. Advertisers determine where to place their advertising budget based on the likelihood that the advertisements will create enough of an increase in sales to justify their expense, in other words, their **return on investment** (**ROI**). Advertisers must be familiar with their market and consumers' media consumption habits in order to be successful in reaching their customers as effectively as possible.

The most basic market segments for advertising are: (1) consumers and (2) trade, which is people within the industry that make decisions affecting the success of your marketing efforts. The first market—consumers—is targeted through radio, television, billboards, direct mail, magazines, newspapers, and the Internet. Consumer advertising is directed toward potential buyers to create a "pull" marketing effect (see Chapter 1). The second target

CONTENTS

market is the *trade* and consists of wholesalers, retailers, and other people who may be influenced by the advertisements and may respond in a way that is favorable for the marketing goals. This creates a "push" marketing effect (again, see Chapter 1). In the recording industry, this would include radio program directors that may be influenced by an advertisement in *Radio & Records* to more favorably consider a particular advertised song for inclusion in the station's weekly playlist. **Trade advertising** is usually done through direct mail, through **trade publications** (usually magazines) aimed at people who work in the industry, and through trade shows. Trade advertising is not common through television, radio, and newspapers because they are too general in nature—not targeted enough to effectively reach the industry.

COMPARISON OF MEDIA OPTIONS FOR ADVERTISING

The most complex issue facing advertisers involves decisions of where to place advertising. The expansion of media has increased the options and complicated the decision. The following chart in Table 11.1 represents a basic understanding of the advantages and disadvantages of the various media options.

Table 11.1	A comparison of media	
Media	**Advantages**	**Disadvantages**
Television	■ Reaches a wide audience, but can also target audiences through use of cable channels ■ Benefit of sight and sound ■ Captures viewers' attention ■ Can create emotional response ■ High information content	■ Short life span (30–60 seconds) ■ High cost ■ Clutter of too many other ads; consumers may avoid exposure ■ Can be expensive
Magazines	■ High quality ads (compared to newspapers) ■ High information content ■ Long life span ■ Can target audience through specialty magazines	■ Long lead time ■ Position in magazine may be uncertain ■ No audio for product sampling (unless a CD is included at considerable expense)
Newspapers	■ Good local coverage ■ Can place quickly (short lead) ■ Can group ads by product class (music in entertainment section) ■ Cost effective ■ Effective for dissemination of information, such as pricing	■ Poor quality presentation ■ Short life span ■ Poor attention-getting ■ No product sampling

Table 11.1	(continued)	
Media	**Advantages**	**Disadvantages**
Radio	■ Is already music oriented ■ Can sample product ■ Short lead, can place quickly ■ High frequency (repetition) ■ High quality audio presentation ■ Can segment geographically, demographically and by music tastes	■ Audio only, no visuals ■ Short attention span ■ Avoidance of ads by listeners ■ Consumer may not remember product details
Billboards	■ High exposure frequency ■ Low cost ■ Can segment geographically	■ Message may be ignored ■ Brevity of message ■ Not targeted except geographically ■ Environmental blight
Direct Mail	■ Best targeting ■ Large info content ■ Not competing with other advertising	■ High cost per contact; must maintain accurate mailing lists ■ Associated with junk mail
Internet	■ Best targeting—can target based on consumer's interests ■ Potential for audio and video sampling; graphics and photos ■ Can be considered point-of-purchase if product available online ■ Can use cost per click instead of impressions for setting rates	■ Slow modem speeds limit quality and speed ■ Effectiveness of this new media still unknown ■ Doesn't reach entire market ■ Internet is vast and adequate coverage is elusive
Mobile	■ Good for "just in time" promos ■ Good for proximity marketing ■ Can reach younger demos	■ Limited message capacity ■ Consumers must opt-in ■ Push strategy only

NEW MEDIA ADVERTISING

Advertisers are beginning to shift their advertising budgets away from traditional media and into new media. New media consists of forms of communication involving computers and portable devices and differs from mass communication in a number of areas: it is interactive, dynamic instead of static, highly targeted and specialized, and multimedia. Advertisers prefer the targeting, the effectiveness, and the interactivity compared to traditional mass media such as magazines, newspapers, television, and radio. Thus, over time we are experiencing a shift in advertising revenue, as illustrated in Figure 11.1.

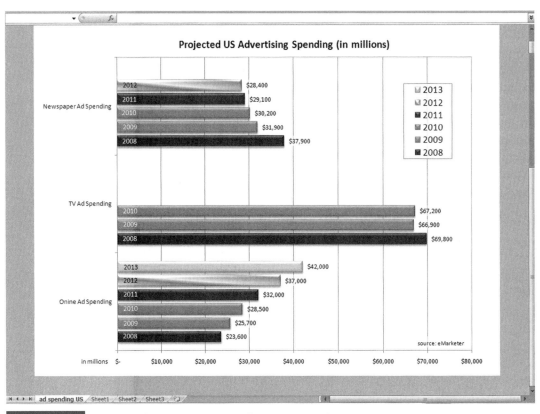

Internet Advertising

Most Internet advertising takes the form of sponsored links and banner ads. A banner ad is basically a graphical advertisement placed across the top or down the side of a web page (also called a sidebar ad) that is linked to the advertiser's web site. Advertisements appear on targeted content sites, such as online versions of magazines, to pitch products available on e-commerce sites. Originally, these ad rates were set based on those in the magazine industry—that is, they were based on the number of impressions. Thus, the advertiser paid based on the number of people who saw the ad. Now, the most popular form of charging for such advertising is "**cost per click**" (**CPC**), also known as pay per click (PPC). With CPC, the advertiser is charged a small fee each time a potential customer clicks on the ad and is taken to the advertiser's web site. This has proved more effective because the advertiser is paying only for those potential customers who respond to the ad by clicking on it and who are directed to the advertiser's site. Google has popularized the use of sponsored links and offers web sites that have similar content the ability to feature Google ads from sponsors who sign up for Google's AdWords.

Here is how Google AdWords and Yahoo! Marketing Solutions work. As stated earlier, advertisers pay on a per-click basis; in other words, they pay a few cents for each time a web visitor clicks on their sponsored link. The advertiser enters in a series of keywords (search terms or words their customers are likely to use in a search engine when looking for a particular product or type of web site). Advertisers place bids to have their ad strategically located in the sponsored links category on search engine results. The advertiser can actually create a list of terms and bid independently on each one. The highest bid for that particular set of search terms has the top spot. You may not want to be the top spot for blues music if you depend on live shows for income because blues fans from all over would be likely to click to your site only to learn that you are not performing in their area. But if the term "blues music" were combined with "East Texas" and you were performing in that area, then perhaps you would want one of the top advertiser spots. Finding the right keywords and combination of keywords may take a bit of trial and error at first.

Contextual Advertising

Contextual advertising is defined as advertising on a web site that is targeted to the specific individual who is visiting the web site, based on the subject matter of the site and featuring products that relate to that subject matter. For example, if the user is viewing a site about playing music and the site uses contextual advertising, the user might see ads for music-related companies such as music stores. Google has added AdSense as a way for web site owners to feature relevant advertising on their sites and share in the CPC revenue generated by sponsors. The source of these ads comes from the AdWords program, so that those who sign up for AdWords can specify whether they want their ad to appear on these related web sites—the **content network**. Google's web site states, "A content network page might be a web site that discusses a product you sell, or a blog or news article on a topic related to your business" (Google, 2009).

Mobile Advertising

The latest form of new media advertising is marketing to mobile phones. As multimedia phone handsets become more popular, and as the proliferation of mobile devices continues to overshadow Internet adoption, marketers are looking to cell phones as the new way to reach consumers who may not respond to more-traditional advertising campaigns. In an article in *AdWeek*, writer Greg Stuarts states, "Mobile has many . . . strengths, including: interactivity; the fact that it's personal (and will hopefully be personalized); its omnipresence (which, while unique, could also be accomplished with media mix); and its ability to do sight, sound and motion, with which marketers tend to be enamored" (Stuart, 2009). Over 4.4 billion people globally are using mobile devices, with more than 40 million in the United States. One mobile advertising company, BuzzCity, reported mobile advertising growth of 47% in the United States in the fourth quarter of 2008.

Skeptics raise the question of consumer resistance to accepting advertising messages on their mobile devices. In a study conducted by *AdWeek*, 76% of respondents said they would refuse to accept advertising messages via their cell phones (Stuart, 2009). The 15% who said "yes" gave a qualified answer—only with their permission. But in Japan, as of 2007, 54% of mobile phone subscribers had already agreed to receive ads on their phones and more than 44% of Japanese mobile phone users actively click on ads they have received (Iacolare, 2007). A study of teens in the United Kingdom by Q Research (2007) revealed that only 32% of teens are willing to receive ads with no enticements on their mobile phones. When asked if they would be willing if they only received push ads on things they were interested in, the positive response jumped to 71%; and when incentives are added in, the response is even higher. In fact, in late 2007, one cellular provider, Blyk, launched a phone service in the United Kingdom that is free to consumers—it is fully advertiser funded. The service targets 16- to 24-year-olds by offering 217 texts and 43 minutes of voice calls every month for free in return for users agreeing to receive six daily push marketing messages.

The results of the Q Research study indicate that push ads must be tightly targeted to the receiver's interests. Nester and Lyall reported that push campaigns have a response rate of about 12%, much higher than other forms of direct marketing. In the study of teens in the United Kingdom, respondents said they preferred to receive advertisements in the form of picture ads, with text ads being the least popular. In the study, only males under age 16 preferred video ads, with other respondents expressing concern over the cost of receiving video ads. Most teens prefer to receive picture ads on their phones, rather than videos or text only.

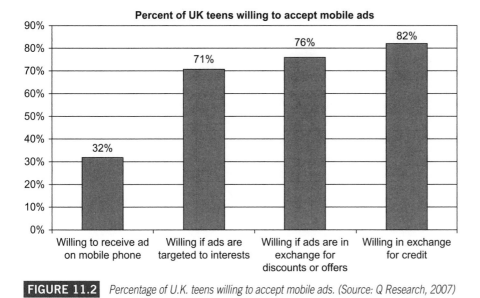

FIGURE 11.2 *Percentage of U.K. teens willing to accept mobile ads. (Source: Q Research, 2007)*

DETERMINING COSTS AND EFFECTIVENESS

Advertisers have many choices when it comes to how to spend the advertising budget to maximize effectiveness, and the options keep increasing. In the 1950s and 1960s, television and magazines were not as fragmented as they are today, and content was more general in nature. **Media fragmentation** is defined as the division of mass media into niche vehicles through specialization of content and segmentation of audiences. In the 1960s, advertisers sought to reach particular groups of potential customers, rather than paying for advertising in general media vehicles that would be wasted on viewers who were not interested in their product.

Magazines became the first medium to go through fragmentation. The popularity of general appeal magazines such as *Life*, *Look*, and *Saturday Evening Post* gave way to the rise in specialty magazines that dealt with everything from fitness to alternative music. Advertisers would rather pay for ads in the particular publications that reach their market than pay the more expensive rates to reach the general population. Fragmentation spread to television with the expansion of cable TV. The Internet offers the ultimate in fragmented and specialized content, or what is now being considered as "media personalization." As a result of this, advertisers can now reach people who are more likely to be potential consumers of the particular products being advertised.

Media Planning

Media planning involves the decisions made in determining in which media to place advertisements. It consists of a series of decisions made to answer the following questions (Sissors and Bumba, 1994):

1. How many potential consumers do I need to reach?
2. In what medium (or media) should I place ads?
3. When and how often should these ads run?
4. What vehicles and in which markets should they run?
5. What choices are most cost-effective?

The development of the media plan must incorporate the following considerations:

1. **Marketing objectives** – What are the goals of the entire marketing plan? To inform the buyer? To motivate the buyer? To change attitudes? To promote repeat purchases? To reinforce a previous purchase? To create a buzz? To emphasize price or value? To create sampling opportunities?

2. **Characteristics of the product** – Some products are more suitable for advertising in one medium over another. For example, perfume samples work well in magazines but not on radio. For music, audio is important but so are visuals.

3. **Pricing strategy** – The high costs associated with some media make media buys impractical.

4. **Channels of distribution** – Media buys should be limited to areas where the product is available for sale and where they would support retailers.

5. **Promotion plans** – This involves determining the relative amount of effort needed for advertising, compared to other marketing aspects, and to complement and support them (Dunn et al., 1990).

How Advertising Effectiveness Is Measured

The basic structure of measuring advertising effectiveness starts with information on reach, frequency, gross impressions, gross rating points (GRP), and cost per thousand (CPM). **Reach** is the number of different persons or households exposed to a particular media vehicle during the specified time period. **Frequency** is the number of times during that period that a prospect or a portion of the population is exposed to the message.

$$\text{Average Frequency} = \text{Total Exposures/Reach}$$

The concept used to measure these two notions together is gross rating points. **Gross rating points (GRP)** are a measure of the total weight of advertising that derives from a particular media buy; reach (expressed as a percentage of the market) times average frequency. It is the sum of the ratings for the individual announcements or programs. In television, GRP is the total weight of a media schedule against TV households. For example, a weekly schedule of five commercials with an average household rating of 20 would yield 100 GRP.

$$\text{GRP} = \text{Reach} \times \text{Average Frequency}$$

Gross impressions (GI) are a measure of total media exposure. It is the sum of audiences of all vehicles used in a media plan. The number represents the message weight of a media plan. The purpose of GI analysis is to get a quick look at the total audience size of one or more media (Sissors and Bumba, 1994). It is good for comparing with other media plans that may include a different group of media vehicles. GI can be calculated by adding the target audience sizes (Audience Size × Number of Exposures) delivered by each media vehicle in the plan.

When determining the effectiveness of an advertising plan, the costs must be factored in. This is done by determining the *cost per rating point* (for electronic media) and cost per thousand (a more general measure). **Cost per rating point (CPP)** is defined as determining what media programs are most cost effective by dividing the cost of advertising by the show's expected rating.

$$\text{CPP} = \text{Cost of Ad Schedule/GRP}$$

Cost per thousand (CPM) is a term describing the cost of reaching 1000 people in the medium's audience—the cost of delivering 1000 gross impressions. It is used in comparing or evaluating the cost efficiency of various media vehicles. The formula varies depending on the media.

As mentioned previously, Internet advertising rates are commonly set based upon a cost per click. In other words, each time a web visitor clicks on an ad that is linked to the advertiser's site, the site that hosted the ad receives a bit of money, paid by the advertiser. For search engine "sponsored link" advertising, the bidding and targeting are based upon keywords—those words or phrases likely to be used by potential customers looking for products in your category. The ad rate, or cost per click, is commonly set using a bidding process, where the advertiser with the highest bid gets the best

Table 11.2	Formulas for print, broadcast, and Internet
Media	**Formula**
Print	Cost of ad × 1000 circulation
Broadcast	Cost of 1 unit of airtime × 1000
	Number of households reached
Internet click-through rate	Number of clicks
	Number of impressions

placement on the web page or the ad receives a greater number of impressions in the ad rotation schedule. Advertisers are able to set a unique bid amount for each keyword or phrase. But they are not charged for impressions, only for clicks.

All other things being equal, you would expect advertisers to look for the lowest cost-per-thousand medium as being the most efficient. But this is seldom the case. Specialized products for narrowly targeted markets may require more scrutiny of each vehicle's ability to reach the target market and the effectiveness of the type of message being delivered. For example, a specialty music magazine such as *Down Beat*, designed to reach jazz fans, may cost more per each thousand people reached than local newspapers, but those who are reached may be more inclined to purchase a new jazz recording. Also, labels want to take advantage of the sampling capabilities of radio, television, and the Internet, rather than rely on print media to sell music.

THE ROLE OF ADVERTISING IN MARKETING RECORDED MUSIC

Advertising is crucial for marketing recorded music just as it is for other products. The primary advertising vehicle in the recording industry is local print, done in conjunction with retail stores to promote pricing of new titles, and is referred to as co-op advertising. But the record industry also relies on magazine, radio, television, outdoor, and Internet advertising. The impact of advertising is not easy to measure because much of its effect is cumulative or in conjunction with other promotional events such as live performances and radio airplay. SoundScan has improved the ability to judge the impact of advertising, but because marketing does not occur in a vacuum, the relative contribution of advertising to sales success remains somewhat of a mystery.

Advertising campaigns usually reflect the imaging of an artist's release. Consistency of artist photos, theme, and even the font of the text create indelible impressions that help consumers make the connection between the advertisement and the actual purchase of the release. Some ads may use photos related to the actual cover art of the CD, but where applicable, the actual CD cover will also be included.

Trade Advertising

The recording industry uses trade publication advertising to set up the release of a single to radio by advertising in publications often read by the radio industry, including *Billboard*, *HITS* magazine, *Friday Morning Quarterback*, and *The Album Network*. Advertising in *Billboard* also reaches the retail industry and may influence retail buyers. Labels may also use direct mail flyers sent to retail accounts to help promote album releases. These flyers are sometimes sent through the distributors or one-stops. Postcard advertisements are also sent to radio stations to generate interest in a single.

Coordinating with the Publicity Department

Any media campaign should be designed in conjunction with the publicity department. A coordinated media campaign is necessary because the vehicles targeted for advertising are the same vehicles targeted for publicity and some synergy may occur. For example, placing an ad in a particular publication may increase the likelihood of getting some editorial coverage. Or at least, if it has been determined that the vehicle reaches the targeted market, those consumers will be receptive to the editorial content of the publication, as well as the advertisement content. For example, if you wanted to sell hiking shoes, you would probably select a hiking, camping or backpacking magazine for your advertising, but you are also more likely to get a product review in those publications than in a parenting or sewing magazine.

Co-Op Advertising

Co-op advertising, or **cooperative advertising**, is generally defined as the joint promotion effort of two or more parties in the selling chain. It offers the potential for synergy between the two parties and is quite common in the recording industry. More specifically, co-op advertising is usually run by a local advertiser (retailer) in conjunction with a national advertiser (manufacturer). The national advertiser usually provides the copy and shares (or bears) the cost with the local retailer. Record labels team up with retail chains to create co-op advertising programs.

Co-op advertising offers a number of advantages to the manufacturer. It gains the retailer's support and endorsement, improves the relationship between the label and the retailer, helps get more products in the store, and utilizes the retailer's knowledge of the local market.

Cooperative advertising programs must follow the guidelines and legal statutes as monitored by the Federal Trade Commission to ensure fairness in trade practices. The Robinson-Patman Act of 1936 forbids price and payment (including payment for advertising) discrimination between suppliers and retailers. In other words, participating in a co-op promotion amounts to the retailer receiving a discount on goods purchased from the manufacturer. The discount comes in the form of a "rebate" paid for providing local advertisements. If the plan is offered to one retail chain over another, it amounts to discriminatory promotional allowances in violation of the Robinson-Patman Act.

Although co-op advertising is usually targeted to consumers of a specific product, the ad dollars exchanged for the exposure will most likely gain added product placement with the retailer. For example, if a mini-album cover of a new release is placed in a retailer's Sunday circular, readers of the insert will then know about its release, what it looks like, the price, and so forth. But for placing the mini in the ad, the record label gains added exposure in-store as well, such as placement of the CD on the best-seller endcap as well as extra product in the bin.

Co-op advertising can be perceived as a double-edged sword, meaning that consumers learn of new releases while the label gains additional exposure in the store. But understand that co-op advertising is usually part of the equation during the initial buy as well as continual support of the release through its lifecycle. Often, co-op ad dollars offset margin because co-op advertising vehicles often feature a release at "sale" price. The retailer can afford to discount the consumer price of the release because co-op ad dollars are aiding the profitability of each unit sold. (Co-op advertising is discussed in greater detail in Chapter 12.)

Consumer Advertising: The Media Buy

When considering a media buy, labels consider many factors:

- Budget
- Target market
- Media
- Timing
- Partners
- Artist relations

Advertising Budgets

Determining the budget sets the parameters and scope of any advertising campaign. Recognizing financial limitations can assist in creating a media buy that is the "best bang for the buck." Using the projected sales of the record plus a profit and loss analysis, labels can create a budget based on forecasted sales. To calculate a ballpark ad budget, published rate cards of advertising costs can be collected to assist in the budgeting process. The chapter regarding the profit and loss statement of a project discusses at length the implication of marketing dollars and advertising.

Target Markets

Focusing on a target market increases the effectiveness of a media buy. Most products have a specific consumer to whom companies aim their promotions. This focus determines the type of media that is purchased. For example, many products created for parents are targeted in women's lifestyle magazines as well as daytime television, with research validating that stay-at-home mothers are consuming these media as well as visiting their branded web sites. When purchasing media, most consultants look at targets demographically and in terms of when these demographics are engaging the media, both physically and virtually. Women ages 25 to 54, and men ages 18 to 24 are examples of profiling the target market. Attention to dayparts when the target is engaged is also relevant. Specifying a particular time of day to increase chances of hitting the target is strategic in creating an effective advertising buy. Evaluating the media by target enables an advertising buyer to focus the ad purchase. Ratings of electronic media outlets are determined by Arbitron and Nielsen. As discussed in Chapter 8, How Radio Works, media outlets operate their businesses to attract a certain consumer so that they can create revenue via the sale of advertising.

Media

Deciding which media to use for advertising can be an agonizing process. Print advertising historically has the lowest cost per thousand (CPM), considering its wide readership. Newspaper advertising is known as the *shotgun* approach—a shotgun uses "shot" instead of a bullet. "Shot" will scatter and possibly hit the target, but will likely hit other areas too. A newspaper is consumed by many demographics, a small portion of which may be the target. Is this the best "bang for the advertising buck"? Depending on the product, it may be most effective. For music, newspaper advertising has an important role in alerting music buyers to new releases. The Sunday circulars of various retailers often feature new releases that are "**streeting**" on the upcoming Tuesday. But some record label executives question, "Can a consumer hear a print ad?"

To hear the product, consumers would have to listen to radio, watch television, or—for a bigger impression—engage the Internet to "test drive" the product. Radio and television usually incur a higher CPM, but the audience is more concentrated, meaning that an advertisement on these outlets has a higher probability of hitting the target.

Labels have shifted much of their paid advertising to *viral marketing* online. With their artists' help, labels have employed fans to become viral marketing armies. Through the use of widgets, which are embedded streaming audio and video playback elements, fans can personalize their social web sites with their favorite artist's latest single and video, thereby creating a "virtual" endcap for that act. In some cases, there is a click-through component that will allow viewers to purchase downloads and merchandise through that portal. Technologies such as these have reduced concerted advertising dollars spent by labels on viral marketing while exponentially increasing the visibility of their artists at the very level of consumption.

Timing

When to purchase the advertising "run" is a critical decision. Most advertising outlets have "lead" times in which they must prepare for the actual advertisement. Note the following rate card information from a well-known magazine. To place a print ad for the July 6, 2009, magazine, ad artwork had to be delivered to the magazine no later than May 18, 2009; if a secondary cover or an insert were wanted, materials had to be delivered by March 23—3½ months ahead of time. For most labels, the lead time needed to place an ad of this nature creates a huge stumbling block because 7 weeks can make or break a single at the top of the charts. And yet, to place an ad means that there has to be a clear commitment from the label that it's "all systems go!"

Radio and television lead time tend to be shorter because their advertising activities often occur in "real time." But what *does* need time is the creation of the actual advertising spot. Both radio and television advertising require prerecorded "spots" in which to air. Creation of these spots includes production issues such as musical content; voice-over ad copy; and, in the case of television, visual images as well. Some labels will preproduce spots with an eye toward the need to make a quick advertising decision.

But labels are leveraging the technology at radio too. Widgets are being embedded on radio web sites as well. When listeners land on favorite radio web sites, they can access new music of the newest releases produced by the label via the playback portal, which could have a click-through component leading to a sales opportunity. This is all a part of the promotional activities set up by the record label promo team as they garner airplay and exposure for their acts.

Rates and Discounts/Audience Profile   PAGE 1

National Editions

Rate Base	Regular Issues 3,450,000	Feature Issues I 3,600,000	Feature Issues II 3,700,000	Feature Issues III 3,800,000	Year End 4,200,000
4-Color Rates:					
Page	$266,780	$279,570	$287,540	$294,810	$325,290
⅔ Page 4C	$227,330	$238,880	$246,120	$252,420	$278,250
½ Page 4C	$180,110	$189,000	$194,250	$199,500	$219,980
⅓ Page 4C	$122,300	$128,100	$131,830	$135,450	$149,100
Cover 2	$333,480	$349,470	$359,420	$368,530	$406,620
Cover 3	$293,460	$307,530	$316,290	$324,300	$357,820
Cover 4	$360,150	$377,430	$388,190	$398,010	$439,170
Black & White Rates:					
Page	$186,750	$195,700	$201,290	$206,380	$227,700
⅔ Page B/W	$159,600	$166,950	$172,200	$176,400	$194,250
½ Page B/W	$126,000	$132,300	$135,980	$139,130	$153,830
⅓ Page B/W	$85,580	$89,780	$92,250	$94,500	$104,370
Insert Cards:					
Supplied	$213,420	$223,660	$230,030	$235,850	$260,232
We-Print	$266,780	$279,570	$287,540	$294,810	$325,290

Additional Information

1) **Insert Cards.** Standard size, 6" x 4¾". Price and space requirements on request. A 5% margin must be allowed in the circulation of insert cards.

2) **Supplied Inserts.** Supplied 4-page signatures, as well as specially designed inserts such as die cuts, etc., are acceptable. Rates, specifications and availabilities on request.

3) **Premium Charges.** Special advertising production premiums are non-commissionable and do not earn any discounts. Rebates received on space charges may not be applied to premium charges.

4) **Non-Standard Colors.** All run-of-book ads that use non-convertible PMS colors or 5 colors will incur $8,400 non-commissionable production premium. If PMS Metallic or Day Glo inks are used, $10,200 non-commissionable production charge will be incurred.

Discounts

1) **Under-the-Card Discount.** In lieu of all other discounts and credits, national full-size advertisements ordered to appear under a publisher insert card earn a 35% discount. Space is limited and subject to availability.

2) **Maximum Discount.** With the exception of the under-the-card discount, an advertiser may not earn more than a 30% discount.

Audience Profile

	% of comp.
Adult Reader	%
Age 18–24	17.0
Age 18–34	37.6
Age 18–49	70.3
Age 25–54	62.5
Age 35–44	21.7
Age 45–54	20.1
Age 55+	20.5
Baby Boomers (40-51)	31.7
Generation X (25-39)	26.2
Attended/Graduated College+	61.3
Dual-Income Households	31.6
Adults with Children in Household	49.0
$60,000+ Household Income	55.4
Live in A & B Counties	76.9
Own Home	67.6
Working Women (of women readers)	45.7
Total Adults	43,528,000
Total Women	29,333,000
Percentage	67.4%
Total Men	14,196,000
Percentage	32.6%
Median Age (Adult)	40.7
Median Household Income	$67,178

Source: MRI Fall 2008
Note: Above audience profile does not reflect special feature issues that deliver increased rate base.

FIGURE 11.3   *Example magazine rate card.*

Closing Dates & Issue Cycles   PAGE 6

	Monday Issue Date	Ad Close (7 weeks)	Friday On-Sale Date	Covers & Inserts Due Date	Issue Cycles Available		Monday Issue Date	Ad Close (7 weeks)	Friday On-Sale Date	Covers & Inserts Due Date	Issue Cycles Available
January	05	No Issue				July	06	05/18	06/26	03/23	N
	12	11/24	01/02	10/6	N, M, M10, M20, S, SP, B		13	05/25	07/03	03/30	N, M, M10, M20, S, SP, B
	19	12/01	01/09	10/13	N, SEL		20	06/01	07/10	04/06	N, SEL
	26	12/08	01/16	10/20	N, M, M10, M20, S, SP		27	06/08	07/17	04/13	N, M, M10, M20, S, SP
February	02	12/15	01/23	10/27	N		03	06/15	07/24	04/20	N
	09	12/22	01/30	11/03	N, M, M10, M20, S, SP, B	August	10	06/22	07/31	04/27	N, M, M10, M20, S, SP
	16	12/29	02/06	11/10	N, SEL		17	06/29	08/07	05/04	N, SEL, SP
	23	01/05	02/13	11/17	N, M, M10, M20, S, SP		24	07/06	08/14	05/11	N, M, M10, M20, S, SP, B
March	02	01/12	02/20	11/24	N		31	07/13	08/21	05/18	N, SP
	09	01/19	02/29	12/01	N, M, M10, M20, S, SP	September	07	07/20	08/28	05/25	N, M, M10, M20, S
	16	01/26	03/06	12/08	N, SEL		14	07/27	09/04	06/01	N, SP, B
	23	02/02	03/13	12/15	N, M, M10, M20, S, SP, B		21	08/03	09/11	06/08	N, SEL
	30	02/09	03/20	12/22	N, SP		28	08/10	09/18	06/15	N, M, M10, M20, S, SP
April	06	02/16	03/27	12/29	N, SEL	October	05	08/17	09/25	06/22	N
	13	02/23	04/03	01/05	N, M, M10, M20, S, SP, B		12	08/24	10/02	06/29	N, M, M10, M20, S, SP
	20	03/02	04/10	01/12	N		19	08/31	10/09	07/06	N, SEL
	27	03/09	04/17	01/19	N, M, M10, M20, S, SP		26	09/07	10/16	07/13	N, M, M10, M20, S, SP, B
May	04	03/16	04/24	01/26	N	November	02	09/14	10/23	07/20	N, M, M10, M20, S
	11	03/23	05/01	01/26	N, M, M10, M20, S, SP		09	09/21	10/30	07/27	N, SP
	18	03/30	05/08	02/02	N, SEL		16	09/28	11/06	08/03	N, M, M10, M20, S, SP, B
	25	04/06	05/15	02/09	N, M, M10, M20, S, SP, B		23	10/05	11/13	08/10	N, SEL
June	01	04/13	05/22	02/16	N		30	10/12	11/20	08/17	N, M, M10, M20, S, SP
	08	04/20	05/29	02/23	N, SP	December	07	10/19	11/27	08/24	N, SP
	15	04/27	06/05	03/02	N, M, M10, M20, S, SP, B		14	10/26	12/04	08/31	N, M, M10, M20, S, B
	22	05/04	05/12	03/09	N, SEL		21	11/02	12/11	09/07	N, SEL
	29	05/11	05/19	03/16	N, M, M10, M20, S, SP		28	11/09	12/18	09/14	N, M, M10, M20, S, SP

Key: M: Metro Editions   M20: Top 20   S: State and Spot Markets
        M10: Top 10   N: National   B: Behavior Scan Availability
        SEL: Selective — Inkjet

        SEL: All inkjet and selective binding insertions require issue flexibility.

FIGURE 11.3 (*Continued*)

In studying lifecycles of releases, most labels recognize the urgency of front-loading the advertising, because the bulk of sales occurs early in the release of a record. Often, labels work diligently to alert consumers to a new release in order to increase the sales response at street date. But on occasion, a record may have a second or third single that is the reactionary song, giving the label an incentive to fuel the sales fire by generating an advertising buy.

Partners

Some advertising buys have a dual purpose. Clearly, the primary objective is to alert consumers to the availability of a release. But in many cases, the advertising purchase enables the label to leverage additional exposure by building in a promotional partner. When purchasing ad time, a record label may consider approaching a retailer to be the destination store tag, such as "Available now at Target." This activity would include logo placement of the retailer on both print and television advertisements. In turn, the retailer may purchase more records of the specified artist as well as prominently place the release in stores, anticipating an increase in sales.

Artist Relations

In an ideal situation, labels and their artists should look at their business relationship as a partnership. The artist delivers the product that the record label is to market and sell. An artist should expect from its label a level of commitment and promotional activity to give the release the best possible shot at success. Many efforts are put forth by record labels to ensure success, and one of these activities is advertising. An artist, with his manager, should review the promotional efforts being created and executed by the label, just as the label reviews the music of the artist.

This television-advertising buy was bolstered by additional ad exposure at radio. These time buys were during the week leading up to the Country Music Association (CMA) Awards Show.

The combined advertising gross rating points (GRPs) for the Atlanta market were 82.5. Although $15,000 was spent at radio, the CPM was dramatically lower than at television. Additionally, many more women ages 25 to 54 were reached because of the focused advertising that radio and its specific targeted demographics can deliver. Generally, radio advertising does not include gross impressions of homes.

With any budget, the greater the GRP, the better the advertising "bang for the buck." An actual advertising campaign would be comprehensive in attempting to reach the target market while maintaining a budget. To do so, buys at many outlets would be secured. The optimal equation to increase the GRP includes varying the combinations of media, dayparts, and number of spots. Depending on the agenda as well as the budget, all advertising should help increase visibility and sales of a specific artist.

Online Advertising

In a search term keyword campaign, various phrases are set up with particular bid amounts for each keyword phrase. The average position indicates where the ad ranks among other competitive ads for impressions when consumers use that particular keyword phrase. The advertiser can then adjust the bid to suit the maximum return on investment. Also indicated in the figure are the number of clicks in a time period, the number of impressions, the ratio of clicks to impressions, and the cost per click charged to the advertiser.

Table 11.3	Sample TV advertising buy								

MARKET	PURCHASED			# of SPOTS	WM: 25–54		HOMES	
DAYPART	PNTS	DOLLARS	CPP		IMPS	CPM	IMPS	CPM
ATLANTA								
PRI	7.5	6000	800	1	75	80	115	52.17
	Market:		**Atlanta**					
PRI	Daypart:		Primetime					
PNTS	GRP:		7.5					
	Cost:		$6000.00					
CPP	Cost/point:		$800.00					
	# of spots:		1					
IMPS	# of impressions of target market–women 25–54:		75,000					
CPM	Cost/thousand on target:		$80.00					
IMPS	# of homes:		115,000					
CPM	Cost/thousand of homes:		$52.17					

Table 11.4	Radio buy								

MARKET	PURCHASED			# of SPOTS	WM: 25–54		HOMES	
DAYPART	PNTS	DOLLARS	CPP		IMPS	CPM	IMPS	CPM
ATLANTA								
AMD	40	8000	200	27	440	18.18		
PMD	35	7000	200	23	375	18.66		
TOTAL	75	15,000	200	50	815	18.4		
	Market:		**Atlanta**					
AMD	Daypart:		AM drive					
PMD			PM drive					
PNTS	GRP:		75					
	Cost:		$15,000.00					
CPP	Cost/point:		$200.00					
	# of spots:		50					
IMPS	# of impressions of target market–women 25–54:		815,000					
CPM	Cost/thousand		$18.40					

+ Add keywords: Quick add | Keyword tool			Edit keywords | Search this list Customize columns						1 - 157 of 157 keywords.	
Pause Unpause Delete	Edit Keyword Settings									
☐ Keyword	Status ⑦		Current Bid Max CPC	Hide Settings Sort by: Bid | URL	Clicks ▼	Impr.	CTR	Avg. CPC	Cost	Avg. Pos
Placements you've targeted ⑦	On		$0.30		2	82,907	0.00% ⑦	$0.16	$0.31	n/a
Other placements on the content network ⑦	Off		Content Auto ⑦ [Edit]		0	0	-	-	-	n/a
Search Total	On		Default $0.45 [Edit]		16	21,716	0.07%	$0.49	$7.87	8.6
Total					18	104,623	0.02%	$0.45	$8.18	4.0
☐ sell music	🔍▸	Active Bid is below first page bid estimate of $0.70	$0.65	▾ Settings $0.65 Max CPC [Edit]	7	7,304	0.10%	$0.55	$3.85	9.6
☐ music business	🔍▸	Active Bid is below first page bid estimate of $1.00	$0.70	▾ Settings $0.70 Max CPC [Edit]	4	4,306	0.09%	$0.52	$2.08	7.9
☐ internet musician	🔍▸	Active	$0.45	▾ Settings Default Max CPC [Edit]	1	1,778	0.06%	$0.25	$0.25	4.7
☐ record distribution	🔍▸	Active	$0.45	▾ Settings Default Max CPC [Edit]	1	373	0.27%	$0.40	$0.40	6.2
☐ get a record deal	🔍▸	Active Bid is below first page bid estimate of $0.80	$0.45	▾ Settings Default Max CPC [Edit]	1	137	0.73%	$0.44	$0.44	11.3

(callout) clicks / impressions / click thru rate / ave cost per click

FIGURE 11.4   *Example of keyword bids in a cost-per-click campaign. (Permission from Google)*

FIGURE 11.5   No Depression *banner ad for artist Steve Earle. (Courtesy of NoDepression.com)*

An Actual Campaign

As mentioned earlier, many labels have opted to use fans as their mouth-piece by giving them tools such as widgets to virally spread news about artists and their new releases. But online advertising campaigns can be directed in a way to maximize dollars while creating exposure and immediate sell-through. Instead of using impressions, as print advertising does, online ad costs are based on PPC, or pay per click, impressions, where sites charge advertisers a fee based on consumers who have "clicked" through

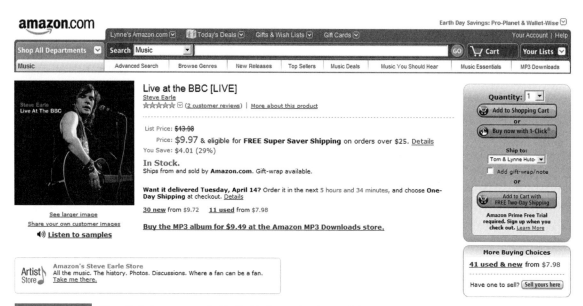

FIGURE 11.6   *Steve Earle Amazon page tied to the* No Depression *banner link.*

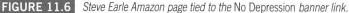

Check out the No Depression community website! It's your roots music playground. The site gives you an opportunity to share your thoughts, photos, videos, and music with other members. Start a discussion in the forum, write a blog about a great show you've seen or a new album you picked up, share photos from a recent festival, promote your events, and invite your friends. www. NoDepression.com.

the actual banner ad. By researching analytics of web site activities, labels can determine the best sites on which to advertise, based on all sorts of great data: gender, age, ethnicity, income, education, and lifestyles. By marrying the target market of the artist with web site analytics, the probability of click-through traffic increases dramatically—and the impression of hearing and selling the artist increases as well.[1]

The online music magazine *No Depression* targets Americana-roots music fans and had, up until summer 2008, been a print, subscription-based magazine with additional newsstand sell-through. But with the downturn in the music industry and the current economic conditions, the owners decided to go online only. Their advertising fees are strictly impression-based, but the actual ads are interactive, meaning that they have click-through and flash capability.[2]

The Steve Earle *Live At The BBC* release was a click-through banner ad that rotated on *No Depression*'s home page. For an Americana artist, this banner ad placement made perfect sense, notifying the target market of the pending release a week prior to street date. By clicking through the banner, the reader could instantly become a consumer, landing on Amazon.com's site where a preorder sale could be made. The reader would also be able to

---

[1]Jenn Harbin, UMG, personal interview, March 5, 2009.
[2]Kyla Fairchild, *No Depression*, personal interview, March 5, 2009.

click through to a specified "Steve Earle" store. Here, catalog releases along with select merchandise could be purchased as well.

There are a myriad of choices when deciding how to alert consumers to a new release from an artist, and as technologies emerge, these options will continue to increase. What is important is to keep the target market at top of mind with the budget close to the vest. Understanding all the "players" in the drama, with their various agendas, will help keep the message from getting garbled. For in the end, the label must do the "right thing"—which is to sell records.

GLOSSARY

cooperative advertising (co-op)—An advertisement where the cost is shared by the manufacturer of the product and the local retail outlet.

content network—A term Google uses to describe the group of web sites grouped together by content or subject matter and targeted for contextual advertising.

contextual advertising—Advertising on a web site that is targeted to the specific individual who is visiting the web site based on the subject matter of the site and then featuring products that relate to that subject matter.

cost per click (CPC)—The advertiser is charged a small fee each time a potential customer clicks on the ad and is taken to the advertiser's web site.

cost per rating point (CPP)—A method for determining what media programs are most cost effective by dividing the cost of advertising by the show's expected rating.

cost per thousand (CPM)—A dollar comparison that shows the relative cost of various media or vehicles; the figure indicates the dollar cost of advertising exposure to a thousand households or individuals.

frequency—The number of times the target audience will be exposed to a message.

gross impressions—The total number of advertising impressions made during a schedule of commercials. GIs are calculated by multiplying the average persons reached in a specific time period by the number of spots in that period of time.

gross rating point (GRP)—In broadcasting/cable, it means the size of the audience during two or more dayparts. GRPs are determined by multiplying the specific rating by the number of spots in that time period.

media fragmentation—The division of mass media into niche vehicles through specialization of content and segmentation of audiences.

media planning—Determining the proper use of advertising media to fulfill the marketing and promotional objectives for a specific product or advertiser.

medium—A class of communication carriers such as television, newspapers, magazines, outdoor, and so on.

opt-in—Consumers choose to be added to a mailing or marketing list.

proximity marketing—The localized wireless distribution of advertising content associated with a particular place.

rating—In TV, the percentage of households in a market that are viewing a station divided by the total number of households with TV in that market. In radio, the total number of people who are listening to a station divided by the total number of people in the market.

rating point (a rating of 1%)—1% of the potential audience; the sum of the ratings of multiple advertising insertions; for example, two advertisements with a rating of 10% each will total 20 rating points.

ROI (return on investment)—Deals with the money you invest in the company and the return you realize on that money based on the net profit of the business.

reach—The total audience that a medium actually reaches; the size of the audience with which a vehicle communicates; the total number of people in an advertising media audience; the total percentage of the target group that is actually covered by an advertising campaign.

streeting—Industry jargon for the date that a record is available to the public at retail outlets.

trade advertising—Advertising aimed specifically for retailers and media gatekeepers.

trade publication—A specialized publication for a specific profession, trade, or industry; another term for some business publications.

vehicle—A particular carrier within a medium, such as *Rolling Stone* magazine or MTV network.

viral marketing—Techniques that use pre-existing social networks to increase brand awareness, increase product sales, or achieve other marketing objectives.

BIBLIOGRAPHY

Bovée, C. L., & Arens, W. F. (1986). *Contemporary advertising* (2nd ed.). Homewood: Irwin.

Dunn, S., et al. (1990). *Advertising: Its role in modern marketing* (7th ed.). Orlando, FL: The Dreyden Press.

eMarketer. Projected U.S. advertising spending. www.emarketer.com. (no longer available).

Fairchild, K. (2009). *No depression* personal interview, March 6.

Google. (2009). Information on AdWords campaigns. https://adwords.google.com/select/snapshot.

Harbin, J. (2009). MCA Nashville new media department, personal interview, March 10.

Iacolare, L. (2007). Mobile mass media creative and advertising models. Mobile phones are the future. July 30. www.webtvwire.com/mobile-mass-media-creative-advertising-models-mobile-phones-are-the-future.

Sissors, J., & Bumba, L. (1994). *Advertising media planning* (4th ed.). Lincolnwood, IL: NTC Business Books.

Stuart, G. (2009). Mobile advertising: maybe next year . . . or the year after that www.adweek.com/aw/content_display/community/columns/other-columns/e3ice 058ab1756ad1650d14d274af7e1ec8. *AdWeek Online, January 12.*

Q Research. (2007). New Survey Data Supports Ad-Supported Mobile Music. http://ad-supported-music.blogspot.com/2007/02/new-survey-data-supports-ad-supported.html.

CHAPTER 12

Music Distribution & Retailing

Amy Macy

INTRODUCTION

Music **distributors** are a vital conduit in getting physical and digital music product from record labels' creative hands into the **brick-and-mortar** and virtual retail environment. Recognize that as the marketplace shifts to a digital environment, distribution companies are reevaluating their value in the food chain and continually developing sales models that reflect direct sales opportunities to music consumers. By the end of 2008, nearly 1 out of every 4 albums purchased were at "nontraditional" outlets, which include digital services, Internet retailers, mail order, "nontraditional stores," and concerts. Whatever the sales channel, distribution companies think of themselves as extensions of the record labels that they represent.

Prior to the current business model, most individual records labels hired their own sales and distribution teams, with sales representatives calling on individual stores to sell music. It took many reps to cover retail, as Figure 12.1 shows. This model shows nine points of contact where each label meets with each retailer. As retailers became chains and business economics evolved, record labels combined sales and distribution forces to take advantage of **economies of scale**, which eventually evolved into the current business model.

Figure 12.2 includes the distribution function and shows the points of contact reduced to six. In today's business model, record label sales executives communicate with distribution as their primary conduit to the marketplace, but labels also have ongoing relationships with retailers. Depending

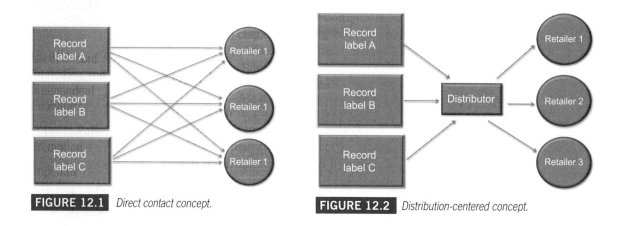

FIGURE 12.1 *Direct contact concept.*   **FIGURE 12.2** *Distribution-centered concept.*

on the importance of the retailer, the label rep will often visit the retailer with a distribution partner so that significant releases and marketing plans can be communicated directly from the label to the retailer. And as deals are struck, both orders and marketing plans can then be implemented by the distributor.

THE BIG 4 AND MORE

Within the last 10 years the **Big 4** have consolidated from the Big 6, reducing their numbers of employees to reflect the ever-decreasing size of the music sales pie. But the mergers began prior to the explosion of file sharing with the amalgamation of Universal with Polygram in 1999, the same year as the emergence of Napster. This merger created Universal Music & Video Distribution (UMVD); in 2006, they became Universal Music Group (UMG) and have maintained their market share position at number 1 since inception.

Most recent was the merger of Sony and BMG in 2004, who consolidated their music conglomerates as well as their distribution workforces but maintained much of the integrity of their imprints. They too gained market share positioning by garnering the number 2 position simply by merging. But as of October 1, 2008, Sony purchased BMG's 50% stake in the Sony/BMG merger and now wholly owns this new venture called Sony Music Entertainment Inc. (SMEI), keeping all the labels that were Sony as well as BMG.

WEA and EMI maintain separate distribution functions as part of Warner Music Group and EMI Group, respectively. Although there has been speculation that the two would consider joining forces to combat the larger entities, no deal has been struck, and EMI was most recently purchased in 2007 by the private equity firm Terra Firma, causing many to rethink the merger/acquisition possibility.

As noted in the market share data, independent labels continue to be a force within the music business, and in 2008 independents held a 16.5% market share. With the burgeoning digital storefronts, any label can now have an instant "sales" point in which to connect with customers directly. But to fulfill physical product, independent labels must reach the brick-and-mortar outlets using traditional methods. Most of the major distributors have created an independent arm of their distribution function within their family. By contracting this function of distribution to independent labels, these "independent distributors" can assist the independent label in placing their music in the mainstream marketplace. But there are many true independent distributors that are *not* tied to the Big 4 that function similarly by placing physical product in stores.

UMVD – Universal Music & Video Distribution

Sample labels that it distributes:

Interscope
Geffen
Island/Def Jam
Universal
UMG Nashville
Hollywood
Disney/Buena Vista

Sony Music Distribution

Sony's RED Distribution represents many independent labels
Sample labels that Sony Music distributes:

Columbia
Epic
Arista/Arista Nashville
J Records
RCA
Jive
LaFace
Razor & Tie
WindUp
RCA Label Group Nashville

WEA

WEA's ADA Distribution represents many independent labels
Sample labels that WEA distributes:

Warner Brothers
Atlantic Records
Bad Boy
Roadrunner
WSM/Rhino
V2 Records
WEA/Fueled By Ramen

EMD

EMD's Caroline Distribution represents many independent labels
Sample labels that EMD distributes:

Blue Note
Capitol Nashville
Capitol Records
Virgin EMI

Vertical integration

Three out of the Big 4 entertainment conglomerates share profit centers that are vertically integrated, creating efficiencies in producing product for the marketplace. To take full advantage of **vertical integration**, labels looking for songs would tap their "owned" publishing company. (Each of the major entertainment conglomerates has a publishing company. If they only recorded songs that were published by their sister company, more of the money would stay in-house.) Once recorded, the records would be manufactured at the "owned" plant. In turn, the pressed CDs would then be sold and distributed into retail—with the money being "paid" for each of the functions staying within the family of the music conglomerate (Figures 12.3 and 12.4).

In addition to the Big 4 companies, again, there are many independent music distributors that are contracted by independent labels to do the same job. Ideally, the distribution function is not only to place music into retailers, but also to assist in the **sell-through** of the product throughout its lifecycle, and independent distributors have developed a niche in marketing unique and diverse products.

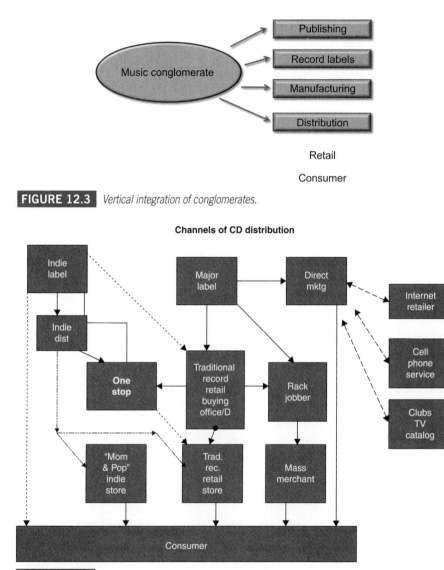

Publishing

Record labels

Manufacturing

Distribution

Retail

Consumer

FIGURE 12.3 *Vertical integration of conglomerates.*

Channels of CD distribution

FIGURE 12.4 *The distribution pathway (Courtesy Dr. Clyde Rolston, Belmont University).*

MUSIC SUPPLY TO RETAILERS

Once in the distributor's hands, music is then marketed and sold into retail. Varying retailers acquire their music from different sources. Most mass merchants are serviced by **rack jobbers**, who maintain the store's music department including inventory management, as well as marketing music to consumers. Retail chain stores are usually their own buying entities, with company-managed purchasing offices and distribution centers (DCs). Many independent music stores are not large enough to open an account directly with the many distributors, but instead work from a *one-stop's* inventory as if it were their own. (**One-stops** are wholesalers who carry releases by a variety of labels for smaller retailers who, for one reason or another, do not deal directly with the major distributors.) Retail chain stores and mass merchants will on occasion use one-stops to do "**fill-in**" business, which is when a store runs out of a specific title and the one-stop supplies that inventory. (Retail is discussed in detail later in this chapter.)

ROLE OF DISTRIBUTION

Most distribution companies have three primary roles: the sale of the music, the physical distribution of the music, and the marketing of the music. Reflecting the marketplace, most distribution companies have similar business structures to those of their customers—the retailers. Oftentimes, the national staff is separated into two divisions: sales and marketing. The sales division is responsible for assisting labels in setting sales goals, determining and setting deal information, and soliciting and taking orders of the product from retail. Additionally, the sales administration department should provide and analyze sales data and trends, and readily share this information with the labels that they distribute.

The marketing division assists labels in the implementation of artists' marketing plans along with adding synergistic components that will enhance sales. For instance, the marketing plan for a holiday release may include a contest at the store level. Distribution marketing personnel would be charged with implementing this sort of activity. But the distribution company may be selling holiday releases from other labels that they represent. The distribution company may create a holiday product display that would feature all the records that fit the theme, adding to the exposure of the individual title.

The physical warehousing of a music product is a huge job. The major conglomerates have very sophisticated inventory management systems where music and its related products are stored. Once a retailer has placed an order, it is the distributor's job to pick, pack, and ship this product to its designated location. These sophisticated systems are automated so that manual picking of product is reduced, and that accuracy of the order placed is enhanced. Shipping is usually managed through third-party transportation companies.

As retailers manage their inventories, they can return music product for a credit. This process is tedious, not only making sure that the retailer receives accurate credit for product returned, but also the music itself has

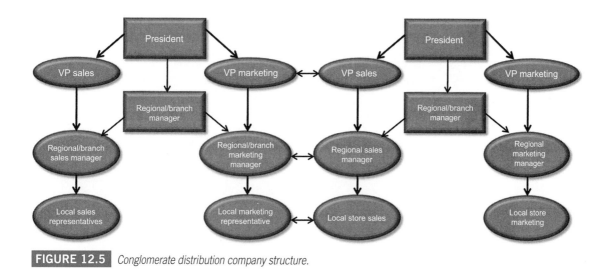

FIGURE 12.5 *Conglomerate distribution company structure.*

to be retrofitted by removing the retailer's stickers and price, shrink wrapping again, and then returning to inventory.

CONGLOMERATE DISTRIBUTION COMPANY STRUCTURE AS IT RELATES TO NATIONAL RETAIL ACCOUNTS

Although there are many variations and nuances to these structures as determined by each distribution company, the basic communication chart in Figure 12.5 still applies. At each level, the companies are communicating with each other. At the national level, very complex business transactions are being discussed, including terms of business as well as national sales and marketing strategies that would affect both entities company-wide. As mandated by the national staff, the regional/branch level is to formulate marketing strategies, either as extensions of label/artist plans or that of a distribution focus. At the local level, implementation of all these plans is the spotlight. But by design, these business structures are in place to create the best possible communication at every level, with an eye on maximizing sales.

National Structure

To optimize communication along with service, distributors need to be close to retailers. Many of the major conglomerates have structured their companies nationally to accommodate the service element of their business. Most distributors have regional territories of management containing

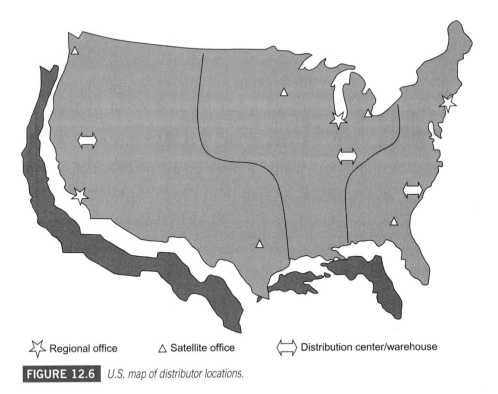

△ Regional office   △ Satellite office   ⇔ Distribution center/warehouse

FIGURE 12.6   *U.S. map of distributor locations.*

core offices and distribution centers (DCs). Each region contains satellite offices, getting one step closer to actual retail stores. As reflected in the structure chart in Figure 12.6, field personnel are on the front line, reporting to satellite offices, who then report to the regional core offices.

This U.S. map shows basic regions for a major distributor with satellite offices and distribution centers. Core regional offices are located in Los Angeles, Chicago, and New York. Regions are naturally eastern, central, and western, with regional offices including Atlanta, Detroit, Minneapolis, Dallas, and Seattle. The DCs are centrally located within each region. In this example, the DCs are located in Sparks, Nevada; Indianapolis, Indiana; and Duncan, South Carolina. No destination is more than a 2-day drive from the DC, making product delivery timely.

Although the core regional offices are centrally located, these offices are primarily marketing teams, executing plans derived by both the labels that they represent and the distribution company. To be clear, over 75% of the music business is purchased and sold through 10 retailers. The locations of these buying offices are key sales sites, and designated distribution personnel are placed near these retailers so that daily, personal interaction can occur. These locations and retailers are shown in Table 12.1.

Table 12.1	Retail chain home office locations	
Bentonville, AR		Wal-Mart
Minneapolis		Best Buy and Target
Miami		Alliance
Albany		TransWorld

Table 12.2	Sales timetable	
Prior to SD	**Activity**	**Example dates**
8 weeks	Sales book copy due to distributor	September 27
6 weeks	Solicitation book mailed to retail buyers	October 11
4 weeks	Solicitation	October 25 to November 5
2 weeks plus	Orders due	November 5
1 week	Orders shipped wholesalers/retail chain	November 16 orders received
5 days	Orders shipped to one stops	November 19 orders received
Street date		November 23

TIMELINE

The communication regarding a new release begins months prior to the street date. Although there are varying deadlines within each distribution company, the ideal timeline is pivotal on the actual street date of a specific release. For most releases, street dates occur on Tuesdays. Working backward in time, to have product on the shelves by a specific Tuesday, product has to ship to retailer's DCs approximately 1 week prior to street date. To process the orders generated by retail, distributors need the orders 1 week prior to shipping. The sales process of specific titles occurs during a period called **solicitation** (Table 12.2). All titles streeting on a particular date are placed in a solicitation book, where details of the release are described. The solicitation page, also known as **sales book** copy, usually includes the following information:

- Artist/title
- Street date
- File under category—where to place record in the store
- Information/history regarding the artist and release
- Marketing elements:
  - Publicity activities
  - Consumer advertising
  - Tour and promotional dates

- Available POP
- Bar code

This information is also available online on the business 2 business (B2B) sites established for the retail buyers.

ONE SHEET: THE SOLICITATION PAGE

On the web site, *MusicDish Industry e-journal*, Christopher Knab of ForeFront Media and Music describes a one sheet as:

> *A Distributor One Sheet is a marketing document created by a record label to summarize, in marketing terms, the credentials of an artist or band. The One Sheet also summarizes the promotion and marketing plans and sales tactics that the label has developed to sell the record. It includes interesting facts about an act's fan base and target audience. The label uses it to help convince a distributor to carry and promote a new release.*

—Knab, 2001

The one sheet typically includes the album logo and artwork, a description of the market, street date, contact info, track listings, accomplishments, and marketing points. The one sheet is designed to pitch to buyers at retail and distribution. The product bar code is also included to assist in buy-plugging the actual release into the inventory code system (Figure 12.7).

CONSOLIDATION AND MARKET SHARE

As consolidation continues, the Big 4 just keep getting bigger, although the indies have been a force to be recognized. Looking at market share date generated by over-the-counter sales of SoundScan, one can view how large these entities have become over time. Note that the piece of the Big 4 pie fluctuates, with independents representing the remaining market share. Over the past 10 years, Polygram and Universal combined and have maintained their strong market leader position, while the Sony/BMG transaction has shown signs of the turbulent times. Remember that during all this time, these pieces of the pie have been shrinking, with music sales down by over 35% since its all-time high in 2000.

Included here is a breakout of 2008 sales showing the power of independent labels. Two of the majors, Warner's ADA Distribution and Sony's Red Distribution, distribute and market independent releases from independent labels and the conglomerates often claim that market share. If broken out independently, the independent labels would have a much larger share of the music market sales—as noted in the 2008 data below.

FIGURE 12.7 *One sheet for Buck 65. (Source: V2 Records)*

DIGITAL DISTRIBUTION

According to SoundScan, year-end 2008 data reveal that digital music accounts for 32% of all music purchased and that there was a 10% increase in overall music purchases exceeding 1.5 billion units. This sounds like

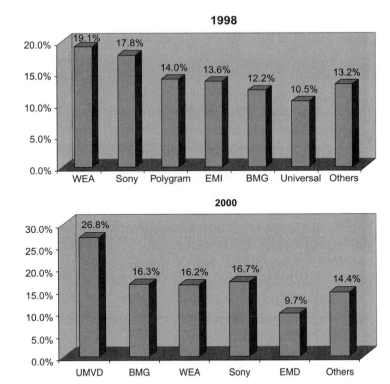

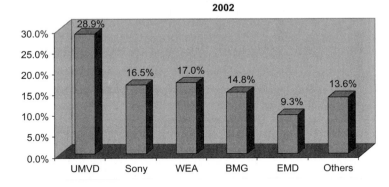

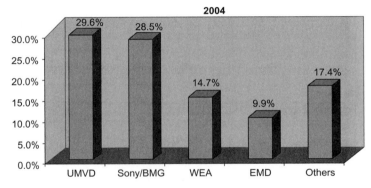

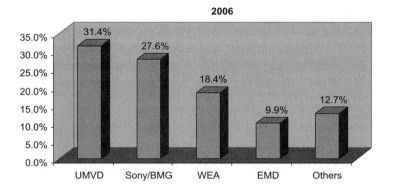

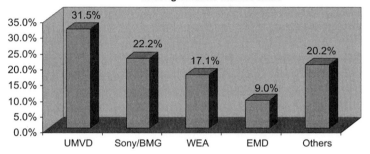

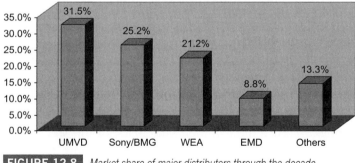

FIGURE 12.8 *Market share of major distributors through the decade.*

good news for the industry, but a comparison of overall album sales including track equivalent album (TEA) sales reflects a decline of −8.5% compared to 2007. Total albums sales alone declined 14% from 2007. The unit sales of digital tracks have increased by 27% since 2007, but did not make up for the loss of album sales, making for another negative year in sales for music.

2008 sales data confirm that consumers are moving to digital consumption. As stated by the Nielsen Company 2008 Year-End Music Industry Report:

- Digital track sales broke the 1 billion sales mark for the first time with more than 1,070,000 digital track sales. The previous record was 844 million digital track purchases during 2007: an increase of 27% over 2007.

- Digital album sales reached an all-time high with more than 65 million sales in 2008; up from 50 million in 2007; an increase of 32% over the previous year.
  - Note that digital album sales accounted for 15% of total album sales compared to 10% in 2007 and 5.5% in 2006.

- In the final reporting week of 2008 the following digital sales records were broken:
  - Digital track sales surpassed 47.7 million. The previous sales record was 42.9 million, week of 12/23–12/30/07.
  - Digital album sales this week broke the two million mark for the first time with sales of 2.4 million sales; breaking the previous record of 1.9 million (12/30/07).
  - The top 200 digital songs for the week posted an all-time high with 13.6 million sales; breaking the previous record of 11.9 million during the last week of 2007.

- 2008 is the first time a digital song broke the 3 million sales mark in a single year. There were 2 songs that achieved this milestone; Leona Lewis' "Bleeding Love" and Lil Wayne's "Lollipop"; with sales of 3.4 and 3.2 million respectively.

- In 2008, there are 19 different digital songs with sales that exceeded 2 million compared to 9 in 2007.

- 71 digital songs exceeded the 1 million sales mark for the year compared to 41 digital songs in 2007, 22 in 2006, and only 2 digital songs in 2005.

—(Nielsen Company 2008 Year-End Music Industry Report)

As a reflection of the shift in consumption, the primary function for music distributors is now split between that of physical distribution of CDs into physical stores and assisting in the relationships between third-party licensing sites. In some cases, many of the major distributors have sites where consumers can buy tracks and albums directly from the conglomerate.

These sites usually promote and sell music from their represented labels only, which makes it difficult for consumers to experience one-stop shopping.

In turn, within most conglomerate families resides a department that licenses music to third parties, which are legal downloading sites such as iTunes, Rhapsody, and Napster, plus the ever-popular ringtone and ringback services that are provided by mobile operators such as Sprint, Verizon, and LG. These licensing departments are critical to the evolution of distribution, positioning them as gatekeepers for the growing downloading environment. At these third-party sites, consumers can purchase from the many sources of music, clearly beyond those of one conglomerate. Currently, the wholesale price of a digital track is approximately $0.68 a license, which includes the artist and producer royalty, mechanical royalty, and revenue to label. (Remember that label revenue is *not* pure profit because the label funded the initial recording process along with any marketing effort.) The remainder of the $0.99 is distributed among the service provider, the distribution affiliate, credit card fees, and bandwidth costs.

Digital Revenues

As consumership of music continues to evolve, so does that of music distribution. Current trends reveal that younger consumers look to purchase tracks, making the classic album less attractive to this buyer. The digital download arena is where these consumers are satiating their needs. In 2008, digital track sales grew 27% to a year-end sales total of 1069 million, and digital album sales accounted for 15.4% of all album sales. So, approximately 85% of albums and 78% of overall music is still purchased in the classic brick-and-mortar environment, but there is a clear trend that digital distribution is having an impact on the traditional music retail environment as hardware devices gain popularity and ease of use.

Future Trends

It's but a snapshot of what 2009 might produce, but music sales continue to slide, with a total album sales decline of −14%. Overall album sales including TEA declined −8.5% compared to 2007. This trend does not mean the end of the music business, but rather a shift of *modus operandi* from the traditional sales of music to a new way on monetizing the recording. Record labels are finding new ways to capture value to consumers who have changed the role of music in their lives—be it single track sales, soundtracks for gaming, streaming while at an advertising site, value-added item for related products, licensing of artists and their images, not to mention the publishing aspect if the conglomerate owns the rights to a hit song—and the list goes on.

An example: To date, Soulja Boy's *Tell'em.com* album has SoundScanned over 850,000 albums, generating estimated revenues for the label in excess of over $8.5 million dollars. His single *Crank That* has digital download sales of 3.3 million at $0.99 a pop, with ringtone sales of over 2 million sold at an estimated $2.49 a download, generating over $8 million in revenue. To look at the money, the single has generated nearly the same amount of revenue as the album, which is a new equation and business model for record labels and how they aggregate their assets.

DISTRIBUTION VALUE

Several distribution companies are exploring ways to add value to the conglomerate equation. Creating distribution-specific marketing campaigns with non-entertainment product lines helps validate distribution's existence, while hopefully enhancing the bottom line. Marketing efforts such as on-pack CDs with cereal, greeting card promotions, and ringtone services add to the branding of the participating artists while increasing overall revenue through licensing and/or sales of primary items. The ultimate value for today's distribution companies is that of consolidator and aggregator. Distributors can consolidate labels to create leverage points within retail. To gain positioning in the retail environment, one must have marketing muscle, and by using the collective power of their various labels' talent, the entire company can raise its market share by coattailing on the larger releases in the family.

THE MUSIC RETAIL ENVIRONMENT

The four P's of product, promotion, price, and placement converge within the music retail environment—this being the last stop prior to music being purchased. This environment is designed to aid consumers in making their purchasing decision. This decision can be influenced in a number of ways, depending on the consumer. For example: Does the store have hard-to-find releases? Do they have the lowest prices? Is it easy for consumers to find what they're looking for? Does the store have good customer service and knowledgeable employees? These questions should be answered, in one way or another, within the confines of the retail environment through the use of the four P's.

NARM

To assist music retailers in determining business strategies, companies look for current sales trends as well as educational and support networks. The

National Association of Recording Merchandisers (NARM) is an organization conceived to be a central communicator of core business issues for the music retailing industry.

Founded in 1958, NARM is an industry trade group that serves the music retailing community in the areas of networking, advocacy, information, education, and promotion. Members include brick-and-mortar, online and **"clicks-and-mortar"** retailers, wholesalers, distributors, content suppliers (primarily music labels, but also video and video game suppliers), suppliers of related products and services, artist managers, consultants, marketers, and educators in the music business field.

Retail members, who operate 7000 storefronts that account for almost 85% of the music sold in the $12 billion U.S. music market, represent the big national and regional chains—from Anderson (who rack all Wal-Mart stores), Best Buy, Borders, Barnes & Noble, Hastings Entertainment, Musicland, Newbury Comics, Target, TransWorld (FYE)—to the small but influential "tastemaker" independent specialty stores such as Waterloo Records in Austin, Texas; Grimey's in Nashville, Tennessee; Ear X-tacy in Louisville, Kentucky; and Amoeba Records in San Francisco, California. Additional members also include online retailers such as Amazon.com, eBay, and iTunes.

Distributor and supplier members account for almost 90% of the music produced for the U.S. marketplace. This includes the four major music companies: Sony, EMI, Universal Music Group, and Warner Music, and many of the labels they represent. A few of the label members are Atlantic, Blue Note, Capitol, Capitol Nashville, Columbia, Curb Records, DreamWorks Records, Elektra Entertainment, Epic Records, Geffen Records, Hollywood Records, Interscope Records, The Island Def Jam Music Group, J Records, Jive Records, Mercury/MCA Nashville/Lost Highway, RCA Label Group, RCA Music Group, RCA Victor Group, Rhino, Sony Classical, Verve Music Group, Walt Disney Records, Warner Bros., Welk Music Group, and Wind-up Records. This list extends even further to include influential independent labels and distributors. For more information regarding NARM and activities, visit www.narm.com.

RETAIL CONSIDERATIONS

The four P's are applicable as an outline for retailer considerations in doing business.

#1 P – Product

How does a retailer learn about new releases? Distribution companies are basically extensions of the labels that they represent. To sell music well, a distribution company needs to be armed with key selling points. This critical information is usually outlined in the marketing plan that is created at the record label level. Record labels spend much time "educating" their distributors about their new releases so that they can sell and distribute their records effectively.

Distribution companies set up meetings with their *accounts*, meaning the retailers. At the retailer's office, the distribution company shares with the **buyer**—that is the person in charge of purchasing product for the retail company—the new releases for a specific release date, as well as the marketing strategies and events that will enhance consumer awareness and create sales. Depending on the size of the retailer, buyers are usually categorized by genre or product type, such as the R&B/hip-hop buyers, or the soundtrack buyers.

Purchasing Music for the Store

When making a purchase of product, the buyer will take into consideration several key marketing elements: radio airplay, media exposure, touring, cross-merchandising events, and—most critical—previous sales history of an artist within the retailers' environment; or if a new artist, current trends within the genre and/or other similar artists. On occasion, the record label representative will accompany the distribution company sales rep with the hopes of enhancing the buyer's knowledge of the new release. The ultimate goal is to increase the total purchase while creating marketing events inside the retailer's environment. Most record labels, along with their distributors, have agreed on a **sales forecast** for a specific release. This forecast, or number of records predicted to sell, is based on similar components that retail buyers consider when purchasing product. Many labels use the following benchmarks when determining forecast:

Initial orders, or *IO* – This number is the initial shipment of music that will be on retailers' shelves or in their inventory at release date.

90-day forecast – Most releases sell the majority of their records within the first 90 days.

Lifetime – Depending on the release, some companies look to this number as when the fiscal year of the release ends, and the release will then roll over into a catalog title. But on occasion, a hit release will predicate that forecasting for that title continues because sales are still brisk.

Inventory Management

Larger music retailers have very sophisticated purchasing programs that profile their stores' sales strengths. Using the forecast, as well as an overall percentage of business specific to the label or genre, the retailer will determine how many units it believes it can sell. This decision is based on historical data of the artist and/or trends of the genre along with the other marketing components.

Keeping track of each release, along with the other products being sold within a store is called **inventory management**. Using **point-of-sale (POS)** data, the store's computer notes when a unit is sold using the **universal product codes (UPC)** bar code. Depending on the inventory management system, a store may have 10 units on hand, which is considered the ideal *maximum* number the store should carry. The ideal *minimum* number may be four units. If the store sells seven units, and drops below the ideal inventory number of four as set in the computer, the store's inventory management system will automatically generate a reorder for that title, up to the maximum number. This **min/max inventory management system** may then download the reorder through an **electronic data interchange (EDI)** to its supplier, and the product is then shipped to the retailer within a few days. To avoid waiting for the product to arrive, a retailer may opt to **drop ship** product directly from the distribution company, avoiding the delay of processing at either the headquarters' distribution center (DC) or the one-stop supplier.

Turns and Returns

A store's success is based on the number of units it sells within a fiscal year. Clearly, the size of the store dictates how much product or inventory it can hold. On average, a 2500-square-foot store may hold 20,000 units. According to NARM 2002 data, the average annual inventory **turn** for music is 4.4 times. As an industry standard, this store would sell: 20,000 units $\times$ 4.4 turns = 88,000 units in a fiscal year. This does not mean that every title sells 4.4 times, but rather the store averages 4.4 sales per year for every unit it is holding.

To manage the real estate within the store, the best-selling product should receive the most space. To keep a store performing well, the inventory management system should notice when certain titles are not selling. Music retailers have an advantage over traditional retailers in that if a product is not selling, they can send it back for a refund, called a **return**. The refund is usually in the form of a credit, and the amount credited is based on when the product is returned along with other considerations such as whether a **discount** has been received. According to NARM 2002 data, the industry return average is 19.5% for music. To put this statistic another way, for every 100 records in the marketplace, nearly 20 units are returned to the distributor.

Loss Prevention

According to the National Security Retail Survey of 2002, the average retailer loses about 1.7% of product to **shrinkage**—or shoplifting. Other research notes that the music retailer can lose between 4% and 4.5% in product. But in any outlet, nearly 50% of the missing product is due to employee theft.

What do music retailers do to protect against shrinkage? Regarding employees, retailers use many screening techniques such as past employment records, criminal checks, multiple interviews, and personal reference checks. As for consumers, there are various techniques to help protect against shoplifting. Loss prevention staff is also employed by many retailers.

Because thieves need privacy, training employees to serve the customer has a dual purpose: to help secure a sale and to let the would-be thief know that the "store" is watching. Reducing clutter and sighting blind spots help keep visibility high. The use of uniformed and plain-clothed detectives is common in larger format stores, with good results; but the use of technology has been a big aid in dissuading the shoplifter. **Source tagging** is the use of embedded security devices placed within the CD packaging. With the use of electronic article surveillance (EAS), stores place security panels at the exit. As a product is sold, the source tag is deactivated. If a product that has not been purchased moves through these security panels, a warning siren occurs, notifying store personnel of the potential shoplifting event.

#2 P – Promotion

Although many music consumers still find out about new music via the radio, social networking as the new "word of mouth," as well as online streaming, is having a huge impact on sales because the listener can become an instant buyer. With the click of a mouse, the computer is now the retail outlet and an impulse buy can quickly be made.

In the past, consumers have been trained to look in Sunday circulars for sales and featured products on almost any item. Major music retailers still use this advertising source as a way to announce new releases that are to street on the following Tuesday, but are finding innovative ways to cross-merchandise not only new music titles, but the store outlet and other featured product.

Once inside the store, promotional efforts to highlight different releases should aid consumers in purchasing decisions. These marketing devices often set the tone and culture of the store's environment. Whether it is listening stations near a coffee bar outlet, or oodles of posters hanging hodge-podge on the wall, a brief encounter within the store's confines will quickly identify the music that the store probably sells.

Featured titles within many retail environments are often dictated from the central buying office. As mentioned earlier, labels want and often do create marketing events that feature a specific title. This is coordinated via the retailer through an advertising vehicle called *cooperative advertising*. **Co-op advertising**, as it is known, is usually the exchange of money from the label to the retailer so that a particular release will be featured. Following are examples of co-op advertising:

Pricing and positioning (P&P) – P&P is when a title is sale-priced and placed in a prominent area within the store.

Endcaps – Usually themed, this area is designated at the end of a row and features titles of a similar genre or idea.

Listening stations – Depending on the store, some releases are placed in an automatic digital feedback system where consumers can listen to almost any title within the store. Other listening stations may be less sophisticated—perhaps as simple as using a freestanding CD player in a designated area. But all playback devices are giving consumers a chance to "test drive" the music before they buy it.

Point-of-purchase (POP) materials – Although many stores will say that they can use POP, including posters, flats, stand-ups, and so on, some retailers have advertising programs where labels can be guaranteed the use of such materials for a specific release.

Print advertising – A primary advertising vehicle; a label can secure a "mini" spot in a retailer's ad (a small picture of the CD cover art), which usually comes with sale pricing and featured positioning (P&P) in-store.

In-store event – Event marketing is a powerful tool in selling records. Creating an event where a hot artist is in-store and signing autographs of her newest release guarantees sales while nurturing a strong relationship with the retailer.

The following grid is a sample forecasting and P&P planning tool used to predict I.O. and initial marketing campaign activities within a retailer's environment.

Rack jobber Anderson purchases over 25% of overall music. They supply music to mass merchant Wal-Mart.

Alliance is the largest one-stop, whose primary business is supplying music to independent music stores, but has also picked up retailers Barnes & Noble and Kmart recently.

Figures are based on a major distributor's overall physical business with these actual accounts. Note that accounts purchase over 80% of all records, which makes them very important in the marketing process. For Target to *not* purchase a title would mean that over 18% of sales would be lost, based on these numbers. This does not mean that a record would sell 18% less over the lifetime of the release, but it does mean that consumers will have to find that title in another store other than Target.

Artist name title selection number				
Account	% of business	Target	Account advertising P&P	Cost
Anderson Merchandisers	26.50%	26,500		
Target	18.00%	18,000		
Best Buy	15.00%	15,000		
Alliance Entertainment, B&N, Kmart	11.05%	11,050		
TransWorld (FYE, Coconuts, Wherehouse, Specs)	4.00%	4000		
Amazon	4.00%	4000		
Borders	3.00%	3000		
Hastings	1.25%	1250		
Costco	0.50%	500		
Top 10 accounts	83.25%	83,250		
All others	16.75%	16,750		
Total	100.00%	100,000		

What is not included in the Top Accounts is the digital business (downloads). For many labels, Apple's iTunes represents over 90% of their digital sales and if ranked within these above accounts, iTunes now claims the number 1 account position—selling the most music of any entity out there. These sales are not based on shipped product but are usually digital single sales that are accounted with TEA equivalency, making this percentage of business more difficult to calculate.

#3 P – Pricing

Although record labels set the suggested retail list price (SRLP) for a release, this is *not* what retailers are required to sell the product for. Most often, the SRLP sets the *wholesale price*, or the cost to the retailer. In negotiating the order, the retailer may ask for a discount off the wholesale price. The retailer may also ask for additional **dating**, meaning that the retailer is asking for an extension on the payment due date. Each distributor has parameters in which this transaction may occur.

Generally, music product comes in **box lots** of 30 units. A retailer will receive a better price on product if it purchases in box lots. For example, a

Artist name
Title
Selection number

Account	% of business	Target	Account advertising P&P	Cost
Anderson Merchandisers	14.6%	14,600		
Best Buy	14.0%	14,000		
Handleman Company	12.3%	12,300		
Target	11.8%	11,800		
Alliance Entertainment	5.3%	5,300		
TransWorld	5.0%	5,000		
Musicland	4.7%	4,700		
Circuit City	4.5%	4,500		
AAFES	3.6%	3,600		
Borders	3.3%	3,300		
Top 10 accounts	79.1%	79,100		
All others	20.9%	20,900		
Total	100.0%	100,000		

FIGURE 12.9 *Sales forecasting grid.*

retailer wants to purchase 1200 units of a new release with a 10% discount and 30 days dating (see Table 12.3).

Normally, the money due for this purchase would be received at the end of the month following the record's release. However, with an extra 30 days dating, the due date is extended, giving the retailer a longer timeframe in which to sell the entire product. Adding extra dating is often a tactic of record labels that want retailers to take a chance on a new artist who may be slower to develop in the marketplace.

The price of product reflects the store's marketing strategy. The major electronic superstores look to music product as the magnet to get customers through their doors, which is why music prices in these environments are often lower than anywhere else. Often, these stores will sell music for less than they purchased it, called **loss leader pricing**. But these stores will also raise the price after a short period of time, usually within first 2 weeks after the street date.

Actual Pricing of Product
Margin and Markup
Margin and **markup** are both calculated using the wholesale purchase price of the product. Percent margin uses the selling price as the denominator,

whereas percent markup uses the purchasing (wholesale) price as the denominator for calculating.

Margin percentage on product is determined with the following calculation:

Formula: Margin % = Dollar Markup/Retail

Program	Program includes	Retail mall # of stores	Retail rural (oc) # of stores	Video/ Music Stores # of stores	Total stores -	Cost per title	Number of titles
Best seller	P&P on the best-seller in-store position	500	250	75	825	$ 22,500	20
New release	P&P on the new release in-store position	500	250	75	825	$ 17,000	16
Music video DVD	P&P on the music DVD in-store position	500	250	75	825	$ 2,700	8
Fast forward	Positioning on the FF in-store position	500	250	75	825	$ 11,700	8
Catalog promotion	P&P on multiple prime in-store positions Supporting in-store signage Supporting promotional levers	500	250	75	825	To be negotiated	TBD
Premium artist package	P&P in all concepts TMG-TV in all concepts Insert (1 cut) (6.5 MM circulation) Magazine direct promotion Magazine RCC (monthly e-newsletter)	500	250	75	825 -	$ 40,000 $ 1,500 $ 3,500 $ 5,500 $ 4,500	3
Total premium package price						$ 55,500	
Genre endcaps							
POP1/RK1	P&P on endcap	500	250	75	825	$ 13,500	12
POP1	P&P on endcap		250	75	325	$ 3,500	2
RK1	P&P on endcap		250	75	325	$ 3,500	2
POP2	P&P on endcap	350	200	75	725	$ 10,000	8
POP3	P&P on endcap	250	75	75	400	$ 7,000	8
RB1/RAP1	P&P on endcap	500	250	75	825	$ 8,500	12
RB1	P&P on endcap		100	75	175	$ 1,600	2
RAP1	P&P on endcap		250	75	325	$ 3,500	2
URBAN 1	P&P on endcap	250	75	75	400	$ 5,100	8
URBAN 2	P&P on endcap	150	41	76	267	$ 3,300	8
OPT1 (open endcap)	P&P on endcap	350	200	75	725	$ 5,000	8
OPT2 (open endcap)	P&P on endcap	250	200	75	400	$ 3,000	8
CNJ (classical new age jazz)	P&P on endcap	133		76	209	$ 1,800	8

FIGURE 12.10 Sample retail chain P&P costs and programs.

Sample rack jobber P&P programs and costs	Country – showboards (one month) $20,000
Super feature: Sunday circular $60,000 (one month P&P)	CCM – showboards (one month) $5,000 GOSPEL – showboards (one month) $2,500
Feature $40,000 (one month)	Various artists – showboards (one month) $10,000
Genre features (one month endcap) urban $5,000 country $5,000 Latin $1,000	Soundtrack – showboards (one month) $10,000 Rising stars developing artist (one month) $9,000 POD – Position out of department (one month endcap) $5,000
Pop – showboards (one month) $24,000 urban – showboards (one month) $12,000	$7,500 Urban direct roto (one week) $1500 single cut $3000 double cut roto – Sunday circular (one week) $10,000 per cut

FIGURE 12.11   *Sample rack jobber P&P programs and costs.*

Table 12.3   Box lot discount

SRLP	$18.98	SRLP	$18.98		
Wholesale	$12.04	Wholesale	$12.04		
×	1200 units	−10% disc.	$1.20		
	$14,448	Total/unit	$10.84	The 10% discount saved the retailer $1440	
		×	1200 units		
			$13,008		

If an SRLP CD of $18.98 is purchased wholesale for $12.04 and the store wants to sell it for $13.98, the margin percentage is $13.98 − $12.04/$13.98 = 13.9%.

Although there are variable ways to calculate margin, most stores use this retail markup calculation because it takes into consideration differing price lines, product extensions, and customer demands in retail value.

Markup uses a similar calculation but divides the dollar markup by the wholesale cost:

$$\text{Formula: Markup } \% = \frac{\text{Dollar Markup}}{\text{Wholesale}}$$

If an SRLP CD of $18.98 is purchased wholesale for $12.04 and the store wants to sell it for $13.98, the markup percentage is $13.98 − $12.04/$12.04 = 16.1%.

There is always an arithmetical relationship between gross margin and markup.

A gross margin of 40% requires a markup of 66.67%, calculated as $\dfrac{40}{(100 - 40)}$.

A gross margin of 60% requires a markup of 150%, calculated as $\dfrac{60}{(100 - 60)}$.

To achieve a target gross margin of 13.9% on the previous example, based on the purchase cost the calculations are as follows:

A gross margin of 13.9% requires a markup of 16.1%, calculated as $\dfrac{13.9}{(100-13.9)} = 16.1\%$

	$
Wholesale CD cost	12.04
Margin 13.9%	× 0.161 = 1.94
Total	13.98

When setting prices, retailers think about markup because they start with costs and work upward. When thinking about profitability, retailers think about margin because these are the funds left over to cover expenses as well as account for profit. Importantly, negotiating for the best discount off the wholesale price improves both markup and margin.

Where the Money Goes

When the consumer plunks down his money at the cash register to purchase a CD, Figure 12.13 shows how that money is divided between all the invested parties.

SRLP	$ 18.98
Wholesale	$ 12.04
Label	$ 5.00
Distr	$ 1.80
Design/manu	$ 1.00
Artist Roy	$ 1.50
Mech Roy	$ 0.80
Rec Cost	$ 1.00
Mkt/promotion	$ 0.90
Retail profit	$ 3.00

LPs non-returnable		
SRLP	Box	Loose
698	4.20	4.33
798	4.70	4.85
898	5.26	5.42
998	4.50	4.50
998A	5.84	6.02
1098	5.00	5.00
1098A	6.70	6.70
1198	7.00	7.22
1298	6.00	6.00
1298A	7.62	7.86
1398	8.26	8.52
1498	8.00	8.00
1598	8.00	8.00
1598A	9.48	9.77
1698	9.00	9.00
1698A	10.48	10.80
1898	11.10	11.44
1998	11.68	12.04
2098	12.11	12.48
2498	18.00	18.00

Video		
SRLP	Box	Loose
598	3.59	3.59
898	5.39	5.39
998	5.99	5.99
1098	6.59	6.59
1298	7.79	7.79
1498	8.99	8.99
1598	9.59	9.59
1798	10.79	10.79
(2) 1998	10.99	10.99
2198	12.99	12.99
2498	14.99	14.99
2998	17.99	17.99

DVD		
SRLP	Box	Loose
798	4.62	4.62
998	5.78	5.78
1198	6.95	6.95
1498	9.45	9.45
1798	10.43	10.43
(3) 1998	11.59	11.59
2098	12.58	12.58
2498	15.75	15.75
2998	17.39	17.39

Compact disc		
SRLP	Box	Loose
398	2.32	2.40
498	3.00	3.09
598	3.60	3.70
698	4.40	4.54
798	4.80	4.95
898	5.39	5.56
998	6.60	6.80
999	6.00	6.18
1099	6.92	7.13
1179	5.99	5.99
1198	7.88	8.12
1298	8.40	8.66
1398	9.14	9.42
1498	9.77	10.07
1598	10.37	10.69
1698	10.79	10.12
1798	11.41	11.76
1898	12.04	12.40
1998	12.99	13.40
2098	13.48	13.90
2198	14.23	14.67
2298	14.88	15.34
2398	15.52	15.99
2498	16.16	16.66
2998	19.39	19.99
3198	20.67	21.31
3498	22.47	23.16
3598	22.97	23.68
3998	25.83	26.63
4498	28.97	29.87
4798	30.71	31.66
4998	31.97	32.96
5498	35.49	36.55
5998	36.10	37.22
7998	52.72	52.72
12998	82.71	85.19

DVD audio		
SRLP	Box	Loose
1798	11.69	12.05
1998	12.98	13.38
2298	14.94	15.39
2498	16.23	16.72

Super audio		
SRLP	Box	Loose
2198	13.95	14.38

Cassette singles		
SRLP	Box	Loose
349	1.93	1.93

CD Single/CD-5		
SRLP	Box	Loose
199	1.24	1.24
298	1.65	1.65
349	2.06	2.06
398	2.65	2.65
498	2.94	2.94
598	3.48	3.59
649	3.80	3.80
698	4.14	4.14
749	4.45	4.45

12" Vinyl/maxi cass returnable		
SRLP	Box	Loose
498	2.75	2.84
598	3.37	3.37
649	3.64	3.64
698	3.91	3.91
898	5.03	5.03
998	5.62	5.62
1198	6.00	6.00
1298	6.50	6.50

Cassette		
SRLP	Box	Loose
398	2.35	2.42
598	3.48	3.59
779	3.99	3.99
798	4.70	4.85
898	5.25	5.41
998	5.84	6.02
1098	6.48	6.68
1198	7.00	7.22
1298	7.62	7.86
1598	9.48	9.77
1998	11.68	12.04

7" Singles-non-returnable		
SRLP	Box	Loose
198	1.28	1.28
259	1.19	1.19
298	1.98	1.98
348	2.28	2.28
398	2.58	2.58

FIGURE 12.12 *Sample distribution pricing schedule.*

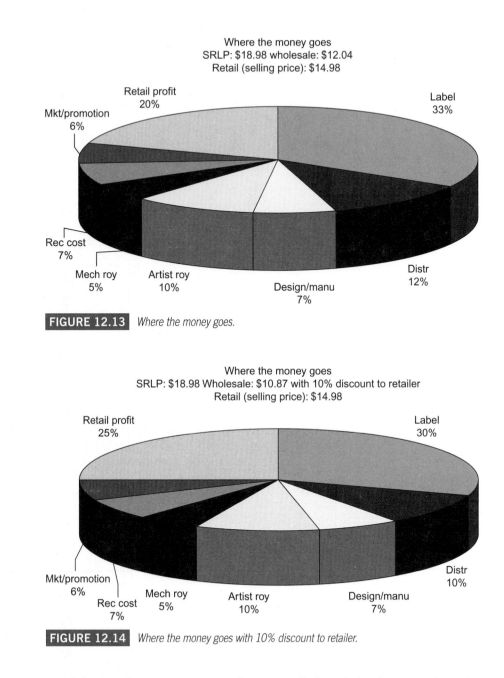

FIGURE 12.13 *Where the money goes.*

FIGURE 12.14 *Where the money goes with 10% discount to retailer.*

If the retailer receives a 10% discount off the wholesale price, then the label gets a reduction of 3% and the distributor a reduction of 2%, while the retailer enjoys a 25% *increase* in margin, from 20% of retail price to 25%, as shown in Figure 12.14.

Free Goods

Many labels use **free goods** to fund co-op advertising. If a retailer has a co-op advertising program that costs $120, instead of exchanging dollars the label may fund the advertising with product. Wholesale for a specific CD may average $12.00, so the label would send 10 CDs to the retailer—being the same dollar value of the advertising program.

This kind of exchange is a win–win scenario for both players. The retailer can then sell those 10 CDs for $14.00/unit, garnering $140 in revenues, which is more than the actual advertising program cost. And, the label value of the CDs on the accounting books is not that of $12.00/unit, but rather $6.00/unit—meaning that in cost to the label, the co-op advertising program was purchased for $60 total.

#4 P – Place

Although promotional activities within the store include the placement of product in designated areas, the choosing of a location or place in the actual store is paramount to the success of the business. Factors to consider are:

Location and visibility – Having built-in traffic helps a store attract more customers. Being in a mall or a high-traffic area is ideal for the store selling mass-appeal product. Additionally, being visible, and easy to find and access can help a store succeed.

Competition – Depending on the store's marketing strategy, having competition nearby can detract from a store's success. Knowledge of area businesses is key to success.

Rent – Besides inventory and employee costs, rent is a big expense that can impact profitability. Again, depending on store strategy, some retailers build their own storefront, while others rent from existing outlets.

Use clauses – A store must be sensitive to non-compete use clauses that can restrict business. Some malls restrict the number of record retail outlets and/or electronic store outlets to help ensure success of existing tenants.

Image – Store image reflects corporate culture along with product for sale.

Floor plans – The store's culture and marketing strategy are best observed in the design and layout of the floor plan. Market research observes that the longer a person is in a store, the higher the probability that the person will purchase an item, based on a marketing concept called **time spent shopping** (**TSS**). On that premise, some stores draw in their customers, creating interesting and interactive displays farther back in the store. Sale items may be placed in the back, or in some cases, there may be multiple floors, all with an eye to keep the shopper in the store longer. Other stores may entice browsers with new releases right inside the front door. Others include the addition of coffee bars to boost TSS.

Genre placement, along with related items, helps define a store. Display placement, including interactive kiosks, listening stations, featured titles, and top-selling charts, helps consumers make purchasing decisions. Larger stores with broader product offerings couple music product with related items. The placement of these elements and traffic flow design can aid consumers and increase sales.

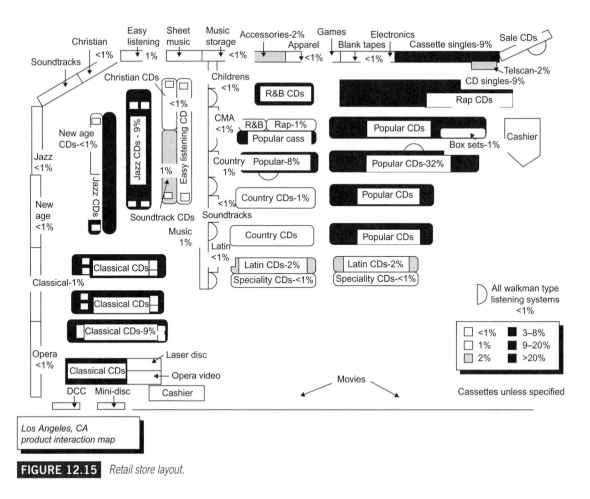

FIGURE 12.15 Retail store layout.

Store Target Market

A music store's target market or consumer generally dictates what kind of retailer it will be. To attract consumers interested in independent music, or to attract folks who are always looking for a bargain, determines the parameters in which a store operates. Music retailers have traditionally been segmented into the following profiles:

- **Independent music retailers** cater to a consumer looking for a specific genre or lifestyle of music. Generally, these types of stores get their music from one-stops. Independent stores are locally or regionally owned and operated, with one or just a few stores under one ownership.

- The **mom and pop retailer** is usually a one-store operation that is owned and operated by the same person. This owner is involved with every aspect of running the business and tends to be very passionate about the particular style of music that the store sells. This passion can be interpreted as being an expert in the knowledge of the genre and can be a unique resource for the consumer looking for the obscure release. Mom and pop storeowners tend to have a personal relationship with their customer base, knowing musical preferences, and keeping the customers informed about upcoming releases and events.

- **Alternative** music stores profile very similarly to mom and pop stores, but with the exception that they tend to be lifestyle-oriented. An electronica music retailer may have many hard-to-find releases along with hardware offerings such as turntable and mixing boards.

- **Chain stores** tend to attract music purchasers who are looking for a deep selection of releases along with assistance from employees who have strong product knowledge. These stores are often in malls and cater to a broad spectrum of purchasers. These stores have been studied and replicated so that entering any store with the same name in any location feels very similar. Often, they have the new major releases up front with many related items for sale, such as blank media and entertainment magazines. Chain stores traditionally buy their music inventory directly from music distributors, with warehousing and price stickering occurring in a central location. Some examples of chain stores are FYE, Hastings, and Musicland/Sam Goody.

- **Electronic superstores** do not make the bulk of their profits from the sale of music but rather use music as an attraction to bring consumers into their store environments. By using loss leader **pricing strategies**, these stores often sell new releases for less than they purchased them, but for a limited time. Meanwhile, they have created traffic to the store in order to make money from the sale of all the other items offered such as electronics, computers, televisions, and so forth. Electronic superstores also buy their music inventory directly from music distributors, with warehousing and price stickering occurring in a central location. Some examples of electronic superstores are Best Buy and Circuit City.

■ **Mass merchants** use the sale of music as event marketing for their stores. Each week, a new release brings customers back to their aisles with the notion that they will purchase something else while there. There is little profit in the sale of music for the mass merchant, but the offering of music is looked at as a service to customers. Often, mass merchants use rack jobbers to supply and maintain music for their stores. It is the rack jobber who initially purchases the music for the mass merchant environment. Some examples of mass merchants are Wal-Mart and Kmart.

Sources of Music

Figure 12.16 shows the basic flow of music as it reaches the consumer level. Recognize that one-stops' primary business is servicing independent record stores but that they also do what is called **fill-in** business for all music retailers.

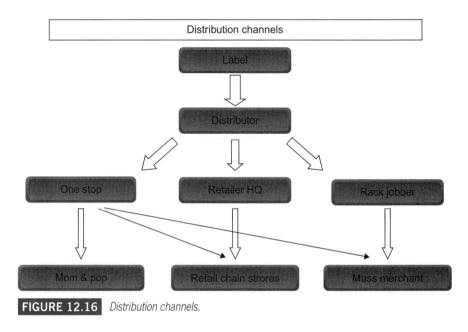

FIGURE 12.16 *Distribution channels.*

Internet Marketing and "Nontraditional" Retailing

Until recently, it was the retailers who had the brand identities that were winning the Internet sales wars. Consumers went to their favorite retail store web sites to browse and purchase music, meaning the actual CD that was to be delivered to the consumer's door. These well-known retailer sites left many start-up web sites with unknown names with little traffic. As noted in the sales overview, downloading and file swapping have become big "business" but were not perceived as a potentially profitable business . . . until now. Downloading activities have hurt not only the labels and their distributors, but the retailers as well. Because of aggressive prosecution and education campaigns to alert downloaders to their illegal practices, consumers are beginning to use legal downloading sites to purchase music. Sites such as iTunes, Napster, MusicMatch, Rhapsody, and—yes—Wal-Mart are all experiencing a high volume of downloads, with more sites coming online soon. Well-known online sites such as Amazon and CDNow have also had an impact on retailing. Plus nontraditional retailers such as Starbucks and Nordstrom have seen great success at selling finished goods to a target market. At the end of 2008, 24% of all albums sold were consumed at a "nontraditional" retailer, the lion's share being digital downloads, followed by Internet sales, then "nontraditional" outlets such as Starbucks, and finally venues and mail order. For more on Internet sales and marketing, refer to Chapter 14 on Internet marketing.

GLOSSARY

alternative—Music stores whose profile is very similarly to mom and pop stores, but with the exception that they tend to be lifestyle-oriented.

Big 4—These are the four music conglomerates that maintain a collective 85% market share of record sales: they are Universal, Sony, Warner, and EMI.

bricks and clicks—The term given to a retailer who has physical stores and an online retail presence. See also *clicks and mortar*.

brick and mortar—The description given to physical store locations when compared to online shopping.

box lot—Purchases made in increments of what comes in full, sealed boxes receive a lower price. (For CDs with normal packaging, usually 30.)

buyers—Agents of retail chains who decide what products to purchase from the suppliers.

chain stores—A group of retail stores under one ownership and selling the same lines of merchandise. Because they purchase product in large quantities from centralized distribution centers, they can command big discounts from record manufacturers (compared to indie stores).

clicks and mortar—The term given to a retailer who has physical stores and an online retail presence. See also *bricks and clicks*.

computerized ordering process—An inventory management system that tracks the sale of product and automatically reorders when inventories fall below a preset level. Reordering is done through an electronic data interchange (EDI) connected to the supplier.

co-op advertising—A co-operative advertising effort by two or more companies sharing in the costs and responsibilities. A common example is where a record label and a record retailer work together to run ads in local newspapers touting the availability of new releases at the retailer's locations.

dating—In creating the "terms" payment for music product, the manufacturer will set a date for when payment will be due. When negotiating, the "date" can be extended, giving the retailer extra time to sell off inventory and allowing the record company more time to develop the artist and single.

discount—The manufacturer offers a discount on orders based on the published SRLP and wholesale price.

distributor—A company that distributes products to retailers. This can be an independent distributor handling products for indie labels or a major record company that distributes its own products and that of others through its branch system.

drop ship—Shipping product quickly and directly to a retail store without going through the normal distribution system.

economies of scale—Producing in large volume often generates economies of scale—the per-unit cost of something goes down with volume. Fixed costs are spread over more units lowering the average cost per unit and offering a competitive price and margin advantage.

electronic data interchange (EDI)—The interfirm computer-to-computer transfer of information, as between or among retailers, wholesalers, and manufacturers. Used for automated reordering.

electronic superstores—Large chain stores such as Best Buy that sell recorded music and videos, in addition to electronic hardware.

endcap—In retail merchandising, a display rack or shelf at the end of a store aisle; a prime store location for stocking product.

fill-in—One-stop music distributors supply product to mass merchants and retailers who have run out of a specific title by "filling in" the hole of inventory for that release.

floor plans—A store layout designed to facilitate store traffic to increase the amount of time spent shopping (TSS).

free goods—Saleable goods offered to retailers at no cost as an incentive to purchase additional products.

independent music retailers—music retailers not affiliated with a retail chain. Usually purchases music from a one-stop.

in-store event—a marketing event to bring consumers into the retail environment, such as an artist signing his or her CD within the store.

inventory management—The process of acquiring and maintaining a proper assortment of merchandise while keeping ordering, shipping, handling, and other related costs in check.

listening station—A device in retail stores allowing the customer to sample music for sale in the store. Usually the devices have headphones and may be free standing or grouped together in a designated section of the store.

loose—The pricing scheme for product sold individually or in increments smaller than a sealed box. When looking at the printed price card of a distributor, "loose" product usually incurs a higher per-unit price than a box lot.

loss leader pricing—The featuring of items priced below cost or at relatively low prices to attract customers to the seller's place of business.

margin—The percentage of revenues left over to cover expenses as well as account for profitability.

markup—The percentage of increase from wholesale price to retail price.

mass merchants—Large discount chain stores that sell a variety of products in all categories, for example, Wal-Mart and Target.

min/max inventory management system—A store may have 10 units on hand, which is considered the ideal *maximum* number the store should carry. The ideal *minimum* number may be 4 units. If the store sells 7 units and drops below the ideal inventory number of 4, as set in the computer, the store's inventory management system will automatically generate a reorder for that title, up to the maximum number.

mom and pop stores—Business entities of a single proprietorship or partnership servicing a smaller music consumer base of usually one or two stores (sometimes known as *indie* stores).

National Association of Recording Merchandisers (NARM)—The organization of record retailers, wholesalers, distributors, and labels.

one-stop—A record wholesaler that stocks product from many different labels and distributors for resale to retailers, rack jobbers, mass merchants (e.g., Kmart), and juke box operators. The prime source of product for small mom and pop retailers.

point-of-purchase (POP)—A marketing technique used to stimulate impulse sales in the store. POP materials are visually positioned to attract customer attention and may include displays, posters, bin cards, banners, window displays, and so forth.

point of sale (POS)—Where the sale is entered into registers. Origination of information for tracking sales, and so on.

pricing and positioning (P&P)—When a title is sale priced and placed in a prominent area within the store.

pricing strategies—A key element in marketing, whereby the price of a product is set to generate the most sales at optimum profits.

print advertising—Can be both trade and consumer advertising efforts in helping influence the purchase of music.

rack jobber—A company that supplies records, cassettes, and CDs to department stores, discount chains, and other outlets, and services ("racks") their music and/or video departments with the right music mix.

returns—Products that do not sell within a reasonable amount of time and are returned to the manufacturer for a refund or credit (less applicable returns fees or expenses).

sales book—Distribution companies compile all their releases for a specific street date into a "sales book," which contains one sheet that outlines the marketing efforts for each title.

sales forecast—An estimate of the dollar or unit sales for a specified future period under a proposed marketing plan or program.

sell-through—Once a title has been released, labels and distributors want to minimize returns and "sell through" as much inventory as possible.

shrinkage—The loss of inventory through shoplifting and employee theft.

solicitation—The sales process of specific titles occurs during a period called *solicitation*. All titles streeting on a particular date are placed in a solicitation book where details of the release are described.

source tagging—The process of using electronic security tags embedded in a product's packaging.

theft protection—Systems in place to reduce shoplifting and employee theft in retail stores. These systems may include electronic surveillance.

time spent shopping (TSS)—A measure of how long a customer spends in the store.

turn—The rate that inventory is sold through, usually expressed in number of units sold per year/inventory capacity on the floor.

vertical integration—The expansion of a business by acquiring or developing businesses engaged in earlier or later stages of marketing a product.

universal product code (UPC)—The bar codes that are used in inventory management and are scanned when product is sold.

BIBLIOGRAPHY

Garrity, B. (2003). Seeking profits at 99¢. *Billboard, July 12.*

Knab, C. (2001). The distributor one sheet. Mar. 25. www.musicdish.com/mag/index.php3?id=3357.

The Nielsen Company 2008 year-end music industry report—Scoop Marketing. (2008). Dec. 31.

Nielsen SoundScan State of the Industry 2007–2008.

Weatherson, J. Vice President and General Manager, Disney Records.

CHAPTER 13

Grassroots Marketing

Paul Allen and Tom Hutchison

GRASSROOTS MARKETING

Consumers are bombarded every day with more commercial messages from more sources than ever before. The competition for our attention ranges from a subtle product placement in our favorite television show or online video to an in-your-face ad from a car salesman on the radio or a tweet from a promoter. Advertisements pop up at us while we are online, they precede theater movies, they appear on our wireless devices, they're on seat backs at stadiums, and they're even in public facilities where graffiti once was posted. It is this competition for the attention of the consumer that has driven marketers to find alternative methods to connect products and services with billfolds and credit cards. This chapter presents some of the less traditional marketing tools and how they may be used to encourage the consumer to buy recorded music and record label products associated with it.

THE POWER OF WORD OF MOUTH

Very few concepts in business are as universally recognized as is the power of word of mouth (WOM) in marketing. Everett Rogers, in the book *Diffusion of Innovations*, talks about the role of "opinion leaders" in the diffusion process (see Chapter 1) and how these trendsetters can be used in facilitating word-of-mouth communication about a new product or other innovation (Rogers, 1995). His ideas are as relevant today as they were when he wrote them in the 1990s. Word of mouth is more effective at closing a deal with

CONTENTS

a consumer than any pitch from any paid medium. Word of mouth is simply someone you know, whose opinion you trust, saying to you, "You gotta try this!" And the companies most effective with this kind of marketing are those who give their consumers reasons to talk about products to others, and then facilitate that communication.

From the company's standpoint, there are several basics to developing a plan to reach new consumers through word of mouth. They are:

- Educating people about your product or services
- Identifying people most likely to share their opinions
- Providing tools that make it easier to share information
- Studying how, where, and when those opinions are being shared
- Looking and listening for those who are detractors, and being prepared to respond

("Word of Mouth 101," 2005)

Among the most popular ways to effectively use word-of-mouth marketing and promotion in the recording industry is through the artist fan club. The club is a network of people who have a passion for the music of the artist and who are willing to be actively involved in promoting the career of the artist. Members of the club are regular chat room visitors, bloggers, video bloggers, message board respondents, YouTube posters, online social networkers, and hosts of discussion groups. Labels often have someone who is responsible for coordinating an artist's promotion with fan clubs, which in many cases has taken the place of the traditional street team. The label coordinator uses her database to give club members updates on tour dates and to initiate contests for the community, built around the artist (Harbin, 2009).

This discussion of word-of-mouth marketing is not intended to be a short course on fan club development, but the elements of it as a catalyst for "spreading the word" are among the best central strategies a label can use to draw from the strength of the idea. Word of mouth is often compared to the work of the evangelist, and fan club members easily fit the definition. Energizing them and arming them with available tools have the potential to spread positive information about the artist in geometric proportions.

Word-of-mouth marketing takes several forms. Among those are:

Table 13.1   Word-of-mouth style

Word-of-mouth style	Elements of the style
Buzz marketing	Using high profile people or journalists to talk about the artist.
Viral marketing	Creating interesting, entertaining, or informative messages electronically by email and passed along to targets.
Community marketing	Forming or supporting fan clubs or similar organizations to share their interest in the artist, provide access to music, videos, tour schedules, etc.
Street team marketing	Organizing and motivating volunteers to work in personal outreach locally.
Evangelist marketing	Cultivating key opinion leaders and tastemakers to spread the word about the artist.
Product seeding	Giving product samples and information about the artist to key groups and individuals.
Influencer marketing	Identifying other artist fan clubs and key members of those clubs to help spread the word about the artist.
Cause marketing	Working with charities and causes that the artist supports.
Conversation creation	Creating catch phrases, highlighting lyric lines, creating emails, or designing fun activities to start the word-of-mouth activity.
Referral programs	Creating tools that help fans refer the artist to their friends.

The importance of word-of-mouth marketing was demonstrated by a U.K. trend-watcher, Haymarket Publishing Services, in its study in 2002 that shows the primary motivators for people to try a new product or a new brand. When applying these principles to promoting a new artist or new music for a record label, Table 13.2 shows the importance of word of mouth. The strength of "advice from friends and family" (60%) and "seeing others using the brand" (27%) suggests that word of mouth can have a powerful influence on the decision of a consumer to try and to buy new music.

Word-of-mouth promotion, as defined by Malcolm Gladwell (2000) in his book *The Tipping Point*, shows the marketer that this seemingly simple strategy can actually have a deep-rooted sophistication. He suggests there are three rules to its effectiveness. (1) There is the law of the few people who are connected to many others, and who can spread the word about a product in ways that have epidemic proportions. (2) The law of stickiness suggests that the idea shared by the connector has to be memorable and must be able to move people to action. And (3) the power of context means that people "are a lot more sensitive to their environment than they may seem" (p. 29) and the effectiveness of word of mouth has a lot to do with

Table 13.2	Factors influencing purchases (Source: Haymarket Publishing Services)

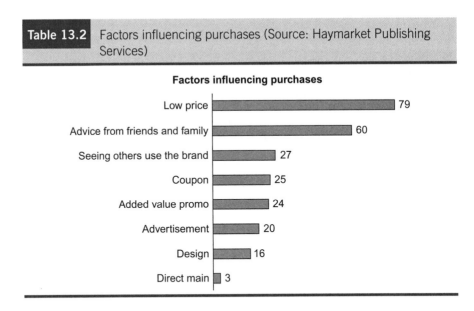

Factors influencing purchases

the "conditions of, and circumstances of, the times and places" (p. 139) where it happens.

A great deal of care must be taken to set up a word-of-mouth promotion to make it effective. The sources of word-of-mouth campaigns will ultimately determine whether the effort was effective, so it is important to keep participants reminded about the ethics of this type of promotion.

For the record label, coordinating the energy and passion for artists through their fans can generate sales. Dave Balter, owner of BzzAgent, a word-of-mouth marketing firm, says, "The key is about harnessing something that's already occurring. We tap into people's passions and help them become product evangelists" (Strahinich, 2005). And that is the essence of the word-of-mouth marketing strategy as it is applied to the recording industry.

STREET TEAMS

The idea of street teams is an adaptation of the strategy of politicians everywhere: energize groups of volunteers to promote you by building crowds, creating local buzz, posting signs, using the strength of the Internet, and rallying voters. Perhaps the iconic street team of the times is the phenomenal success achieved by those supporting Barack Obama's quest for the White House, where literal street teams and e-teams became the core of his fundraising and his campaign. In very practical ways, there isn't a lot of difference between political volunteer groups and artist street teams.

For the artist, they are volunteer ambassadors who quite literally promote personal appearances and music. For the record label, the ultimate rallying point is getting consumers to purchase the artist's music. While many artists use street teams today, they were originally formed in order to promote music that was not radio friendly. Without radio airplay, creators of alternative music sought other ways to connect consumers with their music, and employing street teams became an effective way to do that. And the commissioning of street teams today continues in keeping with their origins—where the Internet has its e-teams, non-mainstream labels like Universal's Lost Highway rely on street teams to promote its roots and alternative artists. And there are occasions when the major labels will hire third-party street team companies as mentioned later in this chapter (Harbin, 2009).

Street teams are local groups of people who use networking on behalf of the artist in order to reach their target market. These team members make up the core of the fan base of the artist and have the deepest passion for the music and message of the artist. Often they are friends and fans of the artist. Some refer to street team members as "marketing representatives" who promote music at events and locations where the target market can be found. These places are tied to the lifestyle of the target market such as at specialty clothing stores, coffee houses, and at concerts of similar acts. A key to effective street teams is for members to understand where to find the target market of the artist and to provide tools and guidance on how to communicate to the target (Tiwary, 2002).

Street teams originally were formed as a way to reach consumers by companies who either did not have the resources for mass-media marketing, and to reach segments of the market that were not as responsive to mass-media messages as they are to peer influence (Holzman, 2009). Now, major marketing and advertising agencies acknowledge street teams as an effective form of youth marketing.

Among the tools that may be provided to the street team members are postcards and flyers, email lists to contact local fans, small prizes for local contests, links for music samples, actual CD samplers to give away at appropriate events, advance information about tour appearances, and release dates for new music. Street team members may engage in sniping—the posting of handbills in areas where the target market is known to congregate. Some labels economize on printing by emailing flyers and posters to local street team coordinators and asking them to arrange for printing.

Street teams require servicing. This means having continuing communication with key members and finding ways to say "thanks" to the street

---

**Case study: Columbia
Records and Switchfoot**

UBI was enlisted to help introduce the band Switch-foot, and their latest album, "The Beautiful Letdown", to the masses. Switchfoot had found success in the Christian Rock format, but their label was interested in creating crossover appeal by targeting college campuses, lifestyle/community locations, clubs/bars, and record stores in 10 major cities across the United States. The goal was to increase awareness of the band, to bring attention to their live performances and to drive sales of their new release.

Universal Buzz Intelligence mobilized their corps of highly trained operatives for an intense 9 week promotional campaign in 10 major markets. The campaign focused on the distribution of promotional materials and other collateral marketing materials throughout college campuses and the communities that surround them. The guerilla marketing effort focused heavily on driving retail store sales for "The Beautiful Letdown" album, as well as promoting the band's live tour dates.

By all accounts the Switchfoot release "The Beautiful Letdown" far exceeded label expectations with over one million copies sold to date (the band sold an aggregate of 100,000 albums for their previous three releases). Of the 10 markets covered by UBI, the band realized 7 sold out performances. In addition, six of their top ten selling markets were markets covered by UBI (which had not previously been strong markets for the band), with the remaining 4 cities in the UBI campaign falling within the bands top 25 selling markets. The bands crossover popularity has bought them video exposure on MTV and The Fuse Network.

Source:
www.universalbuzzintelligence.com/case/switch.htm

FIGURE 13.1   *Universal Buzz case study. (Source: Universal Buzz Intelligence)*

members. Volunteers for causes require a measure of recognition for their effort in order to keep them energized, and it is no different with those working for free on behalf of an artist. The first thing the label coordinator must do is to regularly communicate with core members of the team. Keep them current on the planned activities of the artist, and make them feel they are important. Provide incentives to the extent the budget will allow. Incentives can be CDs, T-shirts, meet-and-greets with the artist, and free tickets.

The hierarchy of street team management begins at the label with a coordinator, who recruits regional street team captains, who then coordinate at the local level. Data captured by artists through their pages on social networking sites as well as on their primary web site become an important building block for street teams. Viral marketing is driven by capturing data about the target market and then effectively using them to meet the needs of the artist's fans for information and entertainment.

While major labels often have in-house grassroots or new media departments to run street teams, there are several independent or outside marketing companies that specialize in street team marketing. The advantage of using one of these companies is that they often promote other lifestyle products, and the opportunities for cross-marketing are increased.

Some of the outside marketing companies used for street marketing are Street Attack, Buzz Intelligence, and Streetwise Concepts and Culture. Street Attack boasts on their web site that their understanding of street psychology and Generation Y mindset makes them specialists in building street teams (www.streetattack.com, 2009). Buzz Intelligence boasts that V2, Virgin, Arista, Interscope, Sony, Universal, Columbia, Geffen, Atlantic, Blue Note, and Warner Bros. are among their clients (www.universal buzzintelligence.com, 2009). Streetwise claims Columbia, Live Nation, Warner Bros., and Universal have used their street marketing services (www.streetwise.com, 2009).

GUERILLA MARKETING

Guerilla marketing is another term closely associated with grassroots marketing. Whereas grassroots marketing refers to a street-level, door-to-door campaign of peer-to-peer (P2P) selling, guerilla marketing refers to low-budget, under-the-radar niche marketing, using both P2P and any inexpensive top-down methods.

Grassroots marketing for large companies may involve a large budget to pay street teams. On the Internet, grassroots marketing includes placing conversational presentations of the label's product or including the artist in discussion forums and message boards, placing self-produced videos and music videos online, and encouraging bloggers to write about your artist. This is done by recruiting fans to be on the artist's street team or e-team.

Large companies hire members of the target market to spread the word to peers about the artist or product. On the Internet, these online street teams seek out web locations where the target market tends to congregate and infiltrates these areas to introduce the marketing message.

Online street team members often use the following activities to promote recording artists:

1. **Posting on message boards** – Street team members who are actually part of the target market visit relevant sites that allow for message posting and casually encourage readers to "check out my new favorite recording at . . . ."

These sites usually have specific rules about the posting of commercial messages and harvesting email addresses (which is universally frowned on by group moderators and system operators). Many groups, however, welcome brief messages from industry insiders notifying interested members of new products. Sometimes street team members do not announce their affiliation with the artist; instead, they engage in communication as if they were just enthusiastic fans. The message postings often are embedded with hot links to the site actually promoting or selling the product. Some user groups require that fans register before they are allowed to post messages.

2. **Visiting and participating in chat rooms** – Chat rooms allow for real-time interaction between members, whereas bulletin boards allow individuals to post messages for others to read and respond to. Because these members or users have a mutual interest in the site topic, user groups offer an excellent way for labels to locate members of the artist's target market.

3. **Blogging or successfully submitting information to bloggers** – Successful bloggers (those with a substantial audience) are opinion leaders, and their message can influence the target market. Online street team members will seek out blogs that discuss music and reach out to the bloggers to check out their artist and write about the music, much the same way a publicist will seek out album reviews in traditional media. Blogging, however, has the potential for street cred(ibility) that is sometimes absent in mainstream media.

4. **Pitching/promoting to online media** – This is much like pitching to bloggers, only more formal. Under these conditions, professional media materials must be supplied to the publication, whether it's an online-only publication or one that also has a traditional media presence. The label may supply completed articles and press releases to busy journalists who don't have time to conduct their own research and whose decision to run an article may be influenced by convenience (see Chapter 10 on publicity for additional strategies to reach bloggers and other online media).

5. **Visiting social networking sites and posting materials including music, artwork, and videos** – Major record labels use interns and young entry-level employees to maintain their artists' presence on social networking sites. These young marketers are responsible for setting up the artist's page on each of these sites, fielding requests from fans to be added to the social group or "friends network,"

providing updated materials (music, news, photos, videos, etc.), and visiting related pages to engage in street team promotion. Some labels have teamed up with these sites to conduct promotions in the form of contests and giveaways, such as encouraging fans to create their own YouTube music video for an artist.

6. **Finding fan-based web sites and asking the site owners to promote the artist** – We address this strategy in the section on fan-based sites in Chapter 14, New Media Marketing. It is often up to online street team members to seek out and identify these fan-based sites. Sometimes they engage in dialogue with the site owner, but often the list of fan-based sites is compiled by the street team member and then passed along to a more senior member of the marketing team. Then these sites are managed by more experienced web marketers who can supply RSS feeds and other technical web assets to the site owner.

7. **Acting as a "clipping service" by scouring the Internet for mentions of the artist** – Google and Yahoo! news alerts have made this task easier by allowing anyone to set up news alerts based on selected keywords. Any mentions of the artist in third-party online sites are identified, and decisions made by the label about how to incorporate them into the overall web marketing plan. Decisions include: Should you create a link to the article? Should you request permission to reprint? Should you rebut negative publicity?

8. **Finding sites that attract the target market and then working with those sites** (see reciprocal links information in Chapter 14) – The following section on turning an artist's competition into partners outlines many of the street team tasks performed in this area. The foremost task is finding and identifying where the target market can be found engaging others on the web—what web sites do they frequently visit? Then the label can make marketing decisions about how to work with those sites, whether through advertising on them or setting up reciprocal links to generate cross traffic.

9. **Submitting the site to search engines and music directories** (see the section on search engines in Chapter 14).

10. **Writing reviews of the artist or album on sites that post fan reviews** – Many retail sites such as Amazon.com allow customers to post product reviews, including music reviews. It is not uncommon for artists to ask their fans to post reviews on these sites or for a label to have street team members post favorable reviews.

Any discussion of guerilla marketing must acknowledge the contribution of Conrad Levinson, author of the best-selling book *Guerilla Marketing*, first published in 1984. Levinson is credited with coining the term, which generally means using nontraditional marketing tools and ideas on a limited budget to reach a target market. In Levinson's words, **guerilla marketing** is "achieving conventional goals, such as profits and joy, with unconventional methods, such as investing energy instead of money" (www.gmarketing.com, 2005).

In the recording industry, much of the activity of street and e-teams is basic guerilla marketing. The list below provides some of the available classic guerilla marketing tactics. While they may not directly apply to promoting music, many of them can be developed and adapted for specific uses by a label to help market recorded music.

1. Marketing plan
2. Marketing calendar
3. Niche/positioning
4. Name of company
5. Identity
6. Logo
7. Theme
8. Stationery
9. Business card
10. Signs inside
11. Signs outside
12. Hours of operation
13. Days of operation
14. Window display
15. Flexibility
16. Word of mouth
17. Community involvement
18. Barter
19. Club/association memberships
20. Partial payment plans
21. Cause-related marketing
22. Telephone demeanor
23. Toll-free phone number
24. Free consultations
25. Free seminars and clinics
26. Free demonstrations
27. Free samples
28. Giver vs. taker stance
29. Fusion marketing
30. Marketing on telephone hold
31. Success stories
32. Employee attire
33. Service
34. Follow-up
35. Yourself and your employees
36. Gifts and ad specialties
37. Catalog
38. Yellow Pages ads
39. Column in a publication
40. Article in a publication
41. Speaker at any club
42. Newsletter
43. All your audiences
44. Benefits list
45. Computer
46. Selection
47. Contact time with customer
48. How you say hello/goodbye
49. Public relations
50. Media contacts
51. Neatness
52. Referral program
53. Sharing with peers
54. Guarantee
55. Telemarketing
56. Gift certificates
57. Brochures
58. Electronic brochures

59. Location
60. Advertising
61. Sales training
62. Networking
63. Quality
64. Reprints and blowups
65. Flipcharts
66. Opportunities to upgrade
67. Contests/sweepstakes
68. Online marketing
69. Classified advertising
70. Newspaper ads
71. Magazine ads
72. Radio spots
73. TV spots
74. Infomercials
75. Movie ads
76. Direct mail letters
77. Direct mail postcards
78. Postcard decks
79. Posters
80. Fax on demand
81. Special events
82. Show display
83. Audiovisual aids
84. Spare time
85. Prospect mailing lists
86. Research studies
87. Competitive advantages
88. Marketing insight
89. Speed
90. Testimonials
91. Reputation
92. Enthusiasm and passion
93. Credibility
94. Spying on yourself and others
95. Being easy to do business with
96. Brand name awareness
97. Designated guerrilla
98. Customer mailing list
99. Competitiveness
100. Satisfied customer

One hundred marketing weapons. Source: Jay Conrad Levinson, www.gmarketing.com/articles.

Guerilla tactics in the marketplace aren't limited to smaller labels with limited resources. When Mike Kraski was vice president of marketing at Sony Nashville, he wanted to find a way to encourage Wal-Mart store associates to become familiar with their new music so that callers to the store could be told the item was available. He announced to the stores he was beginning a promotional campaign where someone from Sony Nashville would occasionally call Wal-Mart record departments and ask about Sony new releases. To those associates who were able to tell the caller about the new music, he gave a limited number of big screen Sony television sets as prizes.

Basic grassroots marketing, whether you call it word of mouth, street teams, e-teams, online teams, peer-to-peer, guerilla marketing, viral marketing, or the latest term du jour, can create an environment that can set a record label's marketing plan and results apart from those of competing artists. One of the biggest challenges of a label-marketing department is to find new and unique ways to present its product to consumers. Including an element of grassroots in the marketing plan can create a plan that steps away from overused templates to add a unique element.

GLOSSARY

bloggers—Short for weblogs, and refers to those who write digital diaries on the Internet.

grassroots marketing—This is a marketing approach using traditional and nontraditional ways to create a bottom-up awareness of an artist. Viral and word of mouth are basic elements of grassroots marketing.

guerilla marketing—This is using nontraditional marketing tools and ideas on a limited budget to reach a target market.

moblogs—Web sites where individuals post still images, video, and text from mobile devices; the services are especially important to time-sensitive information services.

marketing representatives—Another term sometimes used for members of a street team.

sniping—The posting of handbills in areas where the target market is known to congregate.

street teams—Local groups of people who use networking on behalf of the artist in order to reach their target market.

video bloggers—Also known as vloggers, they are the video counterparts to bloggers except the content contains audio and video.

BIBLIOGRAPHY

Gladwell, M. (2000). *Tipping Point*. New York: Little, Brown, and Company.

Harbin, J. (2009). Interview.

Holzman, K. (May 19, 2009). Effective Use of Street Teams. *MusicDish* e-journal. www.musicdish.com/mag/index.php3?id=9909.

Hutchison, T. (2008). *Web Marketing for the Music Business*. Boston: Elsevier/Focal Press.

Rees, J. (November 14, 2004). EMI Clinches Research Deal. Associated Newspapers. (LexisNexis Acacdemic).

Rogers, E. M. (1995). *Diffusion of Innovations* (4th ed.). New York: The Free Press.

Strahinich, John. (January 23, 2005). What's All the Buzz? Word-of-Mouth Advertising Goes Mainstream (LexisNexis Academic). *The Boston Herald*.

Measuring Word of Mouth. (November 8, 2004). *Presentation at Ad-Tech*. New York: Word of Mouth Marketing Association.

Tiwary, V. J. (2002). Starting and Running A Marketing/Street Team. www.starpolish.com/advice/article.asp?id31.

Word of Mouth 101. (2005). Word of Mouth Marketing Association. www.womma.org.

Word of Mouth in the Age of the Web-Fortified Consumer. (2005). Intelliseek. www.intelli-seek.com.

www.gmarketing.com (2005). www.gmarketing.com/what_is_gm.html.

www.streetattack.com (2009). www.streetattack.com/Default.aspx.

www.streetwise.com (2009). www.streetwise.com/indexnew.php.

CHAPTER 14

New Media Marketing

Tom Hutchison

New media has become the new catchphrase for what used to be considered Internet marketing and is now morphing to include content development and marketing for mobile devices. *PC* magazine defines new media as "the forms of communicating in the digital world, which includes electronic publishing on CD-ROM, DVD, digital television and, most significantly, the Internet. It implies the use of desktop and portable computers as well as wireless, hand-held devices" (*PC*, 2008). Record labels now have adapted their Internet marketing departments to include marketing for handheld devices such as the Blackberry and iPhone. This chapter will first address marketing for the Internet and then introduce the concepts of marketing through mobile devices.

INTERNET MARKETING

The Internet has become a powerful force in marketing and commerce and now accounts for an ever-increasing percentage of retail sales. Jupiter Research predicted that by the year 2012, one-quarter of the world's population will be online (Jupiter Research, 2008). By 2008, digital music sales were accounting for 15% of the global music market (IFPI Digital Music Report, 2009). In 2008, PricewaterhouseCoopers predicted that by 2012, half of global music industry revenues would come from digital music sales (in IFPI, 2009). According to the NPD Group (2008), 48% of U.S. teens didn't buy a single CD in 2007, compared to 38% in 2006. In early 2008, Apple claimed that its online retail store iTunes had moved into the top spot as the number 1 U.S. music retailer, beating out Wal-Mart.

In the recording industry, the web has become useful to create awareness and demand for artists and their recordings. But the web is also capable of electronically delivering the product to the customer—a claim that food, beverage, and clothing companies will never be able to make. The Internet provides two opportunities to the music industry: (1) to market and promote artists and their products, and (2) actual distribution of recorded music.

Web marketing should be a part of any marketing plan but should not be the only aspect of the plan. Even though the Internet has become a great tool for selling music, the traditional methods of live performance, radio airplay, advertising, and publicity are still a significant part of marketing and should not be neglected.

> *Rule #1: Don't make the Internet your entire marketing strategy. Internet marketing should not be a substitute for traditional promotion. The two strategies should work together, creating synergy.*

In the music business, it is necessary to build brand awareness. Whether it's with the artist or the record label, you need to create a sense of familiarity in the consumer's mind. This can be done through many tactics available both on and off the Internet and should be designed to lure customers to the artist's web site or drive them into the stores to buy the artist's records.

The web is a great tool to reach a large number of people, with minimum expenses involved. But the web is so vast that unless potential customers are looking for a particular artist, they are unlikely to stumble across an artist's site by chance. Also, many customers are more responsive to traditional marketing methods such as radio airplay and retail store displays. An artist's web site is a great place for his customers to learn more about his products, such as the artist's live shows and recordings, but it is not necessarily the best way to introduce new customers to the artist's products.

> *Rule #2: Build a good web site, but don't expect customers to automatically find it on their own.*

A solid marketing plan incorporates the company web site into every aspect of marketing and promotion, and while the web site will be the cornerstone of Internet marketing efforts, it is just the start. Building a good web site is crucial to marketing success, but the old adage "If you build it, they will come," does not apply to the Internet.

> *Rule #3: Incorporate the artist's web site address into everything you do online and offline.*

Every marketing angle should tie in to the web presence. Internet marketing is more effective when it is conducted in conjunction with other aspects of the plan. Every piece of promotional material should contain the web site address to help direct the potential customer to the web site. Any posters, flyers, postcards, press releases, and so on, should have the artist's web site address prominently displayed. And it goes without saying that CD tray cards should also contain the address. One band had handstamps printed up at the local office supply store, and the stamps had the band's web address on them. The band requested that the bouncers use them to stamp customer's hands as they entered the venue where the act was performing.

This chapter will discuss how Internet marketing and web commerce can become an integral part of the overall marketing plan. Many record labels have an emerging department known as the *new media* department that is in charge of Internet marketing. We will be looking at the following aspects of Internet marketing:

- The artist's web site
- Promoting the artist's web site on the Internet
- Promoting the artist's web site off the Internet
- Promoting the artist's brand on the Internet other than on the artist's web site
- Promoting the artist through mobile networks

THE WEB SITE AS YOUR HOME BASE

The Domain Name

The first aspect of building a web site involves registering a domain name. The **domain name** is the "web address" that your visitors will become familiar with and use to access your site. The **Uniform Resource Locator (URL)** is the means of identifying the web site location on the Internet. An Internet address (for example, http://www.yourname.com) usually consists of the access protocol (http), the domain name (www.yourname), and the top level domain (TLD), such as .com, .org, or .net. The URL will also contain the directory path and file name, such as /myband/index.html. The general default-loading file is named index.html or index.htm. This file name should be the name of your homepage because this is the page that the web browser looks for when your customers enter in the address. The file index. html is served by default if a URL is requested that corresponds to a directory on the server where your web site resides.

It is important for the domain name, or Internet address, to be simple and easy to remember. A long URL will confuse customers and prevent them from finding the web site. There is also more possibility of error when the customer enters in a long name such as www.cheapdomain provider.com/~yourcustomername/personalweb/bandname/index.html.

The domain name is the single most valuable piece of real estate on the Internet. Sometimes bands that have been around since before the Internet, but who were not quick to register their band name when the Internet took off, are forced to use a less than perfect URL address. For example, the band Van Halen does not own the web address www.vanhalen.com. A fan of the band owns that site. The official band web site is www.van-halen.com, and some fans might not know or remember to include the dash. In another example, the official U.S. White House web site is www.whitehouse.gov. A whimsical company that markets political swag (slang for stuff we all get) has set up shop at www.whitehouse.org.

Many services available on the Internet will register the artist's domain name. A quick online check with these providers can determine whether the artist's or band's name is available. The services that sell domain names all have features that will allow you to check the availability of particular domain names and to reference **Whois** (www.whois.net) to determine the domain name owner. Go for the dot-com extension if it is available, and use a domain name that is memorable and simple. This will become your online "brand."

Branding is defined as creating a distinct personality for the product (in this case, the artist, not the label) and telling the world about it. As companies rely more and more on the Internet for promotion, the dot-com name becomes a brand name. So, the URL should reflect the brand that the artist wishes to promote. What if the name you want is already taken? If the artist is not yet well established, it may even be wise to rename the band or change the spelling of the artist's name to accommodate the branding that goes along with having an Internet presence.

As a result of the growth of the Internet, many companies have re-created their brand to reflect the style found on URL addresses, combining words together and using capital letters to distinguish between words. Examples include names such as SunTrust bank instead of Sun Trust. This trend had been extended to include abbreviations commonly used in text messaging and similar to the shorthand found on automobile license plates, where the words "in," "be," "to" and "for" are replaced by the letters and numbers "N," "B," "2," and "4" (e.g., NSYNC, Boyz2Men). Consider all possibilities when acquiring domain names.

It is also important to register derivations of the artist's name, just to cover those visitors who may not know the correct spelling.

Consider the following tips when selecting a domain name:

1. **Make it easy to remember**. Hyphens are hard to remember and hard to communicate verbally when describing the site address to

someone. If you have any peculiar symbols or spellings, be sure to emphasize that in your marketing materials.

2. **Keep it short and simple**. Also, avoid strange spellings, unless you intend to brand your artist with that spelling (such as "z" instead of "s"). Remember that words may be spelled differently outside the United States or in non-English-speaking countries. Some companies and artists like to use acronyms for the domain name (e.g., www.hsx.com for Hollywood Stock Exchange). The problem is that most of the short acronyms are already taken. Dennis Forbes (2006) reported that all of the three-letter acronyms are taken and nearly 80% of four-letter acronyms are already registered. Less than 5% of five-letter acronyms are taken.

3. **Make it descriptive of the site**. It is easier to memorize and identify a site name that relates to the content or subject matter of the site. You can add "band," "sings," "music," "songs," or some other word to the end of your artist's name, especially if the simpler domain name is taken.

4. **Use the dot-com domain, if available**. Most web surfers are accustomed to using the dot-com extension and will default to that unless you make it clear that your TLD is dot-net or dot-org. You may even want to register all variations of your domain name with the different TLDs.

5. **Use a keyword in the name, if possible**. The name of the artist is good, but adding "music" or "band" helps search engines categorize the site based on these keywords. Consider registering the name both ways.

6. **Be consistent**. Even if you register several variations of your domain name to cover all bases, stick with just one for your marketing materials to avoid confusing your customers. It is common to register all variations (www.arstistname.com, www.artist-name.com, www.artistnameband.com) and then forward all of these to the same domain.

7. **Your domain name should be your site name**. This is important because when people think of your web site, they will think of it by name. So, have the domain name match the name of the site that appears in the masthead.

Web Hosting

The host is the physical server in which a web site is stored. If a company has its own server, they can just redirect or forward the artist's URL to their server. If they do not have a **host provider**, there are many commercial companies that offer "space" to host a web site. The free hosting services should be avoided because they tack on banner advertisements and pop-up ads to the site. These advertisements discourage your customers from visiting the site. It is worth the money to pay for a cleaner site without all the ads.

It is important to make sure the hosting site has enough data storage space for all the functions a music-based site needs. Thirty megabytes (MB) of space is probably not enough to hold graphics, photos, and sound files—one MP3 file containing a song of average length can take up to 4 MB. It is necessary to have enough space for expansion. An artist may only need to feature three audio files from her current album, but when subsequent albums are released, the demands on the site will also increase. Up to 5 MB are needed for graphics and web documents. A thorough site should include a bio, photos, band news, tour itinerary, sign-up for a mailing list, press and media coverage (reprints), discography, lyrics, audio files, contests and giveaways, merchandise, links to other favorite sites, video, and contact information. It is vital to include some way for the fans to purchase the artist's recordings—either directly from the web site, or by directing them to an online retailer. All this can add up to quite a bit of storage space. It is recommended that at least 250 MB of storage space and sufficient **bandwidth** be allocated. Bandwidth is the amount of data that can be transmitted in a fixed amount of time. If a server is accessed by a lot of users simultaneously, it requires sufficient bandwidth to transmit the information from the server to the users in a timely fashion.

Before a web site design can begin, it is necessary to outline the goals for the web site. These goals can include creating brand awareness, creating a demand for the products and fulfilling that demand through e-commerce, creating repeat traffic to the site, creating a sense of community among the artist's fans, and enlisting fans to help promote the artist.

Branding is defined as creating a distinct personality for the product (in this case, the artist, not the label) and telling the world about it. Artists who create a well-known brand can parlay that into endorsement deals, an acting career, or becoming a spokesperson for a worthy cause. Artists Madonna and Missy Elliott appeared in TV ads for The Gap; Queen Latifah has moved into films, and Bono of the rock group U2 has become the spokesperson for solving the problems of third-world debt and global trade. So branding should be an important part of any web site. Creating an easily identifiable logo and maintaining consistency in style and design can help support the brand. This extends to the style of the web site, which should convey the image that you want to project for the artist.

The web site should create a demand for the artist's products. The primary products are recorded music and tickets to live shows. Secondary products include posters, bumper stickers, and other souvenirs. Enticing the web visitor to purchase, or creating a desire to purchase the recordings, should be an important aspect of the site. This can be achieved through offering music samples on the site as well as information about the music and the recording. Links to online retailers will encourage the visitor to follow through on purchase intentions, creating the impulse buy. Tour schedules, concert photos, samples of live recordings, and links to sites that sell tickets can create a demand for concert tickets. Maps to the venues also encourage concert attendance. The web site is also a great place for the merchandising of tee shirts and other souvenirs. Photos of these items are important.

Artist web sites are a good place for like-minded people to connect with each other. These fans probably have much in common—especially their interest in the artist. A sense of community can be created through **message boards**, online chats with or without the artist, and by including photos of fans at concerts. These devoted fans can become opinion leaders (the virtual street team), influencing their friends to become fans of the artist and encouraging their friends to visit the web site. MySpace and other social networking sites have provided excellent opportunities for fans to engage in community social behavior centered on a common interest in an artist.

The web site should also provide materials for professional journalists looking for information on the artist, as well as contact information for both journalists and booking agents. Some sites provide print-quality publicity photographs and in-depth biographical information suitable for reprint in mainstream publications.

It is important to provide elements on the web site that will keep visitors coming back as new developments occur in the artist's career. Certain attributes can be included in a web site that encourage repeat visits: including blogs, updated tour itineraries, contests, news, new music samples, and message boards. One entrepreneur even offers a service that provides updated horoscopes that you can post to your web site to encourage repeat traffic (www.tomorrowsedge.net).

Web Site Design

Web design is defined as the creation and arrangement of web pages that make up the site. It's part art and part commerce. A good web site is one that is attractive, uncluttered, quick to load, and easy to navigate. The site must offer something of value to the consumer—information, products, and freebies. It also helps to make the web site fun, refresh content, and give people a reason to return. Effective web sites avoid large flash programs, flashing text, animation, and large graphic files. Half of Internet users do not yet have cable modems, and a web site that is slow to load is sure to fail.

Consider these simple rules of web design:

1. **Design web sites, not pages**. Determine the overall look of the site and the scope of information before attempting to design any one particular page. This will help the designer determine how many pages are needed, what should be on each page, and the overall consistent look of the site.

2. **Keep page size to a reasonable length**. The general rule is to design the page so that it is no longer than twice the length of the computer screen, especially for the homepage. Readers do not like to scroll down and tend to become lost or discouraged by long web pages. The exception to this may be blogging, which tends to be more linear than most other web elements. When long pages are used, be sure to have plenty of bookmarks that can direct the user quickly back to the top of the page. Never use horizontal scrolling.

3. **Use appropriate graphics**. Don't load a lot of photos on one page—it will slow the loading time. Too many photos overload both the page and the eye. Make sure all photos are appropriate for that particular web page. If you want to include an album of photos, relegate it to a special photos page. Use thumbnails to present examples of photographs that can be enlarged at the click of a mouse. Web visitors are now accustomed to thumbnails but should be reminded that they can click for an enlarged version of the photo. Sometimes this is done with a "mouse-over" command, where a text message "click to enlarge" appears as the visitor runs the mouse over the thumbnail.

4. **Clearly specify the purpose of the web site and why the user should visit regularly**. The homepage should have a clear, simple headline description of the site. If that is not suitable, then the masthead and dominant graphic or photo should clearly illustrate the site's purpose or products. Stick to the subject in the web site, keeping your goals in mind. If you want to include something that is not germane to the topic, consider linking to a separate web site. Keeping the site current and changing out content will encourage repeat traffic.

5. **Keep it simple and easy to read**. Nothing screams "amateur" more than a site that is cluttered and hard to read. Avoid backgrounds that

conflict with the text and confuse the eye. Set up a color scheme that is appealing, consistent, and does not induce eye fatigue. Yellow or red text on a white or busy background never works.

6. **Balance the design elements**. Don't go too heavy on either text or graphics. Just like for a magazine layout, white space can be very appropriate when used properly. Basic principles of desktop publishing also apply to web design.

7. **Content is king**. Content should drive the design. Make your design appropriate for the theme and brand of the artist or product. Use content to give visitors a reason to explore the site and to return.

8. **Make navigation easy**. Web site navigation is made easy with the inclusion of a **navigation bar (nav bar)**. The bar usually has prominent placement in the same location on every page. Usually the bar is located across the top, under the banner, or in a column on the left side of the page. The nav bar allows users to be one click away from any place they want to go within the site. It is important to link pages in a consistent, well-planned manner that is intuitive to users. Navigation bars are usually found across the top, under the masthead, or on the left side of the page. Web visitors are used to this design, which was originally created because web pages load from the top and left side, and navigation information was sure to appear on the page regardless of the end user's browser or computer screen. Keep your navigation bar in the same location on every page, with a clearly identified link back to the homepage.

9. **Keep your site up to date**. Visitors will not return to a site that looks abandoned, with old news, outdated information, out-of-style design elements, and cobwebs. If you plan to update on a regular basis, state that intention on the site and then be sure to follow through! If you state you will provide a weekly podcast, then be sure to have a new one available at the same time each week.

10. **Be consistent**. Keep the basic look and theme the same throughout the site. Use the same color scheme, background, and fonts on each page. Do not use many different fonts. For a more consistent look, use the same masthead on every page—with slight variations if necessary to identify individual pages. Stick with standard layouts. The reason the three-column layout is so common is that it works (Kymin, n.d.).

A **splash page** can be used to enhance the image of the web site. This is the first page web surfers are directed to when they access the site. Usually, a splash page will have the logo and elaborate graphical layout, but not much information. The visitor is encouraged to enter the site by clicking on the page. Automatic splash pages direct the visitor to the homepage after a few moments. Do not overload the visitor with large flash programs that take a long time to load. An effective splash page has just a graphic image and the brand logo. Many web designers discourage the use of a splash page because it is just one more step or click required of visitors before they get to the true substance of the web site. If you must use one, Client Help Desk (www.clienthelpdesk.com, 2004) advises you to "consider dropping a **cookie** into each site visitor's computer that automatically skips the splash screen on subsequent visits. Even people with the patience to deal with a splash screen once will be tested upon seeing it repeated each time they return to the site."

The **homepage** identifies your company or brand and extols the benefits of the site and its products. This page should be updated frequently to reflect changes and to notify the visitor of new and interesting developments. Announcements are frequently found on the homepage, although a "What's New" page can also contain news briefs and updates. The homepage should also contain links and navigational tools for the rest of the site. A good web page should balance text with graphics or images and sport a layout that is inviting and interesting.

Web Site Content

An artist's web site should contain the following basic elements:

1. **A description and biography of the artist** – The homepage should have some information or description, but a separate biography page should be created for in-depth information about the artist.

2. **Photos** – Promotional photos, concert photos, and other pictures of interest. This can include shots of the artist that capture everyday life, photos of the fans at concerts, and other photos that reflect the artist's hobbies or interests.

3. **News of the artist** – Press releases, news of upcoming tour dates, record releases, and milestones such as awards. This page should be updated often, and outdated materials should be moved to an archive section.

4. **Information about recordings** – Discography and liner notes from albums to increase interest in the recorded music of the artist;

comments from the artist on the recordings; information on the recording process.

5. **Song information** – Lyrics and perhaps chord charts (again to increase interest in the recorded music).

6. **Audio (and video) files** – These may be located on the purchase page to encourage impulse purchases. They may be 30- to 45-second samples or streaming audio rather than downloadable files, to protect against piracy.

7. **Membership or fan club sign-up page** – Allows visitors to sign up for your newsletter or to access more exclusive areas of the site. This will help you build an email list and allow for more control over content posted on message boards in restricted areas of the site.

8. **Tour information** – Tour dates, set lists, driving directions to venues, photographs/video from live performances, touring equipment list.

9. **E-store** – Merchandise page for selling records, T-shirts, and other swag.

10. **Contests or giveaways** – To increase repeat traffic and motivate fans to visit the site. These can be announced at concerts.

11. **Links** – To other favorite sites, including links to purchase products or concert tickets, venue information, the artist's personal favorites, ezines (online magazines), the artist's MySpace page, and other music sites. Be sure that all your offsite links open in a new window so the visitor can easily return to your site.

12. **Contact information** – For booking agencies, club managers, and the media. This could also include print-quality images for the press.

13. **Message board or chat rooms** – This allows the fans to communicate with one another to create a sense of community. This can be an area restricted to members only.

14. **Blogs** – Recently blogs have become popular on the Internet. A **blog** is simply a journal, usually in chronological order, of an event or a person's experiences. Maintaining a blog of the touring experience is one way to keep fans coming back to the web site to read the most recent updates to the journal. It also gives fans a sense of intimacy with the artist. Video blogging (or vlogging) is also becoming popular.

15. **Printable brochures or press kits** – Electronic versions of any printed materials that the artist uses in press kits and to send out should be made available on the web site. On the link to retrieve these items, be sure to mention whether they are in the PDF format—some older browsers and slower modems lock up when attempting to open a PDF file in a browser window. Sonicbids is an online site that offers users the ability to create professional-looking electronic press kits for artists (www.sonicbid.com).

E-Commerce: The Electronic Storefront

One major decision facing web designers is whether to host an electronic storefront or simply create links to one of the Internet's many online **e-stores**. Setting up a storefront can be complicated because it involves creating a security encryption through a **secure sockets layer (SSL)** and the ability to process credit card transactions. It also means having arrangements in place for product fulfillment: the ability to ship product directly to the customer. Processing credit cards is the most efficient way to handle monetary transactions, but it can be costly. Credit card services usually charge a transaction fee for each order. In addition, the site will need to contain a shopping cart page, wish list database, and order processing and tracking information, all of which can add significantly to web site maintenance.

Some customers are hesitant to give out their credit card numbers to small Internet vendors for fear of fraud. Two companies have emerged to provide liaison services by collecting credit card payments from customers and forwarding payment to the vendor without disclosing the credit card information. PayPal (owned by eBay) is currently the leader in this field but is facing competition from the newer Google Checkout service. Both Google Checkout and PayPal offer the ability for vendors to accept credit card payment online, software to set up the shopping cart page, and assistance with managing shipping.

For handling e-commerce on the web site, it is necessary to follow these recommendations:

1. Provide thorough product descriptions, including graphics.
2. Prominently display the product name and price. If several formats are available, clearly identify the format.
3. Make it easy for the customer to purchase; the fewer clicks, the better. The "buy" button should be obvious.
4. Make sure the customer knows when the order is completed.
5. Once the order is placed, send an immediate email confirmation.

6. Make sure the orders go out as quickly as possible.

7. Make sure the customer knows how long it will take to receive the order.

8. Make sure you have the inventory to fill the orders. If you are out of stock, modify the storefront page immediately to reflect this fact.

The alternative to the do-it-yourself arrangement is to provide links to one of the many online retail stores. For the music business, this includes CDBaby, SNOCAP, AmieStreet, TuneCore, MySpace Music, and Amazon .com. Use of these services does mean giving up a percentage of each product sold but requires little effort on the part of the webmaster. For selling merchandise such as bumper stickers, hats, and T-shirts, www.cafepress.com and www.zazzle.com offer low-cost fulfillment services linked directly from your web site. Links to these e-tailers should direct the shopper to the exact product page and not the homepage of the retail company.

Web Site Promotion on the Site

There are ways to promote a web site and entice visitors to spend more time and visit the site more often. Once visitors have found your site, the most obvious way to have people return to your site is to offer them something of value. This can be as simple as information: tour information, release dates, general industry information, or great links. You can add value to the site by offering such things as screen savers, shareware programs, electronic greeting cards with appealing photographs, jokes, recipes, links, and active message boards.

Contests also lure visitors to return to the site, especially if new contests are offered each month. Successful contests are set up to encourage the visitor to return to or further explore the web site. One example would be to include trivia questions with answers that are easily found on other pages within the web site. It is not wise to offer the main product as contest prizes. If 200 visitors sign up for the contest, they may hesitate on purchasing the product in the hope that they may be one of the winners. It is advisable to offer some other product as prizes.

On the homepage, be sure to remind visitors to bookmark the site so they can easily find it again. In 2002, 46% of web site traffic came through bookmarks and direct navigation.

Tell a Friend

The "tell-a-friend" or "recommend-it" function is a form of viral marketing. **Viral marketing** is "any strategy that encourages people to pass along

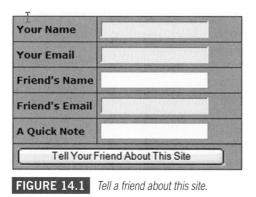

FIGURE 14.1 *Tell a friend about this site.*

Email this page to a friend

Your name:

Your e-mail:

Friend's e-mail:

E-mail link only

Comments:

Send a copy!

FIGURE 14.2 *Email this page to a friend.*

a marketing message to others, creating the potential for exponential growth in the message's exposure and influence" (Wilson, 2000). Visitors to the web site can be encouraged to spread the word and tell others. This is made easier by providing them with a feature to forward the web address to their friends (see Figure 14.1).

This type of marketing is very successful because word of mouth is one of the most credible forms of promotion. Visitors to the site may not be personally interested in what the site has to offer, but they may know someone who is. This makes it very successful in targeting the desired market. This feature can also be used to forward online newsletters, photos, and specific web pages through the "email to a friend" feature that usually contains a subject line such as "(your friend) thought you might be interested in this."

Posting pictures of fans (especially at live performances) is a good way to increase traffic. Those fans that appear in the photos will have an incentive to direct their friends to the site.

Visitor Registration

The most valuable piece of information that can be obtained from consumers who visit the web site is their email address and permission to add them to the "list." A list of email addresses, and permission to contact them, is necessary in this new age of "spamming." Spamming is the activity of sending out unsolicited commercial emails. It is the online equivalent of telemarketing (more on spamming later). An effective up-to-date email list is a valuable marketing tool, and allows for e-newsletters to be sent to fans who have shown enough interest to sign up. Web sites should post their privacy policies to avoid any confusion or legal complications if visitors end up on a mailing list.

When recruiting visitors to sign up, it is more attractive to present this as either a guest book or free membership to the artist's fan club, rather than as just signing up to receive emails. Usually, web sites include a visitor sign-in, registration, or "join now" button leading to a short online form.

When visitors are asked to sign a guest book or register to enter the site, there are two strategies for adding these visitors to the email list: "**opt in**" and "**opt out**." With opt in, the visitor selects a blank check box to be added to the email list. With opt out, the button default is in the checked position and the visitor must uncheck the box to avoid being included in the email list.

Promoting the Web Site on the Internet

Building a web site is not enough. With all the clutter on the Internet, it is necessary to promote the web site by reaching out to web surfers and encouraging them to visit it. It is a commonly held belief among Internet marketers that most of the web site traffic will come from word of mouth, or "word of mouse" in this case.

FIGURE 14.3 *Sauce Boss guest book (Courtesy of Bill Wharton).*

Table 14.1	How Internet users find out about new music (Source: Pew Internet Project: The Internet and Consumer Choice, 2008).		
How Internet users find out about new music	**Ages 18–35**	**Ages 36–50**	**Age 51+**
Going to the web site of an artist, band, or record label	41%	38%	30%
Listening to free streaming samples of songs online	46	25	21
Visiting an online store that sells music	42	32	26
Downloading music files to your computer	42	24	14
Listening to an Internet radio station	29	25	21
Reading online reviews or blogs about songs and artists	28	22	21
Watching music videos online	34	21	16
Going to a MySpace profile of an artist, band, or record label	31	13	8
Receiving an email from a band, artist, or record company	15	9	8
Median number of online music-seeking activities	3	2	1
Number of cases (Internet users)	133	171	190

Search Engines

Once a web site is completed, the URL is submitted to the various **search engines** that scour the Internet when users type in search terms. Search engines can be of some assistance if used correctly, but with so many web sites now on the Internet, it is becoming harder to stand out from the crowd. According to The Pew Center for Research (2008), over 90% of Internet users engage in search engine activities—tied with email for the most popular Internet activity. The most popular search engines on the web are:

1. Google 72%
2. Yahoo! 21.62%
3. MSN Search 9.8%
4. Ask 3.7%

Source: (HitWise, 2008)

Search engines vary in how they find and list sites that match the subjects or search terms entered in by the web user. Some search engines automatically evaluate **HyperText Markup Language (HTML)** documents, looking for particular information used to describe the site (**meta tags**), while others use humans who scour through applications to determine how to rank and categorize submissions.

To ensure adequate listing on these search engines, webmasters will typically submit the web site to the search engine either directly or through services that offer a directory listing. To get listed in Google, click on "About Google." Then, click on "Submit your content to Google." Then, click on "Add your URL." All submissions are evaluated by editors, making this a human-driven search engine. Other search engines use "spiders," or "**bots**." These are automated software programs that constantly roam the Internet, cataloging new web pages they find. They search the site to determine how to categorize information that was submitted to the search engine. Savvy web designers use meta tags to help the spider program define the site and which pages to list.

Search Engine Optimization and Meta Tags

The most common tactics to increase web site visibility involve search engine optimization (SEO). Search engine optimization is the practice of guiding the development or revamping of a web site so that it will naturally attract visitors by gaining top ranking on the major search engines for selected search terms and phrases. You can improve site relevance, which helps determine how prominent your site is in the search engine results, by including a few extras in the site. The goal is to imagine what keywords your customers are likely to use when looking for sites such as yours. There are three aspects of a web site that determine search engine ranking: keywords, content, and inbound links. Search engines look for keywords when sorting and ranking sites. This includes the keywords found in meta tags (see the following paragraph), but also text that appears in the first couple paragraphs of the page to catalog "content-rich" sites. Another characteristic that search engines use to rank keyword results is link popularity: how many other sites think your site is important enough to link to it. The quantity (popularity) and the quality (relevance) of links to your site are used to determine ranking status. Link quality is defined as those links from other sites with high page rankings for relevant search terms. Search engines use this information because they go by the assumption that the most important and relevant sites will have lots of other sites linking to them.

Meta tags are author-generated HTML commands that are placed in the head section of an HTML document. These tags specify which search terms should be used to list the site on search engines. Popular meta tags can affect search engine rankings and are generally listed in the sections "Meta Keywords" and "Meta Description." A meta tag can be generated automatically by the site www.submitcorner.com. The following example tag will let the "bot" know to categorize this web site under "blues music," "slide guitar," and by the artist's other endeavor, "Louisiana-style cooking."

<META NAME="KEYWORDS" content="blues music hot sauce gumbo slide guitar Bill Wharton datil pepper habanero Liquid Summer recipe contest Sauce Boss Jimmy Buffett Parrothead">
<META NAME="DESCRIPTION" content="Hot Sauce and Scorching Slide—Florida Bluesman Bill Wharton makes Liquid Summer Hot Sauce and has made gumbo for over 80,000 people during his high energy concerts with his band—The Ingredients">
(www.sauceboss.com)

The web page title is also very significant and should reflect the nature of the site. The <TITLE> tag is the caption that appears on the title bar

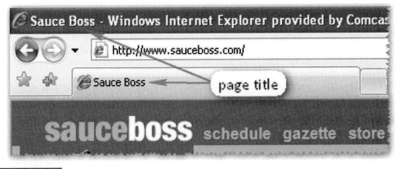

FIGURE 14.4 *Title tag on web browser.*

of your browser and is the name on the clickable link listed in the search engine results (example: <title>Sauce Boss</title>).

Music Directories

Standard **directories** are defined by the I-search Digest (www.led-digest.com, 2004) as "a server or a collection of servers dedicated to indexing Internet web pages and returning lists of pages which match particular queries. Directories (also known as *indexes*) are normally compiled manually, by user submission (such as at www.emap.com), and often involve an editorial selection and/or categorization process." Directories are arranged hierarchically, from general to specific. Because these directories are not compiled automatically by spiders, a site's webmaster must submit the site to the directory. Then their editors will visit the site to determine whether it is appropriate to list in the categories specified in the request.

Music directories are very helpful at directing web surfers to new sites of interest. Some of the more popular directories are: simplythebest.net, the ultimate band list (Internet Underground Music Archive [IUMA]), sonic switchblade.com, music crawl, Listen.com, and StarPolish. Others include:

- Music-Sites.net: Not much of a site but they do offer artist signups. www.Music-sites.net

- MusicXS: An organized alphabetical directory of Internet music links. www.musicdirectory.com

- Simply the Best Music: The SimplytheBest Music Artist database contains composers, groups, lyricists, musicians, songwriters, and vocalists who are looking for management, material, or contacts. The Music Industry database contains web sites of music agents, concert data, music directories, music education, fan clubs, record labels, instrument manufacturers, general music sites, music production, music promotion, music publishers, sheet music publishers, and more. www.simplythebest.net/music/

- Deep Look Music Directory: Deep Look Music Directory is a SEO friendly and human-edited music, art, and entertainment web directory. www.deeplookdirectory.com/music-industry/

- Local Music Directory: Search for bands and musicians worldwide. www.localmusicdirectory.com/

Banner Exchange/Link Exchange/Reciprocal Linking

Exchanging links or banners with another similar web site is a great way to expand traffic to a web site. Exchanges occur when two web sites agree to link to one another, sharing their visitors with the other site. A **banner** is described as a graphical web-advertising unit with a **hot link** to the advertiser's web site. **Banner exchange** occurs when two or more web sites agree to display banners promoting the partners' sites. There are banner exchange services available that help coordinate banner exchanges. Some experts advise against participating in banner exchanges because they may distract visitors and direct them away from the initial site before they have had a chance to explore. Nonetheless, trading links with other web sites that target the same market has been a proven success in increasing web site traffic. The advantages of trading links, according to Insite Media (www.insite-media.com, 2004) are:

1. They bring in a highly targeted audience because the visitor was on a similar web site.

2. Links from other sites create "link popularity" that improves positioning on search engines.

3. It is a more stable form of creating traffic because once these links are set up, it is unlikely that they will change or disappear. As the link partners increase traffic to their own sites, they send more traffic to your site.

Social Networking

Social networking sites are defined as "web sites that allow members to construct a public or semipublic profile and formally articulate their relationship to other users in a way that is visible to anyone who can access their file." According to Hitwise, a company that specializes in online web traffic and marketing, visits to social networking sites accounted for 6.5% of all Internet visits in February 2007. At that time, industry leader MySpace dominated the social networking industry with over 80% of all visits to social networking sites. Contender FaceBook held 10%. By the end of 2008, MySpace was holding 67% of the market, with FaceBook narrowing the gap at 20%. By January of 2009, FaceBook was gaining U.S. market share, with 31% of visits, as MySpace fell to 57% (Hitwise, 2008). While other sites, such as Black Planet, hold a very small share of the market, they may be uniquely targeted to an audience that may be appropriate for particular artists.

Social networking sites are increasingly providing music-based features, such as when MySpace teamed up with SNOCAP to offer digital music sales, and are now offering ecommerce services through MySpace Music. Social networking services are also increasingly offering **widgets**, making it easier to post information from one site to another. For example, band photos loaded up to slide.com can be posted into MySpace fields. When the photos are updated on Slide.com, they are automatically updated on MySpace.

| Number of acts on MySpace (2008) ||
Musical genre	Number of acts
Hip hop	2.5 million
Rap	2.4 million
Rock	1.8 million
R&B	1.6 million
Pop	723,000
Metal	611,000
Punk	468,000
Electronic	413,000
Techno	335,000
Reggae	314,000

Source: MySpace web site, December 2008
Note: Some acts may be registered under more than one category. Figures are rounded.

FIGURE 14.5 *Number of acts on MySpace. (IFPI)*

Web widgets are defined as a portable piece of code that an end user can install and execute within any separate HTML page.

With so many new social networking services cropping up, no one is certain what will be the "next big thing." An astute marketing manager learns to stay abreast of the changing trends in social networking and respond quickly and comprehensively to changes in the marketplace for social interaction. More information on using social networks can be found in Chapter 13 on grassroots marketing.

Viral marketing

Publicityadvisor.com describes viral marketing as a new buzzword for the oldest form of marketing in the world: referral, or word of mouth, which they update to "word of mouse." Viral marketing is any strategy that encourages individuals to pass on a marketing message to others, creating an ever-expanding nexus of Internet users "spreading the word." On the web, this is done through several strategies, including the "tell-a-friend" button mentioned earlier. Incentives can be added by entering each person who "tells a friend" into a contest. You can also add viral tell-a-friend buttons to many of the pages on your web site.

The most common form of viral marketing is through email signature attachments. Hotmail.com has successfully employed this technique by appending their message to the bottom of every email generated by its users. Hotmail offers free email service to users but also places a viral tag at the end of each message that its users generate. If that message is passed along, the advertising tag goes with it.

Ralph F. Wilson, in his article "Six Simple Principles of Viral Marketing," says there are six elements of viral marketing:

1. **Give away something of value**. Giveaways can attract attention. By giving something away up front, the marketer can hope to generate revenues from future transactions. For example, the tag can say "Win free music at www.yourartist.com."

2. **Provide effortless transfer to others**. "The medium that carries your marketing message must be easy to transfer and replicate."

3. **Expand exponentially**. Scalability must be built in so that one message transmitted to 10 people gets passed on to 10 more each, for 100 new messages.

4. **Exploit common motivations and behaviors**. Success relies on the basic urge to communicate and share experiences and knowledge with others.

5. **Utilize existing communication networks**. Learn to place viral messages into existing communications and the message will "rapidly multiply in its dispersion."

6. **Take advantage of others' resources**. A news release reprinted elsewhere will include the viral message and perhaps the link.

Newsgroups/Discussion Groups/Bulletin Board/Chat Rooms

The Internet is full of **discussion groups**, Usenet groups, and special interest web sites that foster the exchange of information and ideas among its members. These topic-oriented web sites usually provide chat rooms; bulletin boards for discussions; and, more recently, blogs that allow for user response and discussion. **Chat** rooms allow for real-time interaction between members, while bulletin boards allow individuals to post messages for others to read and respond to. Because these members or users have a mutual interest in the site topic, user groups are an excellent way to locate members of the target market. Music-related **Usenet** groups are often defined by a common interest in one artist or a genre of music. Internet marketers will often define the target market for their unknown artist by identifying artists with similar appeal. Then marketers will visit sites that cater to those fans.

These sites usually have specific rules about the posting of commercial messages and harvesting email addresses, which is universally frowned upon by group moderators and system operators. Many groups, however, welcome brief messages from industry insiders notifying interested members of new products. These messages often are embedded with hot links to the site actually promoting or selling the product. Some user groups require that you register before participating in their group discussions. Some visitors will engage in "**lurking**" behavior when first introduced to a new users group. Lurking involves observing quietly—invisibly watching and reading before actually participating and making oneself known. Lists of these groups can be found at http://groups.google.com/ and www.topica.com/.

Marketers have also been known to join in chat discussions. Again, targeting the right chat rooms in the right discussion group is important. In street marketing, marketing representatives who are part of the target market pose as fans of the artist and engage in discussions on the merits of the artist and his works (see section on fan-based promotions). Marketing through online chats and message boards is the online counterpart of street teams.

Fan Web Sites

It is not uncommon for fans of an artist to create unofficial web sites or web pages on their personal web site featuring their favorite artists. These sites can be used to spread the word. Most of these dedicated fans welcome receiving press releases and digital photos and will gladly display these on their sites. They usually are also willing to create a link to the artist's official web site. Sometimes these sites are difficult to find because they are not always listed with search engines, but many of the more popular ones can be located through the creative use of search engines or directories.

The marketing goal of the label is to provide these sites with accurate, up-to-date content about the artist. Marketers are now using RSS feeds to accomplish this. An acronym for real simple syndication, **RSS feeds** allow content to be developed in one place, or for one site, and then that content is "fed" to other sites. As the content is updated, it automatically updates that content at all sites that "subscribe" to the feed.

Some grassroots marketers encourage their street teams to create their own "fan" web sites. These members of the target market add information about the artist to their personal web site and generate word-of-mouth promotion from the fan's point of view.

Email Addresses and Spam

Email newsletters have become an effective way to stay in touch with the market on a regular basis. They are the cheapest and most effective way to generate repeat traffic to your web site. The email newsletter (sometimes erroneously called an *ezine* lets the marketer control which messages are sent out to fans, and how often. (**Ezines** are online versions of magazines, generally targeting special interests. There is no shortage of ezines about music and artists. More detailed information on how to submit materials to ezines can be found in Chapter 10 on publicity.)

In his article "7 Reasons to Start an Email Newsletter Today," Rich Brooks (n.d.) extolled the virtues of utilizing this marketing technique, for the following reasons:

1. It complements your web site.

2. It is more cost-effective than print newsletters.

3. It is interactive because it can include links to relevant sites and pages.

4. You can track its effectiveness by monitoring traffic to the web site.

5. It is viral—there is always the possibility that a fan will forward the newsletter to an interested friend.

6. You are preaching to the choir—those people who are already interested in your career.

7. It will help build your subscriber base.

Spam is defined as unsolicited commercial email messages—the equivalent of junk mail or telemarketers. Because of the negativity and problems associated with spam, successful marketers have adopted a code of conduct. The U.S. government and several foreign countries have passed legislation regulating or outlawing the act of spamming. Many special interest sites (discussion groups) strongly discourage "**harvesting**" (collecting from posted messages) email addresses to be used for spamming. The general rules for sending commercial emails are:

1. **Don't send unsolicited messages, period**. Instead, create email lists by encouraging web site visitors to leave their email address so they can receive valuable updates and information.

2. **Send the right messages to the right people**. Touring information is only important to fans in the area of the performances. Leave the comprehensive touring schedule to the web site, and use email for only those fans within driving distance to the venue.

3. **Only send emails when there is something new to report**. Make the message specific and make the best use of the subject field.

4. **Use the email as a teaser, providing enough information to entice the reader into visiting the web site for the rest of the story**. (Be sure to include a hot link or the URL address to your site in the email.) Have a great opening, use short sentences, and focus on the recipient's self-interest.

5. **Don't include everything**. You can link various parts of the email message to specific pages on your web site, so keep your message short.

6. **Don't send out large files full of graphics and attachments**. Instead, rely on the web site to provide the images.

7. **Use the blind courtesy copy (BCC) function**. If you can't send out individual emails, use the BCC function so that recipients cannot view, and thus harvest, other email addresses from your list.

Directing Customers to Online Retailers

The goal of an online newsletter is to either (1) get the fan to revisit the web site or (2) get the fan to purchase the new recording. The fan doesn't necessarily have to revisit the web site to purchase a new release. An effective newsletter announcing the new release should include a direct link to a merchandising web page, whether it is part of the artist's web site or the product page of an online retailer. The idea is to make it easy for the customer to buy or preorder. If the fan would prefer to bypass the artist web site and just purchase the new release, embedded links in the newsletter should facilitate this. Companies such as Amazon.com make it easy to link directly to the product page so the customer doesn't have to sift through storefront pages and menus to find the product. For example, the album *The Best of the Sauce Boss* is available at Amazon.com on the dedicated page at www.amazon.com/exec/obidos/tg/detail/-/B00004SW5Z/qid=1069273304/sr=1-1/ref=sr_1_1/102-6269284-5258519?v=glance&s=music. Instead of dumping customers at the "front door" of www.amazon.com and having them use the search feature, they are directed to the specific page for that release.

Some fans may prefer to shop at a reputable online retailer rather than purchase from the artist's web site. The fan may already have an account with a particular online retailer, making the transaction easier. Creating a link to each of these sites increases the likelihood that the shopper will choose her favorite retailer and complete the purchase. Links to online retailers should also be included on the artist's web site for the same reasons.

Offline retailers, or those with "clicks and mortar" stores are often included in the favorite links area of the artist's web site, thus supporting traditional retailers who often feel overlooked in the world of cyber-promotion. The ezine or newsletter should also mention that the album is available in stores, if indeed it is. Some fans just can't wait for a mail-order copy to arrive. This brief mention in the newsletter may send anxious fans to a physical record store. As mentioned earlier, effective marketing plans integrate offline and online marketing components.

Fan-Based Promotions

Fan-based, or "perceived" fan-based web promotions are very effective and contain the "word of mouth" or "street credibility" that other forms of marketing lack. Marketers have begun to exploit this technique by hiring "street

Table 14.2	Story of the Subway commercial

There was a commercial for Subway sandwiches that ran recently on television. In the commercial, a young man gets on a crowded bus and sees a friend on the bus who works at Subway and is wearing the official uniform. The young man asks his friend what's new at Subway. When the friend describes the new menu offerings, other riders on the bus become interested and head for the nearby Subway restaurant when the bus stops. The two young men then congratulate each other on the effectiveness of their pre-planned tactic. This scheme also is being used in chat rooms to promote artists. One co-conspirator may strike up an online conversation with an accomplice who then displays an interest in the information on the new artist or release. The accomplice will then begin asking questions designed to elicit more promotional information from the initiator. While the ethics of this process may be questionable, the effects are not.

teams," who pose as avid fans of the artist and engage in peer-to-peer promotions. While some of these tactics border on deception, marketers try to include street team members who actually are fans and part of the target market, thus increasing the credibility of their efforts. Nonetheless, these hired e-team members masquerade as fans, and engage in activities much like those of the avid fans they purport to be.

These hired fans typically will find discussion groups and join in the discussions. They will subtly introduce the artist they are promoting and coax participants into visiting an artist web site or previewing the artist's music. They may seek out members of a discussion group most likely to respond to the artist being promoted and send them personal instant messages and emails. They may also post messages on the group message board inviting members to check out a particular web site promoting the artist. Often, these messages take on a tone of fan discovery, such as, "Hey all, I've just discovered a great new artist and wanted to share her (him) with you." Online street teams should be coached or given talking points so that the appropriate messages will be disseminated.

Hired fans also write fan reviews for web sites that post reviews, such as Amazon.com. Often, artists will ask friends and family members to write a fan review for them and post it on Amazon.com or CD Baby, or participate in the "rate this item" feature. It is not considered unethical for an artist to send out an appeal in the newsletter for fans to log on and rate their music or vote for them on some other web site.

Integrating Offline and Online Promotions

To reiterate the theme at the beginning of this chapter, successful marketing plans are those that integrate online marketing with offline promotions.

The two strategies working together create synergy. This starts by incorporating the artist's web address into every aspect of offline marketing. There are three main areas of offline promotion that should be exploited for creating online traffic:

1. Concerts, live performances
2. The press
3. Retail products

Concerts

The artist's URL should be prominently displayed at all live performances. Announcements from the stage and handouts such as decals, bumper stickers, buttons, and postcards (and even temporary tattoos) can help remind fans of the web address. However, promotion of the web site extends beyond merely providing the URL. Concertgoers should be given a reason to visit the site, including contests, photos and music from the performance, product availability, and updates on the next local event.

In addition to supplying fans with the URL, email addresses and perhaps cell phone numbers should be collected from fans at the venue to add to the e-newsletter mailing list. Fans can be encouraged to sign up through contests and the opportunity to be informed of upcoming events and new releases.

The Press

The press kit should contain numerous references to the web site, and the fact that additional materials and updates are available there. All press releases should contain the URL. A special section of the web site can be dedicated to supplying materials normally used by the media to cover the artist, such as print-quality promotional photos, artist history, discography, music samples, and so forth. The press should also be encouraged to include the artist's web address in all articles published.

Retail Products

All music-related merchandise should reference the artist's web site, perhaps encouraging visitation by offering value-added features such as bonus tracks or remixes available only to buyers. The interior liner notes can contain key information for accessing restricted portions of the web site. Again, contests can be geared toward fans that have purchased the music. Fans can also be encouraged to leave feedback or complete a survey. Retail coupons can be offered with arrangements already made for redemption at brick-and-mortar retailers.

MOBILE MEDIA

The wireless handset that we used to refer to as a cell phone is now morphing into a multimedia personal communication device and continues to grow in popularity around the world. Beginning with the third-generation (3G) handsets a few years ago, mobile data networks and increasingly sophisticated handsets have been providing users with a variety of new media offerings. Whereas the Internet and desktop computers have dominated the paradigm shift in the past, the future belongs to wireless mobile technology. Already, the growth of cell phone adoption outstrips Internet adoption. In late 2008, it was estimated that there were over 4.4 billion cell phones in operation globally.

The Pew Internet and American Life Project reported in early 2008 that over 75% of Americans use either a cell phone or a personal digital assistant (PDA). And 62% of all Americans have some experience with mobile access to digital data of some form, whether it is text messaging or music downloading. The most popular alternate use for cell phones is **text messaging**, with 58% of cell phone users stating they have used their device for text messaging. Equally, 58% have used their cell phones to take a picture, whereas only 17% have used their handheld device to play music and only 10% to watch a video. Of interest to the Pew researchers, nonwhites, specifically African Americans and English-speaking Hispanics, were more likely to engage in media activities other than voice phone service than their white counterparts.

Mobile Internet penetration is lower in the United States than many European and Asian countries, with an average of 29% of European Internet users accessing the web on mobile devices, compared to 19% in the United States. The introduction of the iPhone and its competitors in the mid-2000s ushered in a new age of **cloud computing**, mobile web surfing, and mobile marketing.

Mobile Music Sales

Until 2007, the majority of music sales for mobile devices took the form of *mastertones*—short excerpts from an original sound recording that play when a phone rings. The market for ringtones, ringtunes, and the like developed throughout the first half of the decade but saw a decline in sales for the first time in 2007 as consumers moved away from phone personalization features and began to adopt full track downloads to mobile devices (IFPI, 2009). Ringtones accounted for 62% of the **mobile music** market in 2007 but are predicted to account for only 38% of the market by 2012. Ringtone sales fell

FIGURE 14.6 *Sony Ericsson 3G handset with with PlayNow features. (Courtesy Sony Ericsson)*

by 33% in 2008, to 43.8 million units in the United States (SoundScan, 2009). Columnist Colin Gibbs (2008) compares ringtones to the "aural equivalent of bumper stickers" announcing the consumer's message to anyone in earshot, while full-length tracks are to be "consumed."

The global mobile music market is expected to rise to more than $17.5 billion by 2012 according to a Juniper Research study released in early 2008 (Mobile Music Adoption, 2008). This will be driven by subscription music services and full-track downloads. The mobile music market is expected to be more successful in the area of subscription services, unlike the PC-based network, which has seen disappointing numbers for music subscription services. Several European and Asian markets are ahead of the United States in mobile music sales. In 2009, Japan was the leader in mobile music purchases, with 140 million mobile singles sold in 2008.

Mobile service providers and handset manufacturers are introducing value-added features to cell phone subscribers that include music subscription services. Nokia introduced the "Comes with Music" program in 2008–2009, offering a 1-year music subscription service with the purchase of handsets, and the ability to retain music and purchase additional music at the end of the year. Sony Ericsson's PlayNow plus "allows users to download, play and recommend music wherever they are and whenever they like, directly over the mobile network" (IFPI, 2009).

Marketing to Mobile

The emerging area of mobile music creates new challenges for marketing music to mobile devices. The mobile industry continues to test unique marketing ideas to get music consumers more involved with using mobile devices for music discovery and consumption.

Advertising via Cell Phones

Although currently the majority of advertising on cell phones takes place through *text messaging*, marketers are beginning to look at other options including voice message and video ads. More information on mobile advertising can be found in Chapter 11 on advertising.

Voice Messaging

Some artists are beginning to use voice marketing messages to communicate with fans who have opted to receive such messages on their mobile devices. Marketing company Echo Music developed a campaign for an artist that included sending an audio advertising of country artist Dierks

Bentley's voice telling fans about the release of a new album and playing a music sample from the album. Bentley's voice kicked off the message stating, "Hey, dudes. Today's my big day. I'm excited for ya'll to hear it," followed by a snippet from the album. Mobile marketing company Mozes implements voice message marketing plans for major artists. Cameo Carlson (2007), Senior Vice President of Digital Business for Universal Motown Recording Group stated "Mozes' voice message service lets our artists and fans have a two-way conversation that is personal, fresh and real." More than 1 million voice message minutes—personalized voice messages between bands and their fans—crossed the airwaves in the first 6 months after Mozes added voice messaging as a service in October 2007.

Push/Pull and Viral Marketing via Text Messaging

Nielsen reported in late 2008 that their research has shown that over 60% of mobile subscribers use text messaging on a regular basis and that texting is now more prevalent than voice calls. As a result of the popularity of text messaging, cell phone marketers have begun using push messages to consumers' cell phones to market products. A push message is a specially formatted **short message service (SMS)** text message that gives users the option of connecting directly to a particular wireless application protocol (WAP) (mobile Internet) site, using their phones' browser. In other words, the text message has a link; by clicking on the link, the user's cell phone is directed to a special mobile-friendly web site. The majority of mobile campaigns today use SMS messaging. SMS messages are limited to text only and to 160 characters. Despite its limitations, it has proven to be highly effective in generating high response rates.

A pull marketing strategy involves using some other form of media to request that users send a text message in response to the advertiser's call to action. An example of pull marketing would be the *American Idol* TV show that encourages viewers to "text in" their votes.

Viral Marketing

Viral marketing campaigns are new to cell phones and incorporate the notion of recipients passing along the message to others. In the blog "How to Make Viral Marketing Appealing for Phones," JayVee (2006) mentioned three factors for successful viral campaigns on cell phones:

1. **Offer exclusive content** – This should be something that is not available on the Internet but only via cell phones. It could include ringtones, free music files, wallpaper, graphics, and so on.

2. **Make it useful and timely** – For musicians, something centered on a live concert date would be effective, offering something of value related to the concert or the artist.

3. **Clearly define objectives** – The goals should be simple and straightforward, create awareness for the product or service, encourage recipients to pass along the message, and elicit a call to action (log on, download, etc.).

Marketer David Bohan (2009) offers up some examples of mobile marketing campaigns. Paramount Pictures partnered with shopping centers to place interactive touch screens equipped with Bluetooth hot spots to offer free movie previews, screen savers, and coupons to mobile devices. Kroger grocery chain has worked with eight food companies to offer manufacturers' coupons via mobile devices to be used in-store.

Presently, much of the marketing surrounding mobile phone couponing is still tied to Internet access, but **Bluetooth** and other proximity transmission systems will soon be delivering marketing content to willing consumers directly on their mobile devices, including bar code images that can be scanned at the point of sale.

Preparing Your Web Site for Mobi

Cell phone access to the Internet is bringing new challenges to web design. Mobile handsets rely on the **wireless application protocol** (**WAP**) to create web sites specifically formatted for mobile devices. The domain system used for wireless devices is the *.mobi* (top level) domain extension on the end of the URL address.

Just as with an Internet web site, design for a mobi site must be preceded with an understanding of goals for the site. What are you trying to accomplish and provide with a mobile web site? The goals may not be the same as for the computer-based site. An Internet site might focus on more detail, more long-term information, more entertainment, and on creating branding and customer awareness. A mobile site will most likely be used by fans "on the go" who need quick access to information that they can use immediately (sometimes referred to as "just-in-time" information). This may include information on live shows, such as directions to venues, start times, and seating availability. A brief tour schedule may be important if groups of fans are together and making plans to attend an event at a future date. A quick glance at your tour schedule may be what they are looking for. Contests and coupons may be important for fans, as well as access to music downloads. Pull messages that appear on billboards, bus boards, or in newspapers may create an immediate impulse among viewers to respond to the message via cell phone, such as signing up to win something, voting, or downloading new content. With these goals in mind, a separate web site must be developed that allows mobile fans to quickly access what they are looking for without scrolling through a lot of pages and screens. And the site must be updated often to reflect the changing needs of its visitors.

General WAP Formatting Rules

Creating mobile web sites may be as simple as stripping out all standard web formatting to reveal a text-only site similar to the way a search engine spider renders a site, although much of the content must also be reduced for easier browsing. Or better yet, professional "mobi" developers can create a site that dynamically produces formatting from a variety of options, depending on the type of device accessing the site. In other words, the formatting of the site can vary, and for each visitor, the presentation is customized after the server detects what type of handset the visitor is currently using.

Here are some basic rules to consider:

FIGURE 14.7 *Nokia phone with mobi web site for Rihanna. (Courtesy of Nokia)*

- **Be realistic about what will fit on the small screen of a mobile device**. Paring down your information for the screen size is first and foremost.

- **Eliminate all information that is not important to the mobile user**. Less is more on a mobile site. Visitors are usually looking for specific information or completing a specific task.

- **Avoid vertical scrolling**. Web users don't like it, and it's even less popular with mobile users.

- **Reduce the number of clicks**. Don't go deep in page numbers; it's slow going for the mobile user.

- **Keep it clean**. Use readable text on a readable background.

- **Test your design on a variety of handheld devices** to see if it holds up on all of them.

- **Use abbreviations and succinct wording wherever possible**.

■ **Access keys are helpful for speeding up navigation**. (Access keys are those just below the screen on the left and right that can take on a variety of functions, usually with the name of that function on the bottom of the screen just above the key.)

CONCLUSION

In today's marketing climate, new media marketing is a unique area of growth and potential. Web marketing is a cost-effective way to promote music and artists. It offers the ability to provide fans with music samples and artist information. It has the potential to generate new fans, create a sense of community, and transcend geographical boundaries. The Internet is a vast resource. The goals should be to carve out a niche, identify a market, and cultivate current and potential consumers in that market.

The web site serves as the home base for both online and offline promotions. Fans should always know where to look to find updates on their favorite artists. The web site is a dynamic marketing tool and should not be left unattended. While the ultimate marketing goal is to increase the sales base for recorded music and concert tickets, the Internet can be also useful for creating brand awareness and nurturing existing customers.

Mobile media is rapidly becoming a dominant platform for media consumption, especially music and communication. The proliferation of handsets and wireless networks, combined with the evolution of services and software, is bringing about a paradigm change in the way media is purchased, consumed, marketed, and monetized.

GLOSSARY

bandwidth—The amount of data that can be transmitted in a fixed amount of time. Bandwidth is usually expressed in bits per second (bps), or bytes per second.

banner—A graphical web-advertising unit with a hot link to the advertiser's web site.

banner exchange—Two or more web sites agree to display banner advertisements promoting the partner's sites. See also *reciprocal linking*.

blog—A "weblog," or journal that is available on the web, typically updated daily or weekly. Postings are usually arranged in chronological order.

Bluetooth—A low-cost, short-range, low power–consuming wireless networking technology especially popular for communicating between mobile devices and/or accessories.

bots—Automated software programs that constantly roam the Internet, cataloging new web pages they find.

branding—Creating a distinct personality for a product.

chat—An online discussion in real time in which two or more people engage in a written conversation.

cloud computing—Storing files and using software programs from a remote server that can be accessed from anywhere, rather than storing them on the user's personal computer.

cookie—The name for files stored on a hard drive by the web browser that hold information about a user's browsing habits, such as what sites have been visited, which newsgroups have been read, etc.

directories—A server dedicated to indexing Internet web pages and returning lists of pages and links that match particular queries.

discussion groups—A group of Internet users who meet online in a virtual meeting place to discuss a topic of mutual interest. They can use message boards or real-time chats.

domain name—The name given to a host computer or site on the Internet.

e-store—A virtual online store capable of taking orders, collecting money, and shipping product. Also called *e-tailers*.

ezine—An electronic magazine or email organized in a magazine format and sent via email to interested Internet users or made available on a web site.

harvesting—Collecting email addresses by visiting user groups and copying email addresses from their message boards or from distributed email blasts where the addresses are visible.

homepage—The document that is first accessed when visiting a web site, also known as index page.

host provider—The physical place where your web site resides, generally server space rented from an Internet service provider.

hot links—A highlighted or underlined word or a graphic that is linked to another page or web site. The user clicks on that word or image to jump to the new location.

HyperText Markup Language (HTML)—The language used to create links, format documents, and communicate with web browsers to emulate the look intended by the creator.

lurking—The act of observing or viewing a chat room, newsgroup, Usenet, or forum without contributing.

mastertones—Short excerpts from an original sound recording that play when a phone rings.

message boards—Section on some web sites where users post public messages and respond to each others' postings.

meta tag—Author-generated HTML commands that are placed in the head section of an HTML document. These identification tags help search engines identify the content of the document.

mobile music—Music that is downloaded to mobile phones and played by mobile phones. Although many phones play music as ringtones, true "music phones" generally allow users to import audio files from their PCs or download them wirelessly from a content provider.

navigation (nav bar)—The set of buttons on a web site with hot links that, when clicked on, take you to other sections of the site. The nav bar is usually visible on every page and allows the user to jump to other sections of the web site by using the bar.

opt-in/opt-out—The act of explicitly requesting an email distribution by checking a box. For example, when a user signs up to receive an email newsletter, this person is "opting in" to receive it. Opt-out occurs when the check box is automatically checked as the default position. The user must then "unselect" the box to avoid being added to the email list.

push message—A specially formatted cell phone text message that gives you the option of connecting directly to a particular mobile Internet site, using your phone's browser.

reciprocal linking—Trading links with web sites that target the same audience as your own. See also *banner exchange*.

RSS feed—Short for "real simple syndication," this is a form of content distribution in which content is generated at one web site and then is "fed" to and appears on other sites that subscribe to the feed. As the content creator changes out the content, it is automatically updated on all subscriber sites.

search engine—A program located on a web site that acts as a library card catalog for the Internet. Google is an example of a search engine.

secure sockets layer (SSL)—A method of encrypting data as they are transferred between a browser and Internet server. Commonly used to encrypt credit card information for online payments.

short message service (SMS)—Usually refers to wireless alphanumeric text messages sent to a cell phone.

social networking—The practice of interacting with and expanding the number of one's business or social contacts by making connections, typically through social networking web sites such as MySpace and FaceBook.

spamming—The activity of sending out unsolicited commercial emails, the online equivalent of telemarketing or junk mail.

splash page—An introductory first page or front page that is seen on some web sites, usually containing a click-through logo or message, or a fancy graphic.

text messaging—Also called SMS (short message service), it allows short text messages to be sent and received on a mobile phone. Messages can be sent from one phone to another by addressing the message to the recipient's phone number.

Uniform Resource Locator (URL)—An address that identifies the location of any type of Internet resource.

Usenet—The collection of the thousands of bulletin boards residing on the Internet.

viral marketing—Any strategy that encourages people to pass along a marketing message to others, creating the potential for exponential growth in the message's exposure and influence.

wireless application protocol (WAP)—A protocol designed for advanced wireless devices; it allows the easy transmission of data signals, particularly Internet content, to microbrowsers built in to the device's software.

Whois—An Internet utility program that obtains information (such as owner and contact info) about a domain name or IP number from the database of a domain name registry.

widget—A small application that can be ported to and run on different web pages by a simple modification of the web page's HTML.

BIBLIOGRAPHY

Apple. (2008). Press release. iTunes store top music retailer in the US. www.apple.com/pr/library/2008/04/03itunes.html.

Bohan, D. (2009). Don't forget to include mobile marketing www.tennessean.com/article/20090111/BUSINESS01/901110345 (no longer available). The *Tennessean*, Jan. 11.

Brooks, R. (n.d.) 7 reasons to start an email newsletter today. *Flyte New Media*. www.flyte.biz/resources/articles/0303.php.

Carlson, C. (2007). Moses press release. www.mozes.com.

clienthelpdesk.com. (2004). The client help desk (no longer available). www.clienthelpdesk.com.

Gibbs, C. (2008). Finding the marriage of mobile and music www.rcrwireless.com. *RCR Wireless*, Dec. 31.

"Google Nears 72 Percent of U.S. Searches in October 2008." (2008). HitWise. Oct. www.hitwise.com/press-center/hitwiseHS2004/google-searches-nov-08.php.

IFPI Digital Music Report. (2009). www.ifpi.org.

Insite Media. (2004). www.insite-media.com.

I-search Digest. (2004). www.led-digest.com.

Jayvee. (2006). How to make viral marketing appealing for phones. Nov. 28. www.cellphone9.com/how-to-make-viral-marketing-appealing-for-phones.

"JupiterResearch says that one-quarter of the world's population will be online by 2021, totaling 1.8 billion users." (2008). Jupiter Research. Press release, June 16. www.jupiterresearch.com/bin/item.pl/press:press_release/2008/id=08.06.23.economies-will-power%20.html/.

Kymin, J. (n.d.) Basics of web design. http://webdesign.About.com/od/webdesigntutorials/a/aa070504.htm.

Mobile music adoption passes tipping point with revenues set to reach $17.5bn by 2012, whilst ringtones revenues begin rapid decline, according to Juniper Research. (2008). Juniper Research. Press release, Feb. 26. http://juniperresearch.com/shop/viewpressrelease.php?pr=80.

Nielsen SoundScan. (2009). Nielsen.Soundscan.com

Number one real-time intelligence web analytics. (2004). OneStat.com. www.onestat.com.

NPD Group. (2008). The NPD Group: Consumers acquired more music in 2007, but spent less. www.npd.com/press/releases/press_080226a.html.

PC Magazine online. (2008). www.pcmag.com.

Pew Internet and American Life Project. (2008). www.pewinternet.org.

Web Site Design. (2004). www.drtandem.com.

Wilson, R. (2000). The six simple principles of viral marketing Feb. 1. *Web Marketing Today*, 70.

CHAPTER 15

Music Videos

Paul Allen

The value of videos as tools of the music industry is easy to see with the note from promotion company AristoMedia that Americans watch more than 15 billion videos each month, and the number of online video viewers approaches 170 million. Whether they are promotional or entertainment—or both—video usage is becoming a larger piece of the average day for many people.

VIDEO PRODUCTION AND PROMOTIONS

Music videos can play a key part in the success of the well-crafted marketing plan of a record label. They create another dimension to the song and its lyrics by adding a visual element to the musical performance of the artist. It creates strong images to pique the passion of the video viewers and gets them more involved in the music. Ultimately, the marketer's objective is to direct that passion into the need to possess the music in the form of a purchased CD or a download from a music retailer of both the music and the video.

From the point of the record label, the four functions of a music video are:

- Promotion and publicity for recordings and other aspects of the artist's career
- Licensed content for online and wireless entertainment providers
- A product offered for sale
- Programming for video channels

This chapter is dedicated to gaining a better understanding of music videos, how they fit into a marketing plan, and how they are used strategically to stimulate retail sales of a recording, ticketed performances, merchandise sales, and the sale of ancillary digital products such as ringtones.

HISTORY OF THE MUSIC VIDEO

The earliest form of video promotion of music began in 1890. For the next 25 years, **"illustrated songs"** became the rage of thousands of large and small theaters across America, and were credited with selling millions of copies of sheet music. In these theaters, vocalists accompanied by bands or small orchestras would perform songs, while hand-painted glass slides were sequenced and projected on a screen, depicting the story of the lyrics. Often, a vocalist would then lead the audiences in singing the song as the lyrics were projected (PBSKids, 2004).

When radio was introduced in the 1920s, music promotion changed. One of the earliest radio deejays was an announcer named Al Jarvis, who worked in the 1930s at a radio station in Los Angeles. He created a radio show featuring recorded music mixed with chatter. It was also Al Jarvis who went on television in the early 1950s with a program that featured recordings, guests, and information, which made him one of the original **video deejays**. His program was replicated in numerous other television markets and became a staple of many major broadcast companies. By the mid-1950s, video deejay programs were prominent in all top 50 U.S. media markets.

Some of these early television shows featured what was known as a soundie. A **soundie** was a video creation by the Mills Novelty Company that featured performances by bands and was used in coin-operated video machines. In the early 1940s, Mills Novelty created over 2000 promotional soundies. In the early '50s, producer Louis Snader created 750 "**visual records**," and Screen Gems and United Artists began creating their versions of early music videos.

The granddaddy of video deejay television shows was *Bandstand*, which began airing in Philadelphia in 1952 with host Bob Horn. In 1956, Dick Clark replaced Horn, and the show soon became *American Bandstand* on the ABC Television Network.

Television in the 1960s and early '70s featured many shows that included musical performances, but it wasn't until the mid-'70s that artists like Queen and Rod Stewart created videos designed specifically to assist in the promotion of their recorded music projects. But the link of the music video to record promotion caught fire when cable television began to flourish in the early 1980s. Among the first to begin using music videos was a program called *Night Flight* on the USA Network. Soon, scores of cable channels and hundreds of regional and local shows began to regularly feature music videos as part of their entertainment programming (McCourt and Zuberi, 2009).

The use of music videos by cable channels has declined considerably because of on-demand music video capabilities provided by online sources MySpace, YouTube, AOL, Yahoo! and others. Cable programmers realize they are unable to compete with video users who are unwilling to wait for a video to be played on television, because music fans have numerous choices to find any video when they want it and where they want it. Music videos as programmed entertainment are also becoming increasingly irrelevant as high-definition television sets also serve as monitors for computers, which easily create a high-quality visual experience with the music. With the growth of **broadband** and its penetration into more than half of the homes in the United States, the Internet has become a preferred choice for video entertainment for many music fans.

THE MUSIC VIDEO TODAY

Music videos are their own kind of entertainment and are valued far beyond their value of promoting recordings and an artist's career. Unlike in the '80s and '90s, when video cable channels were given free and unlimited use of music videos for their programming, music videos are now licensed content to those who want to use them. Labels that create videos license their use to sites that feature music video entertainment, such as YouTube, Yahoo! MTV, MSN, and AOL. The license is granted in exchange for payments based on the size of the audience that views the videos, or it is based on the advertising that the web site earns from featuring the video. Depending on the online video service, advertising rates associated with music videos generate $3 to $8 for every thousand times the video's page is loaded onto someone's browser (Bruno, 2009).

Video licenses issued by the major record labels are estimated to be worth over $300 million annually, and licensing is one of the fastest growing segments of earnings by the labels. For example, Universal says its earnings in 2008 from the sale and lease of music videos was nearly $100 million, representing an 80% increase from the previous year (Chmielewski, 2008). Artists who are now signed under multiple rights recording contracts earn up to 15% of the net profits generated from the sale and licensing of music video; earlier contracts shared net p`rofits at a rate of 50–50, but earnings potential from music video sales prior to 2004 were negligible.

Nearly every social networking web site provides the opportunity for members to define themselves by the kind of music they enjoy, which has helped create a culture that defines individuals far beyond the answer to the question

"What's your sign?" YouTube and the links to its videos underscore the impact of music video and how it reaches potential buyers of the video, tickets, and related products: 6 of the top 10 videos on the site are music videos, and the site will have shared over 10 billion music video views by the time you read this book. And the growth continues—estimates are that YouTube generates 15 hours of new video content ... every minute (Miller, 2009)!

DECIDING TO PRODUCE THE MUSIC VIDEO

Investing in a music video means that the label views it as an important promotional piece for recordings and for the career of the artist who is under a **multiple rights contract**. It is a very strong promotional device designed to promote the label's interests in an artist, but it also places the image of the artist in front of concert ticket buyers too. There are instances where the music and its images are inseparable, and a video is a necessity. At other times—for whatever reason—the artist may not be camera friendly, and the decision to create a video will include images that do not feature or focus on the image of the artist.

The general kinds of music videos created can be classified as shown in Table 15.1.

Table 15.1	Kinds of music videos (Macy, 2009).
Kinds of music videos	
Scratch video	These videos use coordinated film or digital images that are synchronized with the tempo of the music.
Performance video	This is a concert view of the artist performing and singing.
Abstract video	This video style incorporates surrealistic special effects and sometimes bizarre images.
Storyline video	This video follows the storyline of the song lyrics.

From the record company's perspective, a number of considerations come into the decision matrix to determine whether a video makes business sense for a specific recorded music project. These considerations are detailed in Table 15.2.

Producing the Music Video

Creating a music video involves all of the standard financial considerations that are applied to a label's recorded music project. But, it comes down to an answer to the question: Will the video earn us more than it will cost to create?

Table 15.2	Considerations for making a music video
Is there a story that a video can tell?	Creatives services, the song, producer, the publicist, the artist, the label brand manager, and marketing may all help determine if the song lyrics and the artist can enhance the impact of the recording by creating a video.
Is there a budget available for the video?	The budget needs to be adequate to make the video competitive with those associated with artists of a similar stature.
How much of the production cost is the artist expected to pay?	Traditional artist recording contracts typically require the artist to share costs up to $100,000 and to pay all costs the exceed that amount.
What are the commercial exploitation opportunities for the video?	The label considers income the video will be able to create for the labels as content and as paid downloads.

Creating a budget that stays within available funds is the first step in creating the music video. The criteria for the quality of the production are determined by looking at music videos that fans are viewing. Any music video that the label produces must at least be comparable to production elements used by those that make video for competing artists. We have heard about very low-budget music videos receiving millions of views, but instances like that are anomalies; the label must create a music video that meets the level of creative and technical quality that matches or exceeds those currently being viewed by fans. "Creative" *can* be inexpensive provided that the elements of the video production connect with the target fan of the video. For example, encouraging fans to assemble their own music videos with images and video clips from their wireless devices can energize the community around an artist. When the inclusion of a music video becomes part of the promotional plan for a recorded music project, the label will typically request proposals from video production houses for storyboards and ask these companies to include their suggested concepts for the video. Replies by video companies will also include the budget necessary to produce the video, based on the suggested concept. The label either accepts the idea and its budget, or it may negotiate for a larger or smaller production and accompanying budget.

The Wall Street world of financial accountability has pushed some record companies toward a more conservative approach to music videos by delaying the production decision until it is clear that the single is finding airplay chart success at radio. In the past, this strategy carried risks because music video production required 6 to 8 weeks to take the video from concept to completion. By that time, it may be too late to get a music video promoted to video channels to help the single. This is especially true with rock and urban stations, where time on an airplay chart can be fewer than 10 weeks.

However, it is possible today to request proposals for concepts and then keep the proposals on file until the single achieves chart success. If the video is required, video production companies use current technology to provide rough cuts of the produced video in a week, and have the final version ready for posting on the web and for use in general promotion in less than 2 weeks.

Budgets for music videos produced by labels rarely reach the million-dollar level, though that was relatively common in 2002. Big-budget music videos are now in the area of $300,000, and labels are developing and producing fewer of them (Anderman, 2008). And it is important to understand that professionally produced music videos represent 54% of web streams, but they generate 94% of advertising revenue for web sites—part of which is then shared with the label that produced the video. According to the Diffusion Group, user-generated music videos account for only 4% of a site's advertising (Bruno, 2009).

USES FOR THE MUSIC VIDEO

Promotion

In a perfect world, it would be a simple matter to create a music video and be able to determine the impact it has had in supporting the sale of a recorded music project. Certainly, a bump in sales figures can often be seen in SoundScan data when a music video has been added to web sites and into the rotation of a video channel or local cable program, but there are other opportunities for use of the video to support the project. Often, it is difficult to measure the impact of these other uses for the video, but following are some of those additional opportunities the music video offers the record label.

The publicity department and the independent publicist rely on a video to help them garner media attention for the artist. When a new artist is being pitched to a journalist or a blog for a story, the video is helpful by associating faces and performances with the music. The publicist also uses the video when vying for a performance opportunity on network television shows, including *Saturday Night Live*, *Today Show*, or *Oprah*. This added tool gives the gatekeepers of these television shows a view of what they might expect when the artist appears.

The radio promotion staff may find the video useful, especially if the artist is new. Radio programmers are sent a link to a new artist's video as a way to initially introduce the artist, and then the station is urged to link the video to their own online website when they begin to program the recording on the radio station.

New media departments at record labels use the video to create a presence on the Internet. Many social networking web sites feature music, music samples, and links to videos posted around the Internet. The label's web site includes either the video or a link to the video on the artist's site. Artist management companies that maintain an active web presence will also include links to either the label's or the artist's site to give music fans the experience of the music video.

Sales and Licensing

The marketing department also has specialized uses for the music video by including it as part of a "value added" feature of the release of a physical CD. It is not uncommon for CD jewel cases to display an added bonus for the consumer in the form of video performances as a separate DVD. Packaging music video performances as part of the experience the consumer purchases with a music CD adds value to the product and positions it to better compete with movie DVDs, which regularly include added features.

The Recording Industry of America Association (RIAA) is the industry trade association, and it annually tracks the shipment and value of various configurations of recordings. For 2008, their report shows a dramatic decrease in the interest of consumers of owning videos of music they like, based on how many music videos were shipped to retailers (see Figure 15.1).

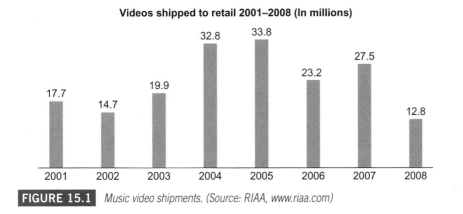

Videos shipped to retail 2001–2008 (In millions)

Year	Value
2001	17.7
2002	14.7
2003	19.9
2004	32.8
2005	33.8
2006	23.2
2007	27.5
2008	12.8

FIGURE 15.1   *Music video shipments. (Source: RIAA, www.riaa.com)*

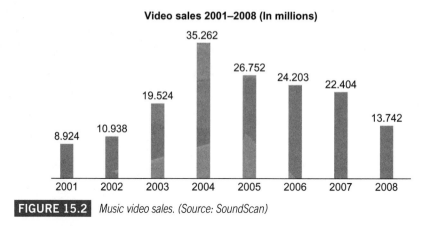

FIGURE 15.2 *Music video sales. (Source: SoundScan)*

The SoundScan numbers report actual sales of videos at retail, whereas the RIAA (2009) reports only shipments. The data shown for these 2 years are a reflection of the new availability of music videos on the Internet embedded within sites that are now paying for their use, for the entertainment of their web site visitors. Music fans no longer need to own a video if they are satisfied with visiting sites featuring the video at no cost but with accompanying advertising displayed on the web page while the video plays. For those who want to own the video, iTunes is the dominant source to download video singles that can be played on wireless devices, game consoles, and computers.

We have already seen how music video licensing is quickly becoming an important source of income for the record label. While the chart above shows the steep decline in sales of music videos, income from licensing within some companies is growing at an annual rate of 80% (Chmielewski, 2008).

Programming

The penetration of music videos into American households through cable and satellite channels is growing each year. Cable and satellite channels that use music videos to program part of their broadcast day continue to thrive, though videos have become an increasingly smaller part of the entertainment they offer to viewers. The proliferation of online and wireless on-demand sources for music videos has reduced television viewer interest in favor of other forms of entertainment. Video channels that program music videos as part of their entertainment offerings are listed in *Billboard* each week along with their charts for videos they are programming.

Table 15.3	Artists programmed on cable and video channels (*R&R*, 2009)
Cable/satellite channel	**Artists programmed**
VH1	The Fray, Kanye West, Taylor Swift, Nickelback, Pink, Kelly Clarkson, Bruce Springsteen, John Legend, Ne-Yo
BET	Keyshia Cole, Ciara, Beyonce, India Arie, Young Jeezy, Guns N' Roses, T.I., Soulja Boy
CMT	Taylor Swift, Toby Keith, Dierks Bentley, Sugarland, Alan Jackson, Carrie Underwood, Brad Paisley
GAC	Dierks Bentley, Keith Urban, Taylor Swift, Carrie Underwood, Brooks & Dunn, Billy Ray Cyrus, Johnny Cash, Trace Adkins
Fuse	Slipknot, Britney Spears, T.I., Kings of Leon, The Fray, Fall Out Boy, T. Pain, Beyonce, The Killers
MTV2	Akon, Kanye West, Stained, T-Pain, N*E*R*D, Gorilla Zoe, Wale, Hoobastank, Soulja Boy
MuchMusic Canada	Coldplay, Kelly Clarkson, Marianas Trench, Jonas Brothers, Danny Fernandes, Lights
CMT Canada	Sugarland, Higgins, Alan Jackson, The Road Hammers, Johnny Reid, Emerson Driver, One More Girl, Dean Brody

Label Staffing

Labels do not do their own music video production, preferring to hire outside firms that are specialists in the field. However, all labels have someone on staff, often in the creative services, who coordinates the creation of music videos as one of her job responsibilities. Depending on the label, video coordination may be a stand-alone job or it could be part of the publicity department. Likewise, many labels hire out video promotion.

Coordination of the creation of a music video involves working with A&R, publicity, marketing, radio promotion, new media, video producers, and creative services to find a common vision for the video. The coordinator seeks to find a story or concept that can enhance the music, and then to find a consensus among the various responsibility centers in the label.

Contracting with an outside music video promotion company can be efficient for many record labels because the video promotion function is not one that is done on a full-time basis at most labels. Video promotion takes on many of the features of radio promotion of a single. Independent music video promotion company AristoMedia (2009) has a 12-week program for its music video promotion clients, which includes:

- Taking new music videos to cable channels and pitches them to the programming staff

- Putting videos on a compilation DVD or file, called a video pool, and services local, regional, and national video outlets

- Following up with video programmers

- Creating tracking sheets so they know the frequency with which the video appears on each of the outlets to which they promote

- Following the charts of each of the music video outlets

- Always promoting for the heaviest rotation possible

Among the services offered by some independent video promoters is a semi-monthly creation of genre-specific compilation discs, sometimes referred to as **video pools**, which are sent to video channels. A promotion like this is fee-based, but it is an efficient method to distribute music videos to hundreds of music video outlets. Independent video promoters also offer new media promotion services for labels that do not have staffing available to create the necessary web presence for their music videos.

GLOSSARY

broadband—High speed Internet access by means other than a 56k modem.

illustrated songs—An early form of promoting music in theaters, featuring live music and projected slides depicting the lyric story line.

multiple rights contract—A recording contract in which an artist agrees to permit the record label to share in the traditional and new revenue streams created by his career in the music business, often referred to as a "360 deal."

soundie—Black-and-white short films of live performances presented in video jukeboxes in the 1940s.

video deejay—Someone who works on a television program that features an announcer, information, and music videos.

video pool—A compilation of music videos that are distributed on a regular basis (semi-monthly and monthly) by independent video promoters to local, cable, and satellite video channels that use this kind of programming.

visual records—Video film productions in the 1950s created to promote sound recordings.

ACKNOWLEDGMENT

A special thanks to Jeff Walker for his input and help with this chapter.

BIBLIOGRAPHY

2007 Year-End Shipment Statistics. (2009). RIAA. www.riaa.org

Anderman, J. (February 17, 2008). *The Itty Bitty Video*. The *Boston Globe* p. N1.

AristoMedia. (2009). Aristomedia.com, "Video Outlet Report," and Jeff Walker.

Bruno, A. (January 24, 2009). Best Bets, Digital: The Big Payback. billboard.com.

Chmielewski, D. (December 23, 2008). Labels, Websites. In: *Video Tussle. Los Angeles Times*, p. C1.

Macy, A. (2009). Personal conversation.

McCourt, T., and Zuberi, N. (2009). Music On Television. http://museum.tv/archives/etv/M/htmlM/musicontele/musicontele.htm.

Miller, M. (January 8, 2009). *Some YouTube Regulars Making Income From Videos. CBS Evening News*.

Music Video 1900 Style. (2004). PBSKids. http://pbskids.org/wayback/tech1900/music/index.

National Airplay Overview. (February 20, 2009). *R&R*. p. 49.

CHAPTER 16

The International Recording Industry

Tom Hutchison

INTERNATIONAL RECORDING INDUSTRY

The sale of recorded music overseas offers challenging, yet potentially rewarding opportunities for future growth of the record business. The entire global music industry has suffered a severe downturn since 2000, when file sharing became one of the most popular forms of music piracy. The downslide continued throughout the decade as other factors such as competition from new entertainment sources and global economic downturn added to the woes.

Countries with well-developed music markets are considered to be **mature markets**, characterized by slow growth and saturation. **Market saturation** is when the quantity of products in use in the marketplace is close to or at its maximum. Newer markets are considered **emerging markets** and offer the potential for rapid expansion. This opportunity is not without complications, the most critical being music piracy.

In this chapter, we will examine how markets are analyzed internationally and what opportunities and challenges the international recording industry faces.

MIDEM CONFERENCE

Each year, Reed MIDEM holds an international music conference in Cannes, France. It is an opportunity for global networking among music industry professionals. The event includes a large trade show and seminars

CONTENTS

Record Label Marketing

347

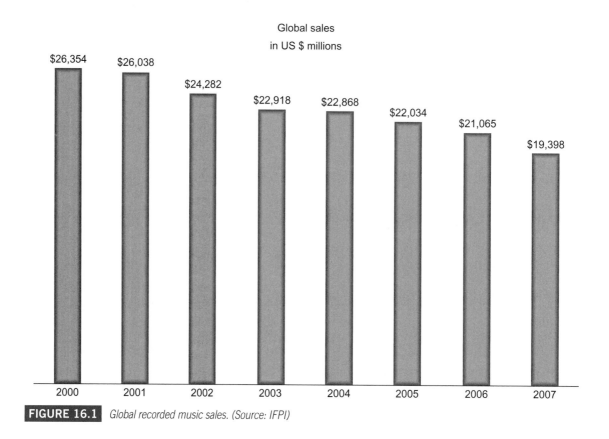

Global sales
in US $ millions

FIGURE 16.1 *Global recorded music sales. (Source: IFPI)*

on international issues in the recording industry. It is considered by many in the U.S. record business to be the most important international conference.

THE IFPI

The predominant organization for monitoring and supporting the recording industry internationally is the **International Federation of Phonographic Industries (IFPI)**. The goals of the IFPI include fighting international piracy, fostering the development of music industries globally, and compiling and reporting economic information about the recording industries.

According to the IFPI, the global sales of recorded music have been falling steadily throughout the first decade of the millennium. Global sales were considered flat for 2004, with the growth of DVD music videos and digital downloads offset by a slight reduction in physical audio sales. In all other years, the global industry has suffered a significant contraction (see

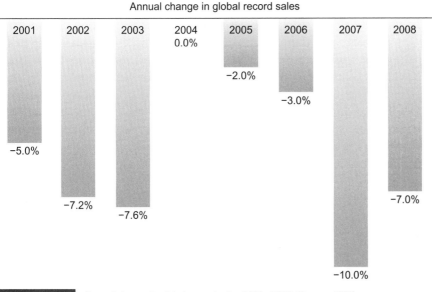

Annual change in global record sales

| 2001 | 2002 | 2003 | 2004 0.0% | 2005 | 2006 | 2007 | 2008 |

−2.0%

−3.0%

−5.0%

−7.2%

−7.6%

−7.0%

−10.0%

FIGURE 16.2 *Annual change in global record sales 2001–2008. (Source: IFPI)*

Figure 16.2) (Pfanner, 2009). As a result, the major record labels have further consolidated and trimmed their payrolls to reduce overhead.

THE NATURE OF OLIGOPOLIES

The recording industry is considered a mature industry. Mature industries are characterized by the presence of a structure called an **oligopoly**. An oligopoly is defined as a situation in an industry where the market is dominated by just a few companies who control most of the market share. By contrast, new, emerging industries are characterized by extreme competition among many small players who are all competing to increase their market share as the industry grows and expands.

The Consolidation Process

Usually at the early stage of marketplace development, the barriers to entry are low, and companies already in the marketplace are not successful in preventing startups from encroaching on their markets. Eventually, the industry is flooded with suppliers and this leads to a shakeout, with some of the more successful small players extending their market share through acquisitions and mergers, and driving out the competition. The motivations for expanding and acquiring other companies include (1) an increase in market share; (2) reduction of costs through economies of scale; (3) more control over the market, including protecting pricing and preventing the entry of upstarts. Eventually, these "chains" or medium-sized companies drive out most of the small entrepreneurs or force them into specialized niche markets. At that point, the medium-sized firms begin to buy up or merge with each other until ultimately, just a few major firms control most of the marketplace—an oligopoly. Examples include the tobacco, automotive, and soft drink industries, which are all dominated by two or three companies controlling over three-fourths of the market share.

Oligopolies

Economists differ in their evaluation of whether an oligopolistic marketplace stifles product diversity or enhances it. Some argue that innovation will more often occur in an oligopolistic market structure because firms have a better ability to finance research and development (Vaknin, 2001). With less competition and larger market share, these companies can better afford the costs and risks associated with developing and introducing new products (new artists in the case of the recording industry) than could a market situation with many smaller players.

Others believe in a negative relationship between industry concentration and product diversity. Their research suggests that complacency and status quo are the norm—that is, as long as these few companies control the marketplace, there is no incentive to invest in research and development (Burnett, 1996). Several scholars have noted that periods of high industry concentration in the recording industry (fewer labels controlling more of the market share) have been accompanied by periods of low product diversity (fewer top-selling titles). This trend started to change in the 1990s when major record labels realized that product development was more effective with smaller business units, such as independent labels. Major labels made great attempts to forge partnerships and partial ownership of the most successful independent labels without completely absorbing them into the larger company. Joint ventures, pressing and distribution deals, vanity labels, and imprints became the norm, thus preserving the innovative edge of the indie label environment.

Pricing by Oligopolies

It is also believed by some scholars that consolidation leads to higher pricing as the few remaining firms have a lock on the marketplace. This notion has been refuted by other economists who have found that mergers do not always drive prices higher. Companies fear losing market share to new entrants and existing competitors. This is especially true in a market where entry by new players is easy and cheap. In the recorded music industry, technology has enabled startup companies to record top-quality products at a fraction of the traditional costs and use the Internet as an inexpensive way to market and distribute those products. In a report commissioned by UNESCO, David Throsby (2002) states ". . . [the majors'] capacity to continue to exert the market power that they have enjoyed in the past is seriously threatened by the fluidity and universal accessibility of the Internet." In 2004, *Wired* journalist Chris Anderson proposed the "Long Tail" theory that demonstrated evidence that the Internet had become a great equalizer, allowing small labels the same opportunities to market globally that the major labels had enjoyed for decades.

The Four Major Labels

As a result of consolidation in the industry throughout the 1980s, 1990s, and into the new millennium, the music industry is now dominated by four major companies who control about 70% to 80% of the world music market. These companies are Warner Music Group, Universal Music Group, Sony, and EMI.

> **Universal Music Group (UMG)** is the largest of the major labels with 28.8% of the market share in 2007. The company controls about

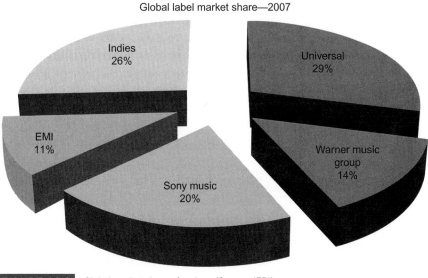

Global label market share—2007

FIGURE 16.3 *Global market share of majors. (Source: IFPI)*

25 record labels, including Interscope, Island, Def Jam, Mercury, Motown, Geffen, and A&M. UMG is a subsidiary of the French telecommunications giant Vivendi Universal, which includes Universal Music Publishing. Universal Music Group has operations in 71 countries. Universal's market share is strongest in the United States and Europe and weaker than other majors in Japan and Latin America.

Entertainment Music Industries (EMI) fell to a global market share of 10.9% for 2007 (IFPI, 2008; Sabbagh, 2008) to rank last among the major labels. Home to labels such as Blue Note, Capitol, and Virgin, EMI was looking to merge with one of the other majors in the mid-2000s, and continues to struggle. The company was acquired by private equity group Terra Firma in 2007.

Warner Music Group (WMG) was purchased from AOL–Time/Warner by a group of private investors including Edgar Bronfman, Jr., and went public in May 2005. With labels such as Atlantic, Elektra, Reprise, and Rhino, WMG has a strong catalog and active A&R department, and has managed to reverse their skid of the early 2000s—increasing their market share to 14.4% by 2007. Warner's share is strongest in North America and Latin America, and weak in Africa and Japan.

Sony Music Entertainment (SME) and Bertelsmann Music Group (BMG) have completed their merger (with Sony buying out BMG) but were finding it difficult to capitalize on the merger, slipping to 20.1% in 2007. Before the merger, SME and BMG had a combined global market share of 25.1%. In an analysis of the merger, David Empinet states that the strength of Sony BMG comes from strong assets brought together from each of the parent firms. "Bertelsmann was and is still today a highly influent international entertainment company at the head of influent media firms like RTL television group, Random House book publishing, Gruner & Jahr magazine publishing, Arvato media services, media clubs like Direct Group, and many more present in over fifty countries" (Empinet, 2009). He went on to state "Sony BMG would be greatly support[ed] by the Sony mother company by affiliating its media services and artists with its position of world's second largest consumer electronics position, better integrating artists music into video games for its Play Station console or into ads for its new high tech products."

The independent sector accounts for just over 25% of global sales.

MARKETS

The IFPI reports industry statistics for regions as well as markets in individual countries. Calculations based on regions include only the most influential markets in each region.

Europe maintains the lead in regional sales but slipped slightly in 2007, while Asia continues to grow as a market for recorded music, with market share increasing from 20% in 2006 to nearly 22% in 2007. Other areas remain consistent and nearly unchanged, with North America accounting for one-third of music sales.

Table 16.1	Countries included in regional analysis (Source: IFPI)
Region	**Countries included in regional analysis**
North America	Canada, U.S.
Europe	Austria, Belgium, Czech Republic, Denmark, Finland, France, Germany, Greece, Hungary, Ireland, Italy, Norway, Netherlands, Poland, Portugal, Spain, Sweden, Switzerland, and the U.K.
Latin America	Argentina, Brazil, Chile, Columbia, and Mexico
Asia	Hong Kong, Indonesia, Japan, Malaysia, Philippines, Singapore, South Korea, Taiwan, and Thailand
Middle East & Africa	South Africa and Mideast
Australasia	Australia and New Zealand

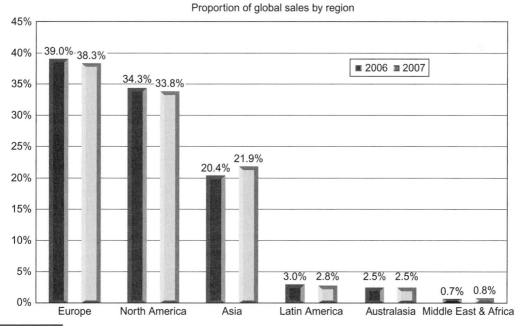

FIGURE 16.4   *Proportion of global sales by region. (IFPI)*

How Markets Are Evaluated

Before one can examine individual geographic markets, it is necessary to understand how the IFPI evaluates national and regional markets. Recorded music sales include an estimate of legitimate retail sales including singles, albums, CDs, digital downloads, and track album equivalents. Final retail value is estimated and converted to U.S. currency values. Price splits are examined—the relative market share of topline product (full), midline product, and budget product. Repertoire origin is examined, which evaluates the relative market share of music originating within the country (domestic) to that which is produced elsewhere (international). World ranking and piracy level are evaluated for each market. The IFPI also looks at sales by genre, retail channels, and per capita music purchases.

The proliferation of consumer electronic devices is seen as an indicator of market potential, and penetration levels of CD players, DVD players, optical recorders, digital music devices, mobile phones, and Internet connections are evaluated by the IFPI for each of the 70 member countries. These are also an indicator of the maturity level of the market. Mature geographic markets are those that exhibit a high penetration level of consumer electronic devices, a higher level of domestic repertoire, a higher proportion of sales in newer formats, lower piracy rates, and higher per capita sales. There is less potential for growth in mature markets, which, by their very nature, are saturated—meaning that consumers are not likely to increase their music purchases from one year to the next.

The use of various music listening devices also shapes the popularity of music formats, from the cassette to the MP3 player. Throughout the 2000s, the industry has experienced a continued decline, but not the termination, of cassette sales, while the CD has given up market share to digital downloads, dominated by single tracks. DVD video had a run through the mid-decade but then exhibited signs of waning.

One emerging trend runs counter to the concept of mature markets exhibiting more innovative behavior and development of communication/entertainment infrastructure: mobile penetration. Rapidly developing countries such as China and India have opted to create wireless data communication systems rather than retrofit a culture with high-speed Internet cable, thus skipping a generation of communication technology that developed in Europe and the United States in the 1980s and '90s. With the market for digital music on the rise, the delivery method for music downloads varies widely from country to country, with Japan leading the way in mobile music downloads (see the section on digital music sales).

Table 16.2   IFPI measures. (Source: IFPI)

IFPI measure	Description
Sales of recorded music	Sales are reported by format, by variable US$ retail value, by fixed US$ retail value, by local currency value, and by annual percentage of change in units. Fixed US$ retail value is a report based on historical local currency values re-stated at the exchange rate at the time the statistics are reported.
Format analysis	Unit sales are reported for singles, LPs, digital tracks, digital albums, CDs, and DVD. Market share (in value) and percent change in value from the previous year are reported.
CD price splits	Based on units sold through retail channels and includes full—based upon full retail price, mid—product priced between 50–75% of full value, and budget—product priced between 35–50% of full price.
Sales by genre	The list of genres varies from country to country and is not included for all countries.
Repertoire origin	Compares domestic and international market share. Sometimes classical or regional are included as additional categories.
Electronic hardware penetration	As an indicator of the market for technology and entertainment software, the total ownership of hardware units is divided by number of households to yield penetration statistics for CD players, DVD players, optical recorders, mobile phones, and digital music players (most likely MP3).
Internet penetration	Digital music sales are not yet compiled by the IFPI, but in anticipation of this, the potential market for Internet delivery of digital music is measured through proportion of households with Internet and broadband connections.
Domestic piracy level	Domestic piracy is a measure of consumption of pirated goods, regardless of the origin of such product. It is measured as a percentage of total (legitimate and pirate) sales.
Per capita sales	Based upon population, per capita unit sales and per capita expenditures

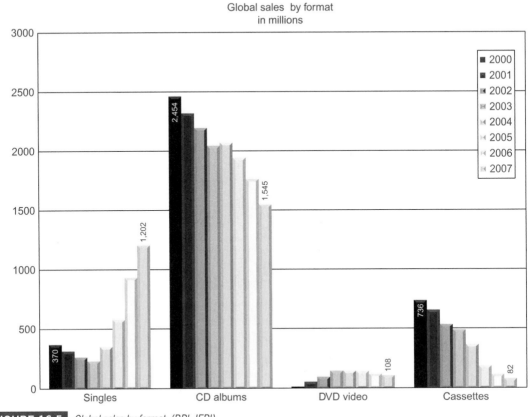

FIGURE 16.5 *Global sales by format. (BPI, IFPI)*

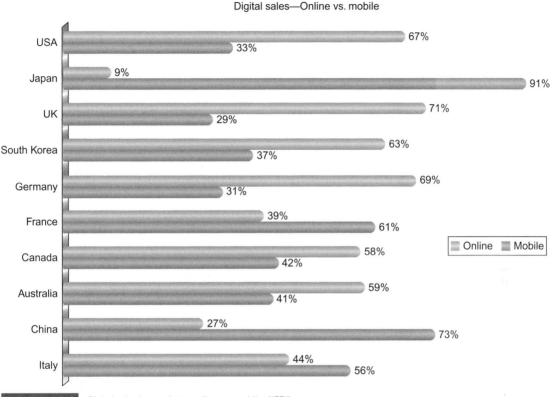

Digital sales—Online vs. mobile

	Online	Mobile
USA	67%	33%
Japan	9%	91%
UK	71%	29%
South Korea	63%	37%
Germany	69%	31%
France	39%	61%
Canada	58%	42%
Australia	59%	41%
China	27%	73%
Italy	44%	56%

FIGURE 16.6 *Digital sales by market—online vs. mobile. (IFPI)*

The Major Markets and Trends

In terms of individual markets (countries), the United States held the largest market share for 2007, with over 30% of the market; Japan was in second place with 18% of the market. These two markets have commanded the top two markets for many years, with third and fourth places rotating between the United Kingdom and Germany. In 2007, the United Kingdom held 10% of the global market share and Germany had just over 8%, a slight growth from 2006.

  Japan – In 2007, Japan suffered a 4% drop in overall sales, representing the ninth year in a row of declining sales. Conversely, digital sales were up 41% from the previous year. Over three-fourths of sales were of domestic origin, as local labels such as Pony Canyon and Avex continue to hold court with the major labels. Per capita unit sales for 2007 were 1.39, with a per capita expenditure on music of the U.S. dollar equivalent of $34.21. For 2007, sales numbers remained flat,

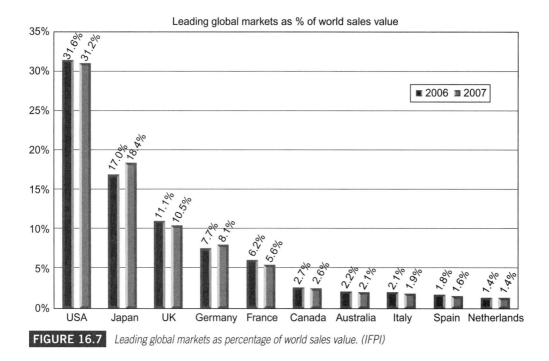

Leading global markets as % of world sales value

FIGURE 16.7 *Leading global markets as percentage of world sales value. (IFPI)*

whereas four of the other five top music markets saw declines. Japan is the second largest digital market, with a 19% market share, and is the world leader in mobile music, with over 90% of all digital music purchases occurring via mobile.

United Kingdom – Britain remains the third largest market for music sales, behind the United States and Japan. The U.K. sales value in 2007 was US$2.97 billion. The United Kingdom's market share decreased in 2007 but remained above 10% of the global market for the fifth straight year. The British Phonographic Industry (BPI) (2008) attributes this to a strong domestic repertoire, one that has found popularity in other English-speaking countries, with artists like Amy Winehouse, Corinne Bailey Rae, Led Zeppelin, and the Beatles. The BPI claims that U.K. artists had 8.5% of the U.S. market share for 2007. Per capita unit sales for 2007 in the United Kingdom were the highest globally, with 2.28 units and a per capita expenditure on music of the U.S. dollar equivalent of $48.87.

Germany – The retail sales value of the German market in 2007 was US$2.27 billion. Germany is the fifth largest digital music market, with 9% of global digital sales. Digital music sales in Germany are showing steady growth. Online single track downloads totaled 37.4 million in 2008, a 22% growth on 2007. Digital album sales increased by 57%, totaling 4.4 million. Per capita sales of all music in Germany for 2007 were 1.38 units, or $27.63.

Digital Music Sales

Despite the overall downturn in music sales since 2000, digital music sales continue to grow—the one bright spot in the otherwise gloomy outlook. In 2007, online music distribution represented about 23% of revenue in the United States and about 15% worldwide. Digital music sales in 2008 grew by an estimated 25% to $3.7 billion in trade value, to account for about 20% of the industry's global recorded music sales (Holton, 2009).

In 2004, digital sales amounted to only $400 million, or 2% of the total industry. By 2006, it was up to 11% as broadband Internet connection became more widespread. And by 2007, the digital music market was worth an estimated $2.9 billion. In 2008, PricewaterhouseCoopers predicted that half of the global music industry revenues would come from digital music by 2012.

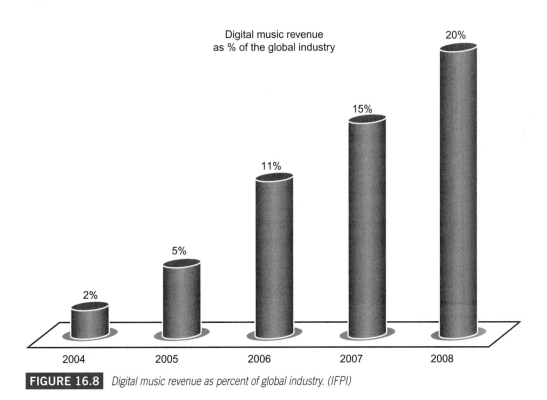

FIGURE 16.8 *Digital music revenue as percent of global industry. (IFPI)*

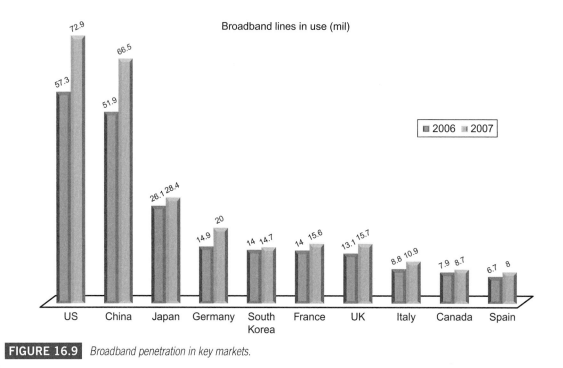

Broadband penetration in key markets.

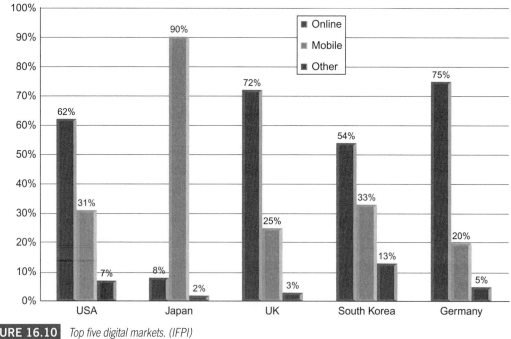

Top five digital markets. (IFPI)

Jupiter Research estimates that the number of worldwide online users will increase 44% between 2007 and 2012, reaching 1.8 billion users. They estimate that by 2012, one-quarter of the worldwide population will access the Internet on a regular basis. According to the report, Brazil, Russia, India, and China will experience some of the highest growth, with China overtaking the United States in number of regular online users by 2011.

Some markets have been quick to adopt the digital music format, with all formats of digital music sales accounting for 30% of the North America market in 2007; in France, however, the digital market represented only 7%. The list of top digital music markets mirrors that of the top music markets with the glaring exception of South Korea, which has developed a strong digital music market. In South Korea, 61% of all music sold in 2007 was delivered as files rather than discs. Almost 80% of South Korea's households had broadband access by 2008, and the country had 44.5 million mobile subscribers in total, or a 90.7% penetration rate.

The digital music market is moving from online digital purchases to music acquisition (of full-length tracks, ringtones, etc.) via mobile devices. In 2008, digital music sales had roughly a 50–50 split between online and mobile sales. As mentioned earlier, nowhere is this move to mobile more evident than in Japan, where over 90% of music is delivered via mobile networks. Full-length song downloads are beginning to overtake mastertones in many countries. In Japan, full-length song downloads were accounting for 60% of mobile music revenues by 2008. Over 140 million mobile singles were sold in Japan in 2008.

Mobile service and mobile hardware providers are beginning to partner with music companies to offer mobile music to their customers. Nokia launched its "Comes with Music" service in the United Kingdom in 2008; however, this service still depends on desktop computers to download the music to the handsets. Sony Ericsson launched a service in Sweden in 2008 that allowed users to download, play, and recommend music directly through the mobile network. At the end of the 6- to 18-month contract, users get to keep up to 300 of their favorite tracks.

Global digital sales by channel—2007

Subscriptions 5%

Mobile 47%

Online 48%

FIGURE 16.11 *Global digital sales by channel. (Source: IFPI)*

Other subscription-based services are finding ways to integrate mobile or portable music devices, using the purchase of a hardware device to activate the subscription for an extended trial period. Music subscription services failed to take off when licensed digital music was first offered to consumers via the Internet. Companies like Rhapsody and Napster failed to capture the interest of music fans who prefer to own music rather than rent it. One of the difficulties stemmed from the fact that music lovers do not like the idea that they must continue to maintain their subscription in order to have access to their music collection. Mobile music consumption has the capability to bypass that concern by offering consumers on-demand music selections that reduce the need for consumers to possess music on their home computers. Other providers have found that bundling subscription services with other mobile communication services is more appealing to customers who have no intention of discontinuing their mobile service, thus causing their collections to become unusable. Others, such as the PlayNow system, reduce consumer skepticism by allowing them to keep a portion of the music acquired during the contract period. These new subscription models bring back the possibility of this revenue stream catching on with consumers and providing more than the 5% of the market they claimed in 2008.

ISSUES AND CHALLENGES TO MARKETING RECORDINGS GLOBALLY

Piracy

The primary threat to the international recording industry is piracy. The proliferation of Internet access and **peer-to-peer** (**P2P**) file-sharing services has created massive headaches for the recording industry that was already struggling to contain physical piracy of CDs. An IFPI report in 2009 claimed that 95% of all music downloads were illegal copies (Smith, 2009). More than 40 billion music tracks were illegally file-shared in 2008, compared to legitimate sales of 1.4 billion.

Physical Piracy

Despite the fact that the IFPI has shifted resources to fighting digital piracy, problems persist with physical pirating of music product. Pirated recordings include factory-pressed CDs, where music albums are copied on blank discs, and pirate cassettes.

In markets where production of factory-pressed pirate CDs is high, the IFPI examines pressing capacity. This is an examination of all the CD factories'

Table 16.3	IFPI enforcement bulletin on piracy

Digital and physical piracy used to be seen as very different beasts. Online piracy would involve technologically savvy people illegally uploading music and sharing it with new kinds of software. Physical piracy ranged from a CD plant churning out fake discs to a man on the street corner selling counterfeit CDs. That distinction is now increasingly outdated.

"There are many examples of how the worlds of physical and digital piracy interlock. Despite the advances in electronic distribution, most record companies still distribute promotional and demo material on physical media. Most of the online pre-release leaks that we in the IAPU investigate originate from someone obtaining (directly or indirectly) a CD and illegally uploading its content to the internet. Conversely, many of the counterfeit CDs on sale on street corners and in markets are burned from illegal online content."

IFPI Enforcement Bulletin – July 2008

manufacturing capacity and a comparison with the legitimate demand for legal discs (of all software) in markets served by those manufacturers. The IFPI claims that overcapacity is a key factor in the spread of disc piracy, including recorded music, movies, and computer software. Various measures are used to monitor production in factories suspected of producing illegal copies (see section on anti-piracy efforts).

Piracy operations have recently been linked with organized crime, as optical discs become a commodity linked with drug trafficking, illegal firearms, money laundering, and even the funding of terrorist activities. A successful Polish police action netting pirated discs and smuggled firearms and cigarettes in 2007 sharply highlighted the links between music piracy and international organized crime. David Wood, Director of Anti-piracy for the BPI stated:

> *Selling copied music, film, games and software on discs is a cash cow for organized gangs, and different gangs control supply and distribution in different regions. Controlling territory is lucrative.*

—BPI, 2008

Some recent raids on illegal music production facilities have netted caches of other contraband. In 2008, authorities raided a Polish syndicate responsible for the sale of over 9 million albums. In November 2008, officials discovered a smuggling ring that was importing CDs from Uruguay to sell in Brazil. Authorities processed an outfit that was selling counterfeit CDs through a web site in Mexico.

Digital Piracy

With the proliferation of digital music tracks, both legal and illegal, the challenges for reigning in piracy have increased. Just when the IFPI was getting aggressive and going after countries that harbored manufacturing plants that produced illegal optical discs, illegal peer-to-peer file sharing began to threaten the industry in ways that were not geographically bound or linked to sovereign nations. The RIAA and other industry trade groups responded by going after the companies that provided a conduit for users to share their music illegally, including Napster, Grokster, LimeWire, Gnutella, BitTorrent, and Kazaa. Success in shutting down one service only led to the creation of a new group of services to replace them. The RIAA started going after infringers, arresting high-profile heavy users of illegal services. This also proved to be unsuccessful in stemming the tide.

In 2005, the IFPI and member organizations turned to Internet service providers (ISPs) to help curtail illegal file-sharing activities. Until this time, the ISPs had adopted a policy of "**net neutrality**." A neutral broadband network is one that is free of restrictions on content, sites, or platforms, on the kinds of equipment that may be attached, and on the modes of communication allowed. Resistance came from advocates of net neutrality who feared the ISPs would engage in data discrimination, block content from competitors, and create a tiered system of access. By 2008, the ISPs were seeing the value in cooperating with the entertainment industries in curtailing the abuse of copyright-protected works. These providers were discovering that infringers were likely to hog bandwidth as they downloaded full-length feature films and other pirated files. The providers also were concerned that some of their own protected content would fall victim to rampant piracy (Warner Communications has both an Internet cable company and entertainment production divisions). ISPs are beginning to develop policies that will identify and sanction repeat offenders. Yet despite all these efforts, the IFPI estimated that in 2008, 95% of all digital music downloads were nonlicensed illegal copies (Smith, 2009).

Anti-Piracy Efforts

The IFPI has been working with government organizations to reduce piracy through the following measures:

1. Encouraging copyright legislation and adequate enforcement, including monitoring borders and educating local law enforcement officials
2. Regulation of optical disc manufacturing facilities, including implementing identification coding and plant inspections
3. Consumer awareness, including effective prosecution and deterrent penalties, education and consumer awareness programs, and making local governments aware of the adverse effects on the development of local artistry

4. Working toward cooperation with Internet service providers (ISPs) to curb illegal digital file sharing

Parallel Imports

Parallel imports are defined as imported genuine copyright goods for resale by an agent other than the local authorized distributor. In other words, the importing of copyright goods into a country that were lawfully made in a foreign country, without obtaining permission of the holder of the copyright for those goods in the recipient country. In the recording industry, this occurs when a particular title is imported from another country when a domestic version of the same physical product is released in that country. Parallel imports are considered a problem when differential pricing policies are in effect. Recorded music is priced based upon what the market will bear, and that varies from country to country. As a result, prices are significantly lower in some countries. Problems occur when manufacturers attempt to maintain pricing policies that are not competitive with the imported version of the same product.

This is currently an issue in the prescription drug industry. Consumers want the ability to import cheaper prescription drugs from Canada. Drug manufacturers are attempting to restrict this practice through government legislation, therefore making the importation practice illegal. This allows the drug companies to maintain higher prices in the United States for the same drugs sold for lower prices in Canada.

Currently, the United States has no such restrictions for the importation of licensed recorded music. As long as the music is a legally licensed copy, it can be imported into the United States and sold alongside the version released for sale in the United States. Until the late 1990s, Australia had laws against the importation of recorded music. As a result, Australian consumers paid more for CDs than consumers in other comparable countries. Consumer groups and retailers petitioned the government and the restrictions were lifted. Also, policies in the European Union call for the gradual abolishment of tariffs and trade restrictions, allowing for parallel imports among member countries. While this practice is good for consumers and retailers, it causes problems for manufacturers.

The primary problem with parallel imports for manufacturers is that it undermines differential pricing policies by leveling the playing field, usually to the lowest available retail price. Parallel imports also create problems with customer support, as local agents are sometimes called upon to offer warranty service on products purchased elsewhere. The imports may also cause problems for local agents who pay for marketing expenses, only to see the consumer demand they have created met with purchases elsewhere. Licensing also becomes less profitable as local agents may hesitate to pay licensing fees for fear that they may not be able to recoup the costs of licensing, manufacturing, and marketing.

Parallel imports are also subject to erratic variations in exchange rates. In the 1990s, when the British pound was low compared to the German mark, German retailers were ordering their CDs from British suppliers, thus forcing TVG-WD, the top German music rack jobber, into bankruptcy. Several years later, when the British pound rose against the mark, these retailers could no longer afford to order CDs from the United Kingdom and found the distribution infrastructure in Germany was now lacking. Now that the European Union has adopted a standard currency, this will no longer be a problem.

The DVD industry has avoided problems with parallel imports by selling systems in Asia that are not compatible with systems sold in the United States. Therefore, DVD movie titles sold in Asia will not work on DVD players owned by U.S. consumers.

The proliferation of online music retailers such as Amazon and CD Baby has caused labels to rethink their record release strategy for major artists—coordinating a simultaneous worldwide release date to minimize on parallel imports and peer-to-peer illegal file sharing among fans.

Licensing

It is still a common practice in the recording industry among the nonmajors to **license recordings** to foreign labels to manufacture and sell in their own territory. Under this arrangement, the original record label grants a foreign record label a license to issue an artist's recording in its local or regional territory, as specified by the terms of the agreement. The licensor (original label) provides the licensee (foreign label) with a master recording and artwork for the licensee to reproduce, distribute, and market. The licensor is paid a percentage of the local **published price to dealer** (**PPD**), which is usually between 15% and 19% of the retail price (Feldman, 2005).

In the absence of parallel imports and other forms of **transshipment**, licensing can be quite profitable for the licensee. However, in the spirit of global trading, many nations have reduced barriers to parallel imports, thus undermining the value of licensing. Also, Internet marketing and digital music downloading services have reduced the effectiveness of territorial licensing by allowing the original record label to distribute globally without relying on other labels internationally to make recordings available on the local level. Licensing is still effective when a physical presence in local retail stores is important.

Currency Exchange Rates

Every nation that has an independent currency usually has one that fluctuates in value relative to the currencies of other nations. Although a few countries officially fix their exchange value to a key currency, the exchange rates between most currencies are primarily determined by market forces and can be erratic when engaging in international commerce. Currency rates are usually cyclical. If the U.S. dollar is strong, American consumers can easily afford to purchase imported goods. But products produced in the United States can become unaffordable overseas. As a result, the demand for U.S. goods is reduced, thus reducing the demand for U.S. dollars. According to the *purchasing power parity theory*, exchange rates will tend to adjust over time so that purchasing power will return to a status quo (The Foreign Exchange Market, 2004).

Not all countries allow their currency values to fluctuate on the open market. Some countries officially fix or peg their currency to that of another country. For example, China has been pegging their currency to the U.S. dollar, for better or worse. This has led to financial complications but allowed China to enjoy the rise of their currency as the U.S. dollar rose in the 1990s and 2000s. The main advantage of a fixed-rate system is that it removes the risks associated with unpredictable fluctuations in exchange rates over time. The disadvantage is that it ignores market demands for currency and requires countries to maintain large reserves of foreign currency to supply the market when there is an excess demand for them at the official exchange rate.

Fluctuations in currency value can interfere with sales expectations and are beyond the control of the marketing efforts of a company (see the section on parallel imports). They can discourage many people from making foreign investments. To overcome this, the European Union has adopted a standard currency, the euro, in an effort to stabilize pricing and encourage commerce among its member nations.

Cultural Restrictions

Global cultural industries that include movies, music, television, books and magazines, and fashions are considered a threat to cultural diversity around the world. Because cultural goods are often mass-produced, there is homogeneity among these goods. As a result, consumers everywhere may wear the same styles and listen to the same music. The underlying assumption is that this will lead to cultural homogenization and a loss of the local culture. The term **cultural imperialism** is used to describe a system where a universal homogenized culture (usually U.S. culture) replaces the genuine local

Table 16.4	Quota action
Country	**Action**
Canada	The Canadian government currently requires that 30% of broadcast material on cable and radio be of Canadian origin.
European Union	Member states must ensure that not only television broadcasters but also on-demand audiovisual media services promote European works.
France	40% of all music aired must be sung in French with 20% of this consisting of new French bands.
Australia	Radio stations that broadcast music have a mandatory obligation to broadcast minimum quotas of music performed by Australian artists. The quotas range from 5% to 25% of music broadcasts.
South Korea	The Korean government reduced the requirements of the quota in 2006, requiring Korean films for only 20% of the time, halved from a previous high of 40%.

culture. To protect local culture, governments may seek to reduce the influence of outside cultural "imperialism" in several ways: (1) providing an alternative, (2) subsidizing local cultural products, (3) banning outside cultural products, and (4) setting quotas.

Examples of this cultural protectionism can be seen in France and China, where government regulation has restricted the importation of American music and films. Canada is an example of a country that has chosen to not only restrict U.S. media content, but to also offer cultural subsidies to local artisans, including musicians, to develop their craft and products. The United Kingdom is an example of a country that has chosen to offer a government-sponsored alternative to "Hollywood" media by subsidizing the British Broadcasting Corporation (BBC). Their stated purpose is "to enrich people's lives with great programmes and services that inform, educate and entertain" (BBC, 2009). The most common forms of cultural protectionism are quotas and censorship.

Quotas

Many nations have imposed quotas or trade barriers on the amount of cultural product imported into their country. Some examples are included in Table 16.4. All are done to preserve the local culture and local cultural industries.

Censorship

Censorship is imposed to protect the moral and cultural values of a society. Censorship may include imported goods but may also apply to entertainment produced locally. There are examples of censorship in the United States, stemming from public protests over broadcast material deemed immoral or indecent by public standards. Public outcry over the 2004 Super Bowl half-time show, in which Janet Jackson had a breast exposed, led to an increase in fines imposed by the Federal Communications Commission for indecency on the public airwaves.

The United States is considered generally more liberal than some other countries in allowing for free expression of ideas. In many parts of the Middle East, North Africa, and China, censorship of music is common (Cloonan and Reebee, 2003). In 1996, officials in Iran arrested 28 teenagers for possessing "obscene" CDs and cassettes. In 1997, South Korean radio station KBS banned teen pop music because of the clothing styles worn by the entertainers. In 2003, heavy metal fans in Morocco were jailed for "acts capable of undermining the faith of a Muslim" and "possessing objects which infringe morals" (Index Online, 2003). And again in 2007, the Iranian government raided a party in Tehran and confiscated 800 "obscene" CDs (Reuters, 2007).

China is one of the most heavily regulated countries with regard to music censorship (Orban, 2003). Nothing can be produced or distributed without permission from Chinese government authorities. Restrictions are placed on all materials that ". . . are against the basic principles of the Constitution . . . that advocate obscenity, superstition or play-up violence or impair social morals and cultural traditions, and those that insult or defame others" (Brenneman, 2003).

OPPORTUNITIES

Opportunities for growth in the international recording industry are coming from two areas: new markets and new technologies. This growth will be fueled by a reduction in trade barriers, improved economic conditions, the availability of new distribution channels, and continued adoption of next-generation technologies (PricewaterhouseCoopers, 2004). Opportunities from new technologies have been outlined in the section on digital music sales.

Geographic Growth Areas

Asia/Pacific is emerging as the key area of growth in the entertainment and media industries, fueled mainly by China and India, both of which are investing in new communication technologies and media infrastructure as well as opening up their markets to international goods. China, with its population of 1.29 billion, and India, with its population of over 1 billion, offer the potential for growth, and both have a low penetration of media hardware at this point in time. Efforts to stem piracy combined with heavy investments in media industries will drive growth in the region.

Trade Agreements

GATT

From 1948 through 1994, the **General Agreement on Tariffs and Trade (GATT)** provided the rules for much of world trade and handled issues dealing with international commerce. Members of GATT were pledged to work together to reduce tariffs and other barriers to international trade, and to eliminate discriminatory treatment in international commerce. The GATT was designed as an interim measure pending the creation of a specialized institution to handle the trade side of international economic cooperation. The first attempt, the International Trade Organization (ITO), failed to materialize when the U.S. Congress refused to ratify the treaty. However, the GATT remained in effect and was the only multilateral instrument governing international trade from 1948 until the World Trade Organization was established in 1995.

WTO

The **World Trade Organization (WTO)** was established as a result of the final round of the (GATT) negotiations, called the Uruguay Round. The WTO is responsible for monitoring international trade policies, handling trade disputes, and enforcing the GATT agreements (www.encyclopedia.com, 2004). The WTO web site describes their mission as follows:

> At the heart of the system—known as the multilateral trading system—are the WTO's agreements, negotiated and signed by a large majority of the world's trading nations, and ratified in their parliaments. These agreements are the legal ground-rules for international commerce. Essentially, they are contracts, guaranteeing member countries important trade rights. They also bind governments to keep their trade policies within agreed limits to everybody's benefit. (World Trade Organization, 2004)

TRIPS

Even before the WTO was formally operational, the Uruguay Round of the GATT negotiations produced the Agreement on **Trade-Related Aspects of Intellectual Property Rights (TRIPS)**, negotiated from 1986–1994. This agreement introduced intellectual property rules into the multilateral trading system for the first time and addressed protection of copyright ownership of recorded music. The agreement establishes a minimum amount of copyright protection for creative works and specifies how countries should protect and enforce these rights.

MARKETING PRODUCTS OVERSEAS

There are several avenues for marketing recorded music overseas: (1) the major labels tap their affiliates in the region; (2) indie labels license the recording to local labels in the territories of interest; (3) products manufactured in the United States are exported through a foreign distributor; and (4) digital delivery to the consumer via the Internet. In all cases, international touring schedules are generally set up to coincide with the release of the album in that territory.

Major Labels and Affiliates

The major labels rely on their branch (affiliate) offices around the world to promote products deemed profitable in those markets. The majors have the lion's share of the market in most countries with a viable recording industry (74%).

Marketing Through an Affiliate

Each regional affiliate of a major label operates independently for the most part, and each branch is responsible for its own marketing. Affiliates generally have the option to accept or decline an opportunity to release a record from an affiliate in another region. For example, if a U.S. branch wants affiliates in other territories to take on one of their releases, they would use a product presentation as an opportunity to pitch the project to affiliates in the targeted territories. If the artist is a major star with international appeal, an international release is customary. If the artist is mid-level or emerging, the U.S. branch may have to convince the affiliates that they should take on the release, even to the extent of paying a portion of the artist's travel expenses. They would also want to demonstrate the marketability of this release.

International affiliates of a major label will gather periodically for product meetings. At these meetings, product presentations by each affiliate feature the artists that they think are most appropriate for international release. The other affiliates at the meeting then take this under consideration, keeping in mind similar acts that they might be working and consumer demand in their territory. The potential releases are then prioritized by label, by album, and by single. Having a great single can move an album up the priority list. If the label doing the pitching does not succeed with their own affiliates, they may consider contracting with an entity outside of the major label system for distribution in that region, but the affiliates get first refusal.

Touring has an impact on the decision of whether to take on a project. An artist, her manager, and her booking agent may work with the label to coordinate an overseas release with a concert tour in that region. At times, a label's clout with one artist may help influence an affiliate to take on projects from other artists on the label. For instance, if an affiliate in Europe wanted to handle the release of a major U.S. artist, that artist's U.S. label may request or require that the affiliate handle other acts on the label as well. An artist's contract with his label may also guarantee that the record will be released internationally. If that artist is still developing, the label may scramble to find foreign territories interested in the release. For a major act, a coordinated worldwide release is normal, with all other affiliates onboard to take advantage of a global marketing campaign. On occasion, a U.S. artist will be released initially overseas to "test market" the artist before investing in a U.S. release.

For the international release of Creed's first album, Jed Hilly, who served as Vice President of Marketing Services, Sony Music International, New York, had this to say:

> When the Creed/Wind-Up deal was announced, I worked with the Wind-Up people and we orchestrated an event. We invited marketing representatives from all over the world—about 20-something people came. Japan, U.K., France—we had guests from all over the world—to Orlando, Florida. We had an afternoon performance where we rented a suite at the local Marriott. We had Julia Darling and Stretch Princess do acoustic sets and then we went to the House of Blues later that night and Finger Eleven warmed up for Creed, for about 2000 people. So, everyone got to go experience that, meet the band members and label personnel. We did that in November and the official Wind-Up launch [internationally] was in January. We had to do this to get people on board.

—Hilly, 2005

Once Creed was established in the marketplace, subsequent albums were released with an international plan in place.

Indie Labels and International Markets

Independent labels must work through arrangements made with labels or distributors in other territories. Arrangements with local labels usually involve licensing and arrangements with distributors involved in the transshipment of products created in the home territory. Again, touring plays a major factor in the success of indie label artists overseas.

Licensing

Licensing was discussed in an earlier section dealing with parallel imports. The label originally releasing an album licenses a label in a foreign territory to release the recording in that territory. The licensor (original label) provides the licensee with a master recording and all related artwork in exchange for a fee and a percentage of sales. The licensee is responsible for manufacturing, distribution, and marketing in the designated territory. Compensation is paid to the licensor at generally 15% to 19% of the local retail list price or computed as a percentage of published price to dealer (PPD), or wholesale price (Feldman, 2005).

When licensing is done, indie labels give up much of the potential for profit, as well as control over manufacturing, sales, and marketing. Generally, the licensor does not place requirements or restrictions on how a record is to be marketed. However, requirements may include the quantity of product for which royalties will be paid.

Some indie labels prefer to avoid licensing if they have the resources to engage in marketing activities on their own or with the assistance of a local distributor. It allows them to gain more control over the marketing and distribution process. However, this may vary depending on the territory. One independent label, Compass records, prefers to use licensing arrangements in most of Asia, while relying on local distributors in other territories (transshipments) (Keim, 2005).

Transshipments/Exports

In this situation, finished recordings are rerouted or exported to a foreign distributor. For the importing territory, they are not considered parallel imports unless a locally licensed version is also available. Indie labels that engage in using foreign distributors must work hard to promote the recordings in those regions. They must contact local press and generally coordinate closely with any touring arrangements. It is also wise to price the product competitively with local pricing rather than pricing as an "import." Indie labels may use different distributors for each territory and perhaps even different distributors within a territory, dependent upon the music genres they are involved with.

The timing of the release date must be taken into consideration for each territory. An artist who does well in Australia may want a fourth-quarter release date. The same artist may not be as well known in the United States and may require a release date in the first quarter of the following year. Thad Keim, Vice President, Sales & Marketing, Compass Records, recalled:

> *When Compass released The Waifs, A Brief History . . . CD in January 2005, it followed the release date in their native Australia by almost 3 months. While the ideal situation would have been a simultaneous release schedule, it's very difficult for an independent label to release a CD in the U.S. during November (and remain competitive) with all the major labels, hit product being released during fourth quarter. Since among other things, marketing costs at retail average three times higher during this period, we decided to wait until January for the U.S. release. The same restrictions don't apply to The Waifs in Australia, as they are a platinum selling act down there.*
>
> —Keim, 2005

Timing is also important to minimize the potential impact of parallel imports.

> *When Compass set up Paul Brady's 2005 release for the U.S. and Europe, we were in a position where our release dates in certain territories had to be within a 1 to 2 week window of one another or we would run the risk of exports from one territory making their way into another and upsetting the domestic release. Exports from the U.S. were of the most concern because the dollar is rather weak against the euro and pound right now. So, in this case, the risk of exports did have some influence on setting our release dates.*
>
> —Keim, 2005

Internet Sales and Distribution

The Internet has created opportunities for smaller record labels to compete internationally by offering products for sale through their web site. Such sales fall into two categories: (1) the shipment of physical products to global consumers, and (2) the commercial sale of downloads. Target marketing is no longer limited to geographic segmentation, as the Internet has eliminated geographic barriers to marketing, sales, and distribution. This is beginning to complicate the issue of territorial rights, differential pricing, and staggered release dates.

Marketing Strategy

The basic marketing aspects of radio, press, tour support, retail promotion, and video are utilized to market overseas, but the relative importance of each may vary from country to country. Many foreign countries have a different radio industry structure from what has been established in the United States. Government-run radio, such as the BBC, dominates many foreign countries that have only recently introduced privately owned radio networks. MTV is popular in Europe and Asia and has been proven successful in marketing, but is beyond the reach of all but the majors. Tour support becomes a more important aspect of many of the U.S. acts promoted overseas.

Music genre classifications vary from country to country, and may be vastly different from what we are used to in the United States. For example, in Europe, country music encompasses folk and Americana.

It is precisely because of these vagaries that marketing in a particular region should be left up to that local affiliate or a distributor with the resources to assist in marketing. They know best what will work in their market. The prerelease setup must be done up to 2 months in advance of the release in order to get press coverage such as magazine covers and media reviews at the time of the release. That can become difficult if the materials are not ready or the label is not willing to release these materials (including an advance copy of the album) overseas in advance of the album release.

With a new artist, an overseas release usually comes after some success, with sales in the United States demonstrating the profitability of the recording. It also allows the artist to wrap-up a U.S. tour before going overseas to support the release with the international tour. For an established artist, a worldwide release date is common to reduce the likelihood of parallel imports and piracy.

CONCLUSION

The international recording industry faces many of the same challenges that face the RIAA and the U.S. recording industry, and to that must be added the increased level of piracy and complications stemming from international trade. The industry is monitored by the IFPI, which provides data on recorded music sales and economic climates of the various member nations. The industry numbers in this chapter were the most current available at the time of printing. The IFPI releases annual global statistics in the fall of each year. By applying the concepts outlined in this chapter to economic indicators in subsequent years, the student of the international recording industry can continue to gain insight into the complex factors influencing international record sales.

New opportunities for music at all levels have been created with the move toward digital music with no geographic limitations. Artists are now able to supply music to any and all fans via the Internet. Global marketing has been made easier because of the Internet, giving potential fans the opportunity to sample and purchase music regardless of where they are. Mobile music distribution will enhance this trend and remove the requirement of hardware and connectivity to acquire new music on the go.

> *This is a global marketplace and you can't approach a pop record and just be thinking about America, because music is one of America's greatest exports. It's the global perspective.*

—Hilly, 2005

GLOSSARY

censorship—The process of banning or deleting content considered unsuitable for reading or viewing.

cultural imperialism—The extension of rule or influence by one society over another.

emerging markets—Countries that are still growing their industrial sector, moving from predominantly agricultural products to manufacturing.

General Agreement on Tariffs and Trade (GATT)—Established in 1948 among member countries to facilitate international trade.

International Federation of Phonographic Industries (IFPI)—The global trade organization for recorded music industries.

licensing of recordings—A grant from the original record label to a foreign record label to issue an artist's recording in the latter's local or regional territory.

market saturation—The quantity of products in use in the marketplace is close to or at its maximum.

mature market—A market that has reached a state of equilibrium marked by the absence of significant growth or innovation.

net neutrality—The principle that data packets on the Internet should be moved impartially, without regard to content, destination, or source.

oligopoly—A market dominated by a small number of participants who are able to collectively exert control over supply and market prices.

parallel imports—The importation into a country of copyright goods that were lawfully made in a foreign country, without obtaining permission of the holder of the copyright for those goods in the recipient country.

peer-to-peer (P2P)—A type of transient Internet network that allows a group of computer users with the same networking program to connect with each other and directly access files from one another's hard drives.

published price to dealer—The wholesale price of recorded works and the price that sets royalty and licensing rates in many countries.

Trade-Related Aspects of Intellectual Property Rights (TRIPS)—An agreement that establishes a minimum amount of copyright protection for creative works and specifies how countries should protect and enforce these rights.

transshipment—The act of sending an exported product through an intermediary before routing it to the country intended to be its final destination.

World Trade Organization (WTO)—An organization responsible for monitoring international trade policies, handling trade disputes, and enforcing the GATT agreements.

BIBLIOGRAPHY

BBC. (2009). www.bbc.co.uk/info/purpose.

Brenneman, E. S. (2003). China: Culture, Legislation and Censorship. Excerpts from Chinese Cultural Laws Regulations and Institutions. www.freemuse.org/sw5220.asp.

BPI Statistical Handbook. (2008). British Phonographic Industry. www.bpi.co.uk.

Burnett, R. (1996). *The Global Jukebox*. New York: Routledge.

Cloonan, M., & Reebee, G. (Eds.), (2003). *Policing Pop*. Philadelphia: Temple University Press.

Empinet, D. (January 11, 2008). *Sony BMG study*. ESCE Paris business school. www.sonybmg-study.com/index.html.

Feldman, L. E. (March 25, 2005). Royalty Agreement. www.leflaw.com/index.php?module=pagemaster&PAGE_user_op=view_page&PAGE_id=131.

Global Music Retail Sales, Including Digital, Flat in 2004. (March 22, 2005). International Federation of the Phonographic Industry (IFPI). Press release. www.ifpi.com.

Global Music Sales Drop 10% in 2007. (January 24, 2008). *Times of India*. www.impactlab.com.

Hilly, J. (Ed.), (2005). Personal interview.

Holton, K. (2009). Global Digital Music Sales up 25 % in 08 – IFPI. Jan. 16. www.reuters.com.

IFPI. (January 2008). Digital Music Report 2008.

IFPI. (January 2009). Digital Music Report 2009.

Index Online. (March 7, 2003). Report: Morocco: Judge Jails Moroccan Heavy Metal Fans. www.indexonline.org/indexindex/20030307_morocco.html.

Keim, T. (2005). Personal interview.

Lathrop, T., & Pettigrew, J., Jr. (1999). *This Business of Music Marketing and Promotion*. New York: BPI Publications.

No Growth in Global Music Sales Before 2005. (2005). The-Infoshop.comPress release. www.the-infoshop.com/press/fi16775_en.shtml.

Orban, M. (2003). Censorship in Music. www.bgsu.edu/departments/tcom/faculty/ha/sp2003/gp9/docs/TCOMPAPERONCENSORSHIP.doc.

Pfanner, E. (January 17, 2009). Global Sales Fell 7% in '08 as CDs Lost Favor www.nytimes.com/2009/01/17/business/media/17music.html. *New York Times*.

PricewaterhouseCoopers. (2004). www.pwcglobal.com/extweb/ncpressrelease.nsf/docid/418FBBF773CAD0A4CA256ECA002F4489.

Reuters News Service. (August 4, 2007). Iran Detains Scores at 'Satanic' Rock Gig. www.reuters.com/article/worldNews/idUSL0423282220070804. (no longer available).

Sabbagh, D. (April 4, 2008). Tough Year Leaves EMI Trailing The Music Majors. www.businesstimesonline.co.uk. (no longer available).

Smith, P. (January 16, 2009). IFPI Report: 95 Percent of Music Downloads Are Illegal. www.paidcontent.co.uk/entry/419-ifpi-report-in-progress.

The Foreign Exchange Market. (2004). www.worldgameofeconomics.com/exchangerates.htmw.

The Industry in Numbers. (2004a). International Federation of the Phonographic Industry (IFPI). www.ifpi.org.

The Recording Industry in Numbers annual report. (2004b). International Federation of the Phonographic Industry (IFPI). www.ifpi.org.

Throsby, D. (2002). *The Music Industry in the New Millennium: Global and Local Perspectives.* Paper prepared for the Global Alliance for Cultural Diversity; Division of Arts and Cultural Enterprise. Paris: UNESCO.

Vaknin, S. (2001). The Benefit of Oligopolies. www/suite101.com/files/topics/6514/files/capitalism.rtf. (no longer available).

World Trade Organization. (2004). www.encyclopedia.com. www.wto.org/english/thewto_e/whatis_e/inbrief_e/inbr00_e.htm.

CHAPTER 17

Tour Support and Promotional Touring

Paul Allen

TOUR SUPPORT

Among the tools the label's marketing department uses to promote a recorded music project is tour support for the artist. Most labels view commercial radio airplay as the most important promotional outlet for new music, but right behind radio are live performances through touring.

Simply stated, tour support is financial aid paid to an artist to offset costs of touring to support the promotion of an album. It is effectively an amount of money that is recoupable, meaning it is a loan from the label that will be paid back to the label from the sale of recordings and other revenue streams shared with the label. Often, tour support comes in the form of money advanced to the artist to cover the losses incurred from being on the road. But it can also be time and energy spent by the label to promote the artist while he is touring, performing, and promoting his new album. This chapter will look at what tour support is and how it figures into the marketing mix as a component of the marketing plan.

The Decision to Support a Tour

Label support for a tour is based entirely on earning money through the sales of recordings by doing "anything it takes to get the artist to perform live on the road" to help promote a new album (Grady, 2005). A decision by a record label to support a tour with either money or services depends a lot on the strength of the proposed tour, whether the tour can be expected to result in significant added sales, and on the caliber of the team that supports the artist. For artists who have multiple rights recording contracts—360 deals—with their labels, the decision to tour also becomes an element of planning for the other aspects of an artist's career.

Among those important team members is a good agent who plans the tour and works with the label to ensure everyone agrees that the markets/cities in the plan are prime locations to sell tickets (to pay for the tour) and benefit from the performances, which will drive sales of the album. Market considerations will include the following:

- Has the artist done well in this market before?
- Are radio stations expected to support the single with airplay?
- Do the local print media write about and support entertainment?
- Does the tour sponsor have a presence in the market?
- Is there a possibility for a media partner in this market for the tour?
- Is this a good retail market for the genre and the artist?

Having a strong artist manager is helpful in the label's decision to provide tour support for the new artist. An experienced manager is an assurance to the label that the tour will be organized and well directed on behalf of everyone.

And finally, the decision to support a tour for an artist may depend on whether the artist's recording contract includes a provision for the label to earn a percentage of ticket and merchandise sales through a multiple rights recording contract. If a label has access to these income streams of an artist, there is an incentive for the label to be more forthcoming with support and reduce the risk that the recording project will not recoup.

Direct Financial Support for a Tour

Tour support can be in the form of financial aid paid to an artist to support her own tour, or to be part of the tour of others. New artists especially have a difficult time being able to fund all of the costs associated with touring, and it becomes necessary to go to the label and ask for a subsidy to cover the losses they will face on the road. Among the touring costs cash from a label helps to support:

- The purchase or lease of a van or bus
- Hotel costs for the artist and support personnel
- *Per diem* (daily food costs) for the artist and band members
- Rental or purchase of equipment for performances
- A tour manager

When a label offers financial support for a tour, it usually requires a proposed budget for the entire tour with an explanation of anticipated losses from touring, up to reasonable limits, often between \$25,000 and \$100,000 (Allen, 2007). The label will advance the money with the expectation that there will be an accounting for the use of the money for the tour and that any excess will be returned to the label when the tour is complete. Because the money is provided to support the tour, which is helping to promote the new and active single and album, tour support in this form is generally recoupable, meaning that the artist must repay the advance as a deduction from royalties earned from the sale of recordings. Also, it is not unusual for an artist who has purchased equipment or a vehicle for the tour from advance money, to return them to the record label when the tour is over.

Most advances in the recording industry are immediately commissionable to the artist manager, meaning the manager takes part of it right away. The artist-manager contract usually requires that a percentage of the artist's earnings—including advances—be paid to the manager. However, in most instances, advances from labels for tour support are offered to an artist with the requirement that they are not commissionable by the artist manager. The logic behind this is that the tour is trying to find a breakeven point with the label's support, and the manager should be paid when the artist is actually making money from the tour.

Major labels have considerably more financial ability to support new artists with tour support in both cash and services. Labels are willing to support recorded music projects in this way because of the large numbers of unit sales the company is expecting from the marketing plan. If a label does not see a return in the form of earnings for the company from touring, they will not make the investment.

Tours that help promote a major release of a new artist require considerable coordination and cooperation by the stakeholders in the project. The label is often willing to provide publicity support for the tour when it is initially announced, and then will continue the media blitz with each stop of the tour. This means setting up each performance in the press in advance, and following up tour stops with post-event press. The artist accommodates the local media at each stop with press interviews, visits the local radio stations that play (or should be playing) the single from the album, and also must be willing to appear in local record stores to help promote the new album. The artist manager must agree to travel with the artist or have someone support the artist on the road in the role of a surrogate for the manager, meaning a tour manager, personal assistant, or road manager.

Tour support is also provided for new artists who open concerts for established acts as a way to share a similar fan base (read: target market). If an opening act has music that is active in the market and is selling, the headliner's management may choose to share earnings from ticket sales because the new act is helping to sell the tickets for the show. However, if an artist is new and trying to build exposure to both fans and to radio, the artist will be required to buy in for a place on the show as the opener. Buying in means that the artist is paying for the opportunity to open for the headliner. In this instance, a label gives the artist financial tour support because he will not be making money for his performances to promote the company's album.

Artists signed to independent labels find that tour support in the form of cash is very limited, often nonexistent. The economics of independent labels prevents them from having enough resources to financially support touring artists with anything but the smallest amounts of cash. Independent labels rely on grassroots elements to support tours of their artists. Touring for the independent label is especially critical because very few receive airplay on commercial radio, which limits the exposure of their artist's music to the public. On a nationwide basis in the United States, major labels can expect as high as 100 million weekly exposures of a hit recording to individual listeners to commercial radio, as noted elsewhere in this book. Independents can't compete at that level, so they must rely on live performance and touring as the best—and sometimes the only—way to present the music of their artists.

The position of the independent label is expressed best by David Haley of Compass Records. He says, "If you're not touring and not present in the market, there's no interest . . . there's no retail interest, there's no consumer interest, there's no radio interest, no print interest, no nothing" (Haley, 2005). He says that without an artist continually going back to perform in markets that work, it becomes difficult for independent labels to stimulate interest in the record company's recordings. Touring keeps fans energized to find and buy an artist's music at the myriad retail locations online as well as at physical retail locations. Having online access to new music by independent artists assures them that the excitement they build on the road can be instantly turned into revenue by wireless and wired music players.

Tour support for many independent labels comes in the form of mobilizing street teams and e-teams through the media via the label's publicity specialist and by using many of the tools outlined in Chapter 13 on grassroots marketing. Haley says dealing with gatekeepers causes his label's staff to become more and more creative. "Unfortunately, the way the system is set up, there are a ton of gatekeepers and our job is to deal with those people while we figure other ways around them" (Haley, 2005). In some ways, this is not much different from the music business in general.

Tour Sponsors

Sometimes labels can assist with indirect tour support by working with artist management to find compatible sponsors for tours. Sponsors pay money to the tour in exchange for having a presence at live performances, matching brands with artists and their music. Products and service companies often link themselves to specific tours because they can connect with the similar target markets delivered by certain touring acts.

An example is the decision of the Ford division of Ford Motor Company to enlist singer Chris Brown to promote its vehicles at his 23-city concert tour in 2008. Brown appeared in three video commercials that were shown at each of his shows and radio commercials featuring him for Ford aired in cities prior to his appearances. The car company was wanting the 18- to 25-year-olds at Brown's performances to "know that Ford is not just a brand for your mom and dad, but that Ford is a brand for you as well" (Connelly, 2007). Brown's pre-Grammy show arrest in 2009, however, strained the sponsorship relationship.

Another example was the decision of Coors to sponsor Brooks & Dunn's Neon Circus Tour in 2002. The Coors logo was integrated into the Brooks & Dunn logo for the tour and was featured in all promotional materials much like a title sponsor. For their sponsorship, they provided 1230 cases of Coors beer (tba Entertainment, 2002) plus $2 million in sponsor support. The sponsorship helped fund the 41-date tour that required 140 people working with the tour on the road.

Some artists fear their fan base may view them as "selling out" for the promise of sponsor money, but Marcus Petersell of GMR marketing notes that an artist taking on a sponsor that embraces her perceived lifestyle will find out that fans will be okay with it. A prime example is when Cruzan Rum approached Kenny Chesney with $1 million to take their tour sponsorship with the plan to move from being the number 7 rum in the United States to number 4. Within 90 days of their sponsorship, Cruzan Rum became the number 3 rum in the United States. The Caribbean/country lifestyle of Chesney blended well with the sponsorship and his fan base.

Petersell goes on to urge labels to encourage their artists to be passionate about products they use and then seek sponsorships. It is much easier to place sponsorships if the artist already uses the product. He also suggests labels use the artist's existing fan base to pursue sponsorships. Approachable sponsors could easily be within the artist's database, and letting fans know about the artist's interest in securing sponsorships could result in additional income (Petersell, 2009).

The promise of negotiating tour sponsorships is a key element of the agreements Live Nation has with several artists. The company has agreed to provide an array of music-related services to the megastars on their roster by licensing multiple rights (360 deals) for each, which means they are the exclusive licensing agent to corporate entities that want to associate themselves with the artist. The company expects to recoup $50 million of its promised $120 million to Madonna from tickets, merchandising, and sponsorships during the first three tour cycles of its 12-year contract with her. Live Nation is pursuing similar earnings from artists on their roster such as Jay-Z, U2, and Shakira (Sherwin, 2008).

The Payoff for Record Labels

Touring can contribute to record sales, especially in the absence of any mass media presence. Indie artists who lack airplay, television exposure, and a limited online presence depend on touring for income, and their labels depend on touring to sell recordings. But major acts who tour can also reap the benefits in the sale of recordings. The chart in Figure 17.1 shows how sales in local markets tended to rise just before and after each local performance date for the artist U2.

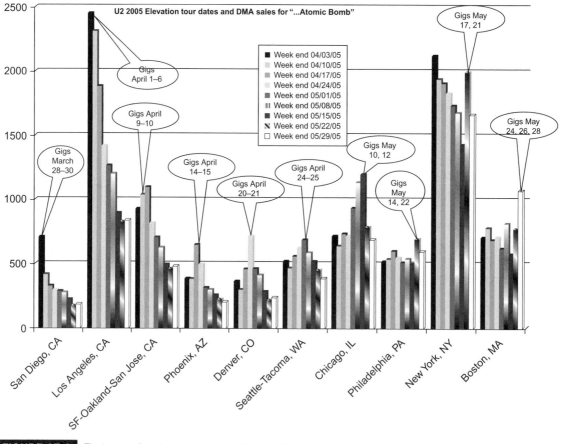

FIGURE 17.1   *The impact of touring on music sales. (Source: Thomas Hutchison, SoundScan)*

Promotional Touring

Promotional touring is somewhat different from a pure concert tour. With a promotional tour, the label is taking the time to introduce an artist to radio prior to the release of a single and ultimately an album. This kind of promotion also is the kind that includes visits by the artist to big retail accounts where possible and to the press. The difference between a promotional tour and touring in general is that a promotional tour is not funded by public appearances and it is designed to promote the artist as a commercial product to the industry before he is promoted to the consumer.

By the nature of a promotional tour, it occurs well before the release of an album, but just before the release of a single into the marketplace. Some labels prefer to travel with artists to key radio stations to introduce them to program directors and music directors. Others prefer to schedule a regional showcase where programmers are flown in for a private performance to introduce the artist. While the latter sounds like the more expensive option, an on-the-road visit to radio stations by a group plus the label entourage for 4 or 5 weeks can be more expensive than a showcase. Promo tours to retail are very similar to radio and can be the very same event. If a label goes through the expense to create an event showcasing an act, it makes sense for as many decision makers as possible to be there, taking in the same event. Labels have artists visit large retail accounts at their corporate offices prior to the initial order date. Meeting key personnel and playing music for the office staff creates good will and connectivity with the artist and her release, and hopefully will generate support through the distribution channel to retail.

Promotional touring aimed at retail, referred to as in-stores, are usually targeted at launch date, often to boost first-week sales at SoundScan and iTunes to create buzz and reorders.

GLOSSARY

advances—Cash given to the artist in the form of an interest-free loan that will be repaid as a recoupable expense, meaning it will be deducted from royalties.

agent—Someone who negotiates agreements for live concert performances with promoters, on behalf of an artist.

media partner—A local newspaper, radio station, or television station that helps promote a concert in exchange for a promotional presence in advertising and at the event.

recoupable expense—An item of cost paid by the record label that the artist agrees to pay back to the label in the form of a deduction from earned royalties.

tour manager—An individual who coordinates an artist's live performances while on the road and who works directly for the artist manager.

tour support—Money or services provided by a record company to offset the costs of touring.

BIBLIOGRAPHY

Allen, P. (2007). *Artist management for the music business*. Boston: Elsevier/Focal Press.

Connelly, M. (2007). Ford sells sync with soul. *Automotive News, Dec. 10*, 24A.

Grady, J. (2005). Personal interview.

Haley, D. (2005). Personal interview.

Petersell, M. (2009). Panel discussion at the leadership music digital summit, March 24, Nashville, TN.

Sherwin, A. (2008). Live nation to stick with star strategy and exclusive tour deals despite leadership split. *The Times (London), June 30*, 42.

tba Entertainment. (2002). *Neon circus program book*. tba Entertainment.

GENERAL REFERENCES

Berkman, B. (2002). www.starpolish.com./advice/.

Passman, D. (2006). *All you need to know about the music business*. New York: Simon and Schuster.

Waddell, R., Barnet, R., & Berry, J. (2007). *This business of concert promotion and touring*. New York, NY: Watson-Guptill.

CHAPTER 18

Strategic Marketing and Special Products

Amy Macy

CONTENTS

In an era when retail floor space dedicated to music CDs is diminishing, when the single—specifically the file—is fast becoming the music format of choice, and consumers have so many options when it comes to spending their entertainment dollars, record labels are looking to monetize their assets in as many ways as possible. The "recording," along with the artist who produced it, has become marketable in so many more ways, via the use of Internet, licensing agreements, cross-merchandising, and using the artist's other talents to magnify and exploit all money-making opportunities, with dedicated label staff to mine these possibilities. In the past, these gratuitous ventures had been called "Special Products and Special Markets," meaning that their profits were considered "gravy" to the label and not projected funds to the bottom line. But with record sales still on the decline, labels look for these products and markets to make up for revenue that has since been lost in the traditional marketplace.

LIFESTYLE

With the emergence of lifestyle and specialty product retailers, a new outlet for music has evolved. When grabbing a cup of coffee from Starbucks, it's hard to ignore the "cool" music playing and CD display right under your nose. A visit to Pottery Barn or Restoration Hardware elicits more than just home décor: an ambience that is enhanced by the music playing—which you can replicate in your own home through the purchase of the "branded" CD.

Joss Stone

With the purchase of $60.00 or more, Gap consumers would receive a gift compilation CD containing 8 songs along with a director's cut of the television commercials and exclusive behind-the-scenes footage of the making of the advertisement (The Gap, 2005).

FIGURE 18.1 *Joss Stone.*

More and more retailers are creating "branded" CDs as product extensions of their retail offerings. Pottery Barn has offered over 70 CD **compilations** that have sold millions of units (Rock River, n.d.), all of which are an "easy way to put lifestyle in a home," according to Patricia Sellman, marketing vice-president at Williams-Sonoma (Strasburg, 2003). The Gap Fashion Retailer wanted to drive sales of their merchandise through the use of music. In-store play as well as exclusive music and video CDs and DVD offers were enhanced by using the artists as part of the overall advertising campaigns. The initial Madonna and Missy Elliott combo spawned several other creations that included Jason Mraz, John Legend, Joss Stone, and Keith Urban—all of which included high-visibility promotions with TV and print ad campaigns that Gap attributes to a 30% rise in sales for their fashion line alone (Rock River, n.d.).

Starbucks promotes music at another level. Not only does the retailer sell music CDs through their retail coffee stores, but also through the web site www.hearmusic.com, Starbucks logs the playlists of songs being aired in its stores nationwide. If a consumer hears a song she likes, she can go to the web site and find out the title and performer, with additional links to the performer's catalog and online sales site (Finn, 2002).

In an unprecedented event, Starbucks established itself as a significant music retailer. Coinciding with the national movie release of *Ray*, Ray Charles's final recording *Genius Loves Company* debuted on the *Billboard* Top 200 charts at number 2. Starbuck's sold more than 44,000 units, which was approximately 22% of total album sales. This percentage outsold both traditional music retailers and mass merchants—making Starbucks the number 1 seller during the CD's first week of sales. "Starbucks continues to be thoughtful and strategic as we extend our reach into the music and entertainment arena," said Ken Lombard, president, Starbucks Entertainment. "Our expertise in music through Hear Music brand, coupled with our unparalleled distribution and consumer reach, created what we hope to be the first of many successful music collaborations. It is truly gratifying to see our customers embracing the fabulous CD and purchasing it in our stores" (*Business Wire*, 2004).

And their beat goes on. Starbucks signed Paul McCartney to an exclusive artist deal on their Hear Music record label and released *Memory Almost Full* on June 5, 2007. Starbucks Stores hosted a global listening event on the day of release that coincided with the artist's birthday. Of the 10,000 international stores participating, 10 Starbucks locations invited fans to participate in a unique video tribute that would premiere on June 18 via the Internet. The Starbucks XM Channel also created special programming featuring McCartney and his new release, and by 2009, Starbucks had sold over 300,000 units of the 620,000 records purchased—meaning that over 50% of McCartney's sales occurred in Starbucks (Cavarocchi, 2007).

So what is going on here? Why are music consumers purchasing their CDs from the most unlikely **brick-and-mortar stores**? Starbucks looked at the tumultuous music retailing environment as "the perfect storm." As retailers began their pricing wars with the $9.99 CD, music became devalued and the actual purchasing experience less than sexy. Coupled with the consolidation of radio, which lessened the importance of disc jockeys and their abilities to educate listeners to new music, Starbucks has capitalized on these forces which "have fractured the traditional music industry and diminished the sense of discovery many consumers used to feel about music, giving non-traditional retailers an 'in' to capture those consumers" (Yerak, 2004).

The Benefits

Why, then, are artists aligning themselves with retailers? What is the benefit to them and their existing record label partners? Initial response would be money. The licensing of existing recordings from labels can be very lucrative. Because labels never know the value of their recordings long term, record companies have found it hard to calculate the value of licensed product. Historically, the licensing of music has been considered "gravy"—monies generated beyond the "meat and potatoes" of the business. But to address the burgeoning market, many conglomerates have created strategic marketing departments devoted to mining the catalog of their holdings and finding new, innovative homes beyond traditional music channels. Additionally, private marketing entities have emerged as potent messengers for products, including an innovative organization called Rock River, a leader in **branded CD** production.

Artists also win in the exchange of **private label** products that carry their music and image. Hallmark Gold Crown stores have partnered with several artists; James Taylor's *A Christmas Album* sold over 1 million units during the 2004 holiday season. Taylor's previous release, *October Road*, released in 2002, was the final recording for his Columbia Records contract, and he had not resigned with any other label. "Mr. Taylor is thrilled that so many people are getting to hear the music and that he had an opportunity to record this album," said Gary Borman, who manages Mr. Taylor and stars like Faith Hill for Borman Entertainment. "There were no financial incentive issues to worry about and no artistic issues either. It was a very fair sharing of the pie. It made it simple" (Ogunnaike, 2004). With the purchase of three Hallmark cards, consumers could buy James Taylor's *A Christmas Album* for $6.95, or could buy it outright for $10.95.

Hallmark released a Valentine's Day–themed CD in 2005. Using market research, "We went to our core consumers and asked what artists they're connected with. Martina (McBride) was on the top of the list. It's to drive new consumers in, Martina McBride fans that may not currently shop in the store," says Ann Herrick, integrated marketing manager for Hallmark (Ebenkamp and Wasserman, 2004).

In the case of James Taylor, Hallmark offered an after-burn market for the artist and his music, capitalizing on name value and music, as well as tapping older consumers familiar with his work who are no longer active music purchasers. Martina McBride may find new fans, because current fans with similar buying practices already frequent Hallmark, making the alignment a win–win for all three parties: Hallmark, Martina McBride, and RCA, her record label.

Since these landmark releases, Hallmark has continued their product extensions with releases from George Strait, Michael McDonald, Barry Manilow, Michael Bolton, and Elvis. Along with these artist titles, the Hallmark Music collection has increased to include 32 exclusive everyday titles such as *Sleep Tight for Baby*; *In the Mood for Love*, a collection of romantic standards from jazz legends; and *Sing Along*, a collection of children's classics.

And for Christmas 2008, Sheryl Crow released exclusively with *Hallmark Home for Christmas*, which included 10 holiday favorites—Sheryl style— with an original Christmas tune: *There Is A Star That Shines Tonight*. She performed this song on The Jay Leno Show December 15, 2008. The success of this "private label" venture is unprecedented, with both James Taylor and George Strait earning Platinum status and Michael McDonald and Barry Manilow mining Gold. This shows how a specified product line such as greeting cards can enhance their bottom line with a product extension of music utilizing the same target market—and the artists win too.

Enhancing Careers

"Matching music to an automobile's image is essential to developing a consistent and credible message," says Rich Stoddart, Ford Division's marketing communications manager. In recent years, car advertisers have paired Led Zeppelin with the new generation of Cadillac, Aerosmith's *Dream On* with Buick LaCrosse, and Jimi Hendrix's rendition of the *Star-Spangled Banner* with the 2005 Ford Mustang. While music "won't sell the car," says Gordan Wangers, president of AMCI Inc., an automotive consulting firm in Marina Del Rey, California. "It is capable of grabbing the attention of the right audience to say, 'Here's my new car and here's my message'" (LaReau, 2004).

Ford uses country music to market its F-series pickups. The theme song *Ford Truck Man* was written and performed by country superstar Toby Keith. Clearly, the attention of the right audience has been grabbed. Research validates that pickup truck owners are country music listeners, and Ford is trying to reach this broad audience where it taps its foot . . . and gas pedal. Aggressive advertising including radio, television, and print prominently displays the artist's image along with the pickup—to ensure a connection between the messenger and the message to the consumer.

But in the case of Mitsubishi, the music can get in the way of the auto-maker's marketing message. Their TV advertising campaign featured a band called Dirty Vegas, showing young people dancing in their cars to the group's techno-music. "Mitsubishi sold a ton of records for Dirty Vegas but not a lot of cars. It was a marketing debacle," stated market analyst Gordan Wangers (LaReau, 2004). The 2002 release by Dirty Vegas has scanned over 550,000 units, with the bulk of those sales during the Mitsubishi advertising push. Since then, Dirty Vegas has released several more albums, with none of them scanning over 10,000 units.

In recent years, television shows have helped to highlight the talents of rising stars. *Grey's Anatomy* started the phenomena and expanded its brand with a CD compilation of its own, featuring artists from the music of the show. With three volumes released since 2005, artists such as KT Tunstall, The Chalets, Snow Patrol, John Legend, and many more, have received mainstream visibility that has enhanced their careers. *The OC* has copied this blueprint but has a sophisticated purchasing model via the web where fans can visit the "Music from The OC" web site (www.musicfromtheoc.com) and can click through on any one of the six volumes of music derived from the seasons and buy it, created by Warner Brothers in affiliation with Fox Television.

Product Extensions

Under the description of "special products," many labels are attempting to capture more of the same consumer's dollar by creating product extensions of a specific artist's release. The concept is to cover the losses being generated by file-swapping and burning by creating "supersize" versions of the bigger releases and targeting hardcore fans that are willing to dig deeper into their pockets in order to have the "special" release. According to research done by the Handleman Company, a former rack jobber to mass merchants Wal-Mart and Kmart, "23% of music buyers account for an estimated 62% of album sales, buying an average single CD every month" (Leeds, 2004). Though sales vary, a deluxe edition commonly accounts for 10% to 20% of an album's overall volume. And they usually carry a much higher profit margin.

Who can forget the "special" release of Interscope's landmark release of U2's *How to Dismantle an Atomic Bomb*? This record had three versions: the 11 track, $10 basic version, or for $32, a fan could purchase the "collector's edition" including a DVD and 50-page hardcover book. The mid-priced version had the DVD but no book. But if a consumer bought a new "U2" Apple iPod, he got the preloaded version of *How to Dismantle an Atomic Bomb* for free.

Beyonce's alter-ego release of fall 2008 *I Am . . . Sasha Fierce* was released in a two-CD version that sold for $4 to $5 more than the single-disc version, according to Mathew Knowles, her manager and father. He said that the less expensive edition was aimed at meeting the demand for CDs priced under $10 and expected sales to split 50–50 between the two versions (Willman, 2008). The release topped the Top 200 Album chart, selling 482,428 units the first week and remained in the Top 10 for 3 months. Interestingly, in that period, of the 1.8 million that sold, only 7% were digital album sales, which is far below the national average of approximately 30% (SoundScan, 2009).

In the case of the Bruce Springsteen album released in conjunction with his Super Bowl XLIII performance in January 2009, the deluxe version contains content not available with the download, but in comparison, the $14 deluxe iTunes version had some of the video content of the extra DVD. According to SoundScan (2009), *Working on a Dream* debuted at number one the first week and scanned 223,741 units, with 20% of the sales being digital. And the next release from U2 *No Line on the Horizon*, was made available in five different versions in retail and two versions by download, including a bonus track only available from iTunes (McElhearn, 2009).

"The concept of holding on to a fan was eroding for a lot of reasons— other interests in the digital world," said Tom Whalley, chairman of Warner Brothers Records. "We've sold the CD as if there's one type of buyer for 15 years. There are different kinds of buyers." Warner Brothers Records released three versions of the November 2004 Josh Groban album *Closer*, including a "fan edition" that was sold only through the singer's web site and fan club. The basic CD included 13 songs; the limited edition included two extra songs and a DVD; and the fan version, listed for $29.98, included two songs more and the DVD. Total sales include over 5.7 million copies for the regular CD to about 15,000 for the topline version (Leeds, 2004).

Labels have continued to find the limit of their CD offerings, and they "found that there's not a ceiling," said Jennifer Schaidler, vice president of music for Best Buy, the electronics retailer based in Richfield, Minnesota. "If you put the right value to the consumer, you can charge more" (Leeds, 2004).

Compilations

Since moving pictures became "talkies," music has been an integral element to the success of the movie business. Early in the evolution of movies, the industry would hire actual singers to play a part, integrating music and story onscreen. Often, the music from these movies would become nationwide hits and the popularity of the singing stars would explode, making many of them classic voices of the genre. Movie soundtracks have continued to evolve, with the 1960s delivering the Beatles *Hard Day's Night*, and *The Graduate* with its hits *Mrs. Robinson* and *Scarborough Fair* by Simon and Garfunkel. The '70s included *Urban Cowboy, Grease,* and *Saturday Night Fever*, all containing a common thread—one very hot actor, John Travolta. The '80s brought *Footloose, Dirty Dancing,* and Roy Orbison's famed *Pretty Woman*. The '90s elevated Whitney Houston to a new level of superstardom with her film debut in *The Bodyguard*. And Celine Dion reached new heights with her megahit *My Heart Will Go On* from the epic *Titanic*. All of these soundtracks contained songs that would eventually become huge pop hits, selling millions of units and igniting a new kind of consumer to purchase music. And in some cases, a single song has helped to launch a career and create a megastar.

Soundtracks reflect lessons learned from the past as well as the purchasing practices of today's music consumers. With file-sharing, many music purchasers have become their own disc jockeys, collecting and burning their "personal soundtracks" containing the individual songs that they want to hear. By compiling songs that capture this spirit while representing the theme of a movie, soundtrack managers have fashioned a "genre" of music that has produced sales.

Overall sales for the genre have slipped in recent years, but the numbers are a direct reflection of the business in general. Between 2006 and 2008, soundtracks maintained over a 4.5% market share of all albums sold, according to SoundScan (2009). The genre can hang its star on the point of Disney with its branded movie and television series *High School Musical (HSM)* and *Hannah Montana*, which have flooded the preteen and teen marketplace not just with music products, but with every conceivable licensed product imaginable, from Corn Flakes with free download opportunities to the musical toothbrushes that will eradicate the breakfast item from your kid's enamel. Not only are these marketing opportunities that create in-store impressions, but the licensing of the music generates additional revenue for the label and, in Disney's case, for the product's branding department associated with the icon such as *HSM*.

2008 created a resurgence in the music of Abba through the movie *Mamma Mia*, which followed the popular international tour of the stage musical by the same name. With 1.4 million units in U.S. sales alone, this release speaks to the power of the baby-boomers who seeks music that they can relate to and want more of. Other highlights include the dark teen hit *Twilight*, which featured a collection mostly of new rockers and also sold over 1.1 million. Bigger soundtracks for 2007 outside of the Disney imprint included older musical remakes including *Hairspray* with 921,000 units sold and *Dreamgirls* with 704,000 units sold.

Record label A&R reps along with music publishers mine the film and television opportunities aggressively. Placing a recording and/or song on a movie soundtrack or television score can only positively affect the bottom line of a company. Developing relationships with music supervisors for film and television is key to securing these coveted positions, even utilizing password-protected web sites that aid in the connection of song and artist with opportunity. *The Hollywood Reporter* along with *Variety* magazine keep industry watchers in the know as to upcoming productions, cueing label and publishing reps of opportunities. In addition to in-house representation, both labels and publishers are using the knowledge and connection of **specialty brokers**. Similar to independent radio promoters, these brokers will represent a catalog in the marketplace and look to be rewarded in the success of the movie or television placement by receiving a percentage of the royalties.

There is a new concept in which movie makers are contracting prominent musicians and artists to conceive the music and create an original score. In this case, the movie production company has created a "work for hire" scenario, where the film company then owns the music. And, of course, if the movie and/or soundtrack gains popularity, it's the movie production company that reaps all the benefits.

Brand Partnerships

Taking a page out of the soundtrack playbook, nonmovie compilations have emerged as powerful sales items as well as marketing tools. The *Now!* series is considered the most successful compilation collection in the history of music sales. The U.S. participating collaborators include EMI Recorded Music, Universal Music Group, Sony Music Entertainment, and the Zomba Group, with the albums rotating among Sony, Universal, and EMI for marketing and distribution. The U.S. series has released 29 compilations selling approximately 62 million U.S. units since its inception in 1998. Key to its success has been the hits—since the brand relies on the top singles to drive sales.

Intangible is the impact on sales of participating artists. Someone purchasing a *Now!* CD might know three or four of the artists on the package, but the consumer most likely is introduced to a new act that he might also enjoy. The residual purchase of the full-length product of that new act is the result of this **cross-promotional** item.

The new CDs allow access to a web site that features acts below the usual superstar status and offers extra promotion songs and ringtones from *Now!* artists. Also in the works is a potential television show that would highlight *Now!* acts as well, which is an effort to try and compensate for overall lost sales volume in the marketplace. An interesting observation by *Now!* executives is what they perceive as its "bifurcated" demographic, meaning that the releases' target market tends to be girls ages 8 to 13 who are too young to purchase music online or women ages 24 to 30 who buy the music for themselves, not their children (Sisario, 2008).

Products in the Music

Busta Rhymes invented the modern-day product placement with his massive single *Pass the Courvoisier*. It was a hit: at both radio and consumers of Allied Domecq brand of cognac. Although the rapper was not paid to write the top-shelf liquor into his song, the artist wrote the hit after trying the drink, compliments of hip-hop entrepreneur Russell Simmons, who instigated the relationship between the brand and the artist and has continued all sorts of brand extensions, including a clothing line that included Motorola partnering with the attire division to position its phones as a fashion accessory.

Maven Strategies, a marketing agency that specializes in placing corporate brands into songs, believes that more artists will consider product placement deals since hip hop specifically sells product. To minimize risk for the corporate partners, Maven Strategies have created compensation programs that reward the hits that receive heavy airplay, while songs that do not do well on the radio receive less. Deals have been structured where artist/songwriters would receive $1 to $5 each time their song played on the radio (Graser, 2005). "We value what the hip-hop community brings to the table, but we deal in an environment of unknowns. It's important to structure agreements that fairly compensate the artists for their capabilities while minimizing the investment risk for our clients," says Tony Rome, president of Maven Strategies (Banerjee, 2004).

Not originally written into the song, Chris Brown's *Forever* borrowed lyrics from the perennial Wrigley's Double Mint gum theme: "Double your pleasure/Double your fun." It was a natural for Wrigley to "double back" and approach the burgeoning hip-hop youth to remake his hit and borrow imagery from his *Forever* video to launch a national television/Internet advertising campaign, utilizing his music and artistry, with Brown as the star.

Actual labels are now creating imprints to accommodate corporate requests. Island Def Jam has created a private label with Proctor & Gamble that will sign and release albums by new hip-hop acts and is named after a brand of body spray owned by the company. Hip-hop and R&B producer Jermaine Dupri, a music-industry veteran, will head up this joint venture and musically accommodate this P&G who will fund the production, promotion, and then distribute the product themselves.

Nike has gotten into the game too, and the music was so strong that it was nominated for a Grammy Award for best rap performance by a duo or group. Kanye West featuring Nas, Rakim, and KRS-One wrote a song themed and titled *Better Than I've Ever Been* that does mention Nike sneakers (along with key "graffiti" graphics in the video). The caveat is that profits went to the Force4Change Fund, a charity for youth leadership program—but the promotional benefits were all Nike's (Levine, 2008).

Retail Exclusives

To lure consumers into a specific store, many music retailers are negotiating **exclusive** product for their customers. This **value-added** product is called by the industry "superior" versions of an album, and the independent music store is "in the position of selling an inferior product. It has a corrosive effect," says Mike Dreese, co-owner of the 27-store Newbury Comics chain in New England (Mansfield, 2007). For the 2008 holiday selling season, Target released its very own Christina Aguilera hits album *Keeps Gettin' Better* complete with new cut, video, and holiday commercial for not just the release, but Target imaging too. Best Buy's reissue of Pearl Jam's 1991 best-seller *Ten* includes a token which is redeemable for three bonus tracks for the game "Rock Band" on both platforms Xbox 360 and Playstation 3 when purchasing the deluxe version of the album. John Mellencamp's release *Freedom's Road* included something for everyone: a 4-song bonus track at Best Buy, a DVD at Wal-Mart and downloads at Wal-Mart.com and Circuit City, and a video and two rough mixes at iTunes (Mansfield, 2007).

But the ultimate "exclusives" within the recent past must be awarded to Wal-Mart and their arrangements with the Eagles and AC/DC. The Eagles *Long Road Out Of Eden* was a double CD with the low price-point of $11.88 and released on October 30, 2007, just in time for the holiday shopping season. With over 3 million units sold to date, this release paved the way for the next private label signing and release: AC/DC's *Black Ice* streeting on October 2, 2008, and it could not be missed when entering the store. The men's clothing department became an AC/DC merchandising mecca with T-shirts, signage, and purchasing opportunities from the moment shoppers entered the store to when they checked out. Over 2 million units have sold to date.

Not to be outdone, the "nontraditional" music stores have endless offerings as well. Amazon's catalog of exclusives includes a *15 Years of Veggie Tales Box Set*, an *Elvis Costello Limited Edition Collectors Box Set*, a remastered *The Distance* by Bob Seger, and many, many more. And the ultimate digital store, iTunes, has paved the way with its hardware/software combination of the iPod and iPhone that dominate the marketplace. Its music store contains over 10 million songs, 40,000 TV episodes, and 4000 films that include over 1000 in high definition. So it would only make sense (and cents) for this outlet to be the exclusive gateway of selling the performances of television's most watched talent show *American Idol*. Through Fox's web site myidol.com, streaming video of the previous day's performance has a click-through icon where consumers are forwarded to the iTunes store and can purchase the latest rendition by their favorite contestant (The Futon Critic, 2009). iTunes even has a service where consumers can be alerted to new exclusives that are coming online, since they are time-sensitive, such as Metallica's digital box set. Exclusively released to iTunes on March 31, 2009, other digital outlets were selling on April 28 (Ruggieri, 2009).

FIGURE 18.2 *iTunes homepage. (Courtesy of Apple)*

Who wins in these exclusive offerings but the consumers who choose the "right" store at which to purchase their new music? But some independent stores look at the **superior product** practice to be anticompetitive and hostile, and have decided to apply sanctions within their store environments. Pricing and positioning advertising along with awareness programs will not be enforced, and the stores will refuse to report sales or chart positions for any artist and product released by the affected label. Although these stores can be instrumental in launching the careers of developing acts, the overriding power of the larger merchants usually undermines these types of policies, and the practice continues.

The trade-off for giving away exclusive tracks is the gaining of valuable marketing real estate and promotional efforts put forth by the retailer. Specifically in the holiday season, marketing programs such as front window displays, product dumps, and in-store positioning can double in price. But at any time of year, marketing efforts with national retailers are expensive, so leveraging the assets of the record company—a single track of music that may have never otherwise seen the light of day—can be smart business and a profitable venture.

Tour Sponsorship and More . . .

Not to be confused with label funded tour support, **tour sponsorships** are considered those ever important financial relationships between product and service providers with artists who can emote their message. Beyond tour sponsorship and sync licensing, many record labels wanting to tap the deeper pockets of consumer brands have positioned their acts for lucrative partnerships. Activities such as product placement, DVD underwriting, and digital tie-ins are being explored, with specific target markets to be exploited.

Traditionally, tour sponsorship was monopolized by "sin" products because there were so many restrictions in ad placement and targeted audiences. But new entries into the tour sponsorship business reflect star power and the need to sell . . . anything and everything.

Talk about a target, AARP sponsored Tony Bennet in a 20-city tour in the spring of 2007 for $900,000. "Tony is the perfect musical ambassador for AARP," says Shereen Remez, AARP group executive officer for member value. In exchange, Tony appeared at several AARP events and even sang at a sweepstake winner's home,[1] all in an effort to create more visibility to younger prospective AARP members as well as increase visibility to other AARP offerings.

---

[1] www.todaysseniorsnetwork.com/tony_teams_with_aarp.htm.

For $1.25 million, Fergie forged a complicated tie-in with Candie's shoes and apparel along with Kohl's department stores where consumers were encouraged to try on Candie's product in exchange for bubble gum that would reveal an "instant win" message. Winners would receive tickets to an upcoming tour event, one of a 20-city tour, where they would become onstage models during the concert, styled by Fergie's tour stylist. In addition, concert and event footage would then be uploaded to Candies.com for additional viewing and promotion (*Billboard*, 2007).

Philanthropic in nature, Build-A-Bear Workshop—in conjunction with Disneymania Concerts—sponsored "make your own stuffed animal" events at regional zoos where $1 of the proceeds was donated to the World Wildlife Conservationists, and the Disney act T-Squad performed. Although small in comparison, this "sponsorship" cost $175,000 for a 20-zoo tour.

As Blackberry launched its new Curve multitasking phone, it wanted to reach a younger demographic to show the capabilities of Blackberry beyond the workplace. They chose John Mayer and his 2007 Tour to deliver this message, and although dollar value of this tour sponsor and exchange was not disclosed, most industry analysts believe that it far exceeded the $1 million mark (Waddell, 2007).

These deals are secured in various ways. Sometimes the record label and its conglomerate will help broker the agreement by being the intermediary for artist and product manager. Additionally, artist managers can be approached directly from product representatives. And on occasion, a third-party music marketing company will help with a hook-up between a product and artist, assisting in the navigation of very different corporate cultures and agendas.

The Digital Age

Buy a Coke, win a free download. Purchase gasoline, win a free download. Click on a pop-up ad, win a free download. . . . The ability to transfer music, via a digital file, with such ease and casualness has caused a bonanza in "free" downloads, with the idea that it's a "prize." But what has that done to the perceived value of the music? Have we not created at least a generation, if not an entire marketplace, that does not value the music anymore—but sees it merely as the dispensable Happy Meal toy in the box? The digital element is a great sampling tool, no doubt—but it's so easy to fill up on the free appetizers that consumers seem not to want to buy the real thing. Here are a couple of digital marketing push strategies.

Via social networks such as MySpace and FaceBook, fans can use Musicbox to embed ad-driven video players that are artist specific of Sony BMG acts and label web sites. Through these viral word-of-mouth efforts, not only do artists gain secondary and tertiary exposure, but the brand partnerships that include AT&T, Honda, Chase, Puma, Starwood, MSN, and L'Oreal gain cool cache as they rub up against the pop culture of the day. In the giveaway category, Burger King created a "microsite" where restaurant goers could obtain a code with purchase to then sample and download a free song from EMI. Cover Girl Cosmetics created a similar deal with Rihanna but used the artist as the spokesperson as well as the musical magnet (IFPI, 2008).

No doubt that the marketing "punch" is attractive when cutting these deals, but the hope is that the sampling of the music will entice a purchase. But looking at the numbers straight up, the technology seems to be the very reason why some are not buying. However, we cannot forget that there are so many other variables that are contributing to the decline of music sales.

As the discretionary income of consumers continues to fragment to the many entertainment choices, music makers are becoming very creative in connecting with buyers. As the "perfect storm" continues to rage, with file sharing, price wars, radio consolidation, and conglomerate mergers, record labels and their artists will have to continually mine the marketplace for willing partners and entertainment synergy. No longer are "special" markets and products considered incremental business to the bottom line, but this evolving revenue source is essential to the survival of the industry.

GLOSSARY

branded CDs—A CD sponsored by and sporting the brand logo or mention of a company not normally associated with the release of recorded music. Examples include product designed for and sold at Pottery Barn, Pier 1 Imports, and Victoria's Secret.

brick-and-mortar stores—Businesses that have physical (rather than virtual or online) presences, in other words, stores (built of physical material such as bricks and mortar) that you can drive to and enter physically to see, touch, and purchase merchandise.

CIMS—The Coalition of Independent Music Stores.

compilations—A collection of previously released songs sold as a one album unit, or a collection of new material, either by single or multiple performers, sold as a collaborative effort on one musical recording.

cross-promotion—Using one product to sell another product, or to reach the market of the other product.

exclusives—Retail exclusive marketing programs that are not offered to other retailers, but arranged specifically through one retail chain.

private label—A label unique to a specific retailer.

tour sponsorship—A brand or company "sponsors" a concert tour by providing some of the tour expenses in exchange for product exposure at the events.

superior products—A value-added version of a product that is sold in the "regular" version elsewhere.

value-adds—To lure consumers into a specific store, music retailers offer exclusive product with added or "bonus" content or material for their customers.

BIBLIOGRAPHY

American Idol performances available exclusively on the iTunes Store. (2009). The futon critic, Feb. 17. www.thefutoncritic.com/news.aspx?id=20090217fox01.

Banerjee, S. (Nov. 6, 2004). Billboard bows ringtone chart. *Billboard*, *116*(45), 47.

Cavarocchi, D. (2007). *Starbucks connects music fans worldwide to McCartney's new album through unprecedented global listening event*. Starbucks Press Room May 17.

Ebenkamp, B., & Wasserman, T. (Sept. 20, 2004). The biz: McBride (Hall)marks v-day. *Brandweek*, *45*(33), 10.

Finn, T. (2002). *The Kansas City Star*, May 9.

Graser, M. (2005). McDonald's buying into hip-hop song lyrics. *Advertising Age*, *Mar. 23*.

IEG/Billboard tour sponsorship. (Feb. 3, 2007). *Billboard*, *119*(5), 19.

IFPI Digital Music Report. (2008), p. 17. www.ifpi.org/content/library/DMR2008.pdf.

Jeans and music—The perfect fit at Gap this fall. (2005). The Gap. Press Release, Aug. 22. www.prnewswire.com/mnr/gap/22461/.

LaReau, J. (Dec. 20, 2004). *Automotive News*, *79*(6126), 22.

Leeds, J. (2004). $10 for a plain CD or $32 with the extras www.nytimes.com/2004/12/27/business/media/27music.html. The *New York Times*, *Dec. 27*.

Levine, R. (2008). It's American brandstand: Marketers underwrite performers www.nytimes.com/2008/07/07/business/media/07music.html. *The New York Times*, *July 7*.

Mansfield, B. (2007). Exclusives aim to pull music fans into stores www.usatoday.com/life/music/news/2007-03-01-exclusive-music_N.htm. *USA Today*, *Mar. 1*.

McElhearn, K. (2009). The future of CDs www.macworld.com/article/138861/future_cd.html. *Macworld*, *Feb. 17*.

Ogunnaike, L. (2004). The media business: Advertising. *The New York Times*, *Dec. 17*.

Rock river music. (n.d.). The Pottery Barn. www.rockrivermusic.com/case-studies/pottery-barn.html.

Ruggieri, M. (2009). Music notes: Jason Mraz's lost Grammy speech www.timesdispatch.com/rtd/entertainment/music/article/W-RIFF19_20090218181702/209954/. *Richmond Times-Dispatch*, *Feb. 19*.

Sisario, B. (2008). Now that's what I call marketing: Pop hits and more www.nytimes.com/2008/11/12/arts/music/12now.html. The *New York Times*, *Nov. 12*.

SoundScan data—2009.

Starbucks leads unprecedented change in music industry, and helps take Ray Charles to the top. (2004). *Business Wire*, Sept. 9.

Strasburg, J. (2003). Bands to fit the brand: S.F. company's CDs sell for likes of Pottery Barn www.sfgate.com/cgi-bin/article.cgi?f=/c/a/2003/07/25/BU267694. DTL. *SFGate.com, July 25.*

Waddell, R. (June 2, 2007). BlackBerry juiced; unprecedented sponsorship of John Mayer tour will help launch curve phone. *Billboard, 119*(22), 10.

Willman, C. (2008). Split personality for new Beyoncé album http://hollywoodin sider.ew.com/2008/10/split-personali.html. *Hollywood Insider, Oct. 15.*

Yerak, B. (2004). Merchants range far beyond their staples. *Chicago Tribune, Nov. 11.*

CHAPTER 19

The Recording Industry of the Future

Tom Hutchison

The record label of the future is expected to be very different from what has been in place for the past 50 years. The current business model is still structured around music as a physical product as the primary source of income, sold through brick-and-mortar retail outlets, and is just now extending to the sale of digital product through online retailers. The record label of the future may be shifting away from that of product provider to more of a service provider, much like television networks are today. The labels' role may be limited to developing *content* and then marketing that content to consumers. This represents a drastic paradigm shift for record labels that have resisted moving into the digital age. Labels are also moving to capitalize on all income sources for the "brands" they finance and develop under a concept called the 360 deal.

Before unraveling the mystery of the future of recorded music, we must first examine the basic principle behind recorded music and how it is consumed. Then we must examine how, through the course of time, new innovative methods replace old methods of accomplishing this task. In a nutshell, recorded music allows the consumer to possess a captured musical performance and replay that performance whenever and wherever desired. The music business for the past 100 years has been supplying that to consumers in the form of a physical product. But many things have changed, especially in the fields of entertainment and communication. To understand the future, we must move away from the concept of supplying consumers with a physical product. After all, the goal is for the consumer to replay the recording at will, and modern technology does not mandate that it must be in a physical form. Instead, we should look at the record

CONTENTS

Table 19.1	Communication flow				
Source		**Medium – channel**		**Receiver (hardware)**	
Labels – product provider	→	Cylinder	→	Phonograph, graphophone	
		→	Vinyl disc	→	Gramophone, Victrola
Service provider	→	Radio broadcasts	→	Radio	
Labels – product provider	→	8-track/cassette	→	Tape player	
		→	CD	→	CD player
		→	MP3 via Internet	→	Computer
Service provider	→	Digital wireless	→	Universal mobile device	

Adapted from C. E. Shannon and W. Weaver, The Mathematical Theory of Communication *(1949).*

business as *music delivery systems*, because the goal is to get consumers to pay for the convenience of listening to their favorite music on demand.

The new model for record labels as service providers involves a drastic overhaul of the label system. The chart above illustrates how within the past 100 years, with the exception of radio, all new music delivery systems have followed the same basic format of creating a physical product. The one exception, radio, is not an on-demand system capable of being user controlled; and at this point, terrestrial radio does not provide direct income for the labels—although changes in this policy are being debated at the time of publication of this book.

LABEL PARTNERSHIPS

Changes in music delivery systems have at times driven changes in record label ownership and partnerships. When radio was first developed in the 1930s, radio networks sought to own record labels. RCA, with its NBC Radio Network, purchased Victor records in 1929, and CBS purchased Columbia and Okeh Records in 1938 (Sony, 2005). From the 1970s through the 1980s, entertainment hardware companies sought alignment with labels. Philips electronics created PolyGram Records in 1972 (Hoover's, 2005), and Thorn Electrical Industries purchased EMI in 1979. Sony purchased CBS records in 1988, and Matsushita bought MCA Records in 1990. This alignment started to change in the 1990s when telecom and utility companies took an interest in record labels. Vivendi bought Universal in 2000, and AOL merged with Time Warner (parent to the Warner Label Group) in 2001. Bertelsmann AG, parent company to RCA and BMG, ventured into the broadcast media industry (PBS, 2002). These mergers and relationships were designed to create *synergy*, defined as a mutually advantageous conjunction or compatibility of distinct business participants or elements (as resources or efforts) (Merriam-Webster Online, 2005). Synergy allows for a company to use one branch to support the activities of another branch. Disney is an example of a company that has used synergy to create a "superbrand" of some of their assets, using a collaborative effort among their film, animation, music, theme park, and product merchandising branches to push a brand into the marketplace. In the 1980s, MCA President Irving Azoff managed to use Universal's film division to propel his artists' careers (Dannen, 1991). BMG has the potential to use their magazine and television affiliates to promote an artist.

Under the old label/receiver relationship model of the 1980s, electronics companies sought to promote new consumer entertainment hardware through the acquisition of music and movie catalogs. After Sony lost the format debate on VCRs,[1] they pursued the purchase of CBS Records, and Columbia and Tri-Star Pictures. The idea was that they could then provide prerecorded software in a format that supported their new hardware. The Sony MiniDisc is an example. Sony attempted to make the MiniDisc into the next recorded music platform by releasing prerecorded titles from the CBS catalog. At one time, five of the six major labels were affiliated with electronic hardware companies.

As the promise of digital delivery of movies and music loomed, companies began to realign themselves in partnerships involving content (source) and service providers (channels). The expectation was that the synergy created by aligning record companies with online and cable service providers would attract customers and increase sales. America Online merged with Time Warner, gaining access to vast amounts of content to fill the AOL "channel." Bertelsmann AG, parent company to RCA and Arista, began to increase its presence in the European media landscape (*Frontline*: Merchants

Table 19.2	The label/receiver relationship model	
Source	**Channel**	**Receiver**
Sony/CBS Records		Sony Electronics
MCA Records		Matsushita
RCA		General Electric
WEA	Warner Cable and TV	
EMI		Thorne Electronics
PolyGram		Philips Electronics

[1]A more recent example would be the format battle between HD-DVD and Blu-ray, with Blue-ray the ultimate winner when software companies (movie studios) decided to embrace Blu-ray and dispense with offering titles on both formats.

of Cool, 2001a). Vivendi purchased Universal in 2000, combining "Vivendi's telecommunications assets with Seagram's film, television, and music holdings (including Universal Studios) and Canal+'s programming and broadcast capacity" (*Frontline*: Merchants of Cool, 2001b). Universal had previously purchased Polygram under the leadership of Edgar Bronfman, Jr.

But these companies failed to capitalize on the synergy they hoped to create. The labels and other divisions of these media giants were criticized as being too entrenched in their business practices to open up to innovative synergistic opportunities. AOL divested itself of Warner Music Group, and BMG and Sony Music Entertainment further consolidated the industry by joining forces but have since failed to capitalize on perceived advantages of the merger. Vivendi tried unsuccessfully to sell Universal Music but ultimately sold the film and video division to General Electric, who also owns the National Broadcasting Network (NBC).

NEW MEDIA CONCEPTS

The study of new media, including the recording industry, requires an understanding of what constitutes *new media* and how it affects the future of music delivery. To understand how media is evolving, it is necessary to look back at earlier technology waves and apply them to the context of diffusion theory. We have previously discussed the evolution of the fixed-music business relationship, noting that radio is the only historical innovation to deliver music in a format other than physical product. And this arrangement failed to directly bring in revenue for record labels. But the future holds more promise in the area of providing music as a service rather than a product, especially one that generates revenue for the record labels. Add to this the fact that satellite and Internet radio do pay sound recording licensing fees to record labels and their artists, and current legislation is pushing to extend this to FM and AM radio.

Table 19.3 The label/channel relationship model

Source	Channel	Receiver
Sony/CBS Records		Sony Electronics
MCA/Universal Records	Vivendi	
RCA/BMG	Bertelsmann AG	
WEA	AOL	
EMI		
PolyGram	Part of Vivendi/Universal	

The Development of Online Services

The telephone can be considered the first streaming (or online) communication device to proliferate into U.S. households throughout the 20th century. The public utility business structure made it possible to supply most households with a telephone line at a reasonable cost, and adoption became widespread. The basic limitation of the 20th-century telephone line was limited **bandwidth**. Bandwidth can be described as the capacity, or the maximum data transfer rate, of an electronic communications system. The telephone line is limited and restrictive. When community antenna television (CATV) became popular in the 1970s and '80s, it became necessary to install an additional cable in the home—a **coaxial cable** (a transmission line of high-bandwidth television signals that consists of a tube of electrically conducting material surrounding a central conductor held in place by insulators). The problem with initial coaxial cables from the CATV providers was that the cable only allowed for one-way communication—from the source to the viewer—it was not interactive. Interactivity is described as a two-way interaction between sender and receiver. While the standard telephone line is interactive, it is not capable of rapid multimedia transmission due to the analog signal and limited bandwidth.

By the 1980s, homes were wired with one cable and one telephone line, each for separate tasks. However, advances in both areas have caused convergence of media in the home: interactive cable and broadband telephone lines. This convergence brings a higher level of service into the home, but telecom and cable companies have been wrangling over who will be allowed to provide what services. This may have actually delayed the onset of comprehensive media and communication services. By 2009, AT&T and Comcast were competing in many markets to be the single source of communication and entertainment services to households.

Today's market: high-speed coaxial cable and fiber optics lines are capable of delivering television programming, radio programming, Internet access, and telephone services. But all this high-speed communication is still dependent on wires and is mostly limited to residential service for entertainment. Add to that the growing market for third-generation

Table 19.4	Phone lines vs. cable	
	Telephone line	**Cable TV**
Advantage	Interactive, two-way communication	Wide bandwidth, fast transmission rate
Disadvantage	Slow data transfer speed	Not originally capable of interactivity

wireless transmission services, satellite Internet and entertainment, and the plethora of wireless (**WiFi**) "hot spots."

Online and Offline Content

We receive our entertainment and information through two types of channels: **offline** and **online**. Online is considered any type of transmission that is streaming in real time, and includes airwaves and wireless communication transmission (radio and cell phones). We also receive information and entertainment through fixed media such as CDs, videocassettes, DVDs, and newspapers. Under those circumstances, we are considered offline. Reception of offline content often requires some physical product, and the inconvenience of driving to the video rental store or music store, or ordering through the mail. The fixed-media content business has been allowed to thrive mainly because the streaming media infrastructure was not in place to provide the same media products at a reasonable cost and convenience (of both hardware and service) to the consumer.

This trend has been changing dramatically as high-speed Internet connections and wireless data services become more ubiquitous. Computer programs that were offered only as stand-alone downloads or through software discs are now being made available online, with the program residing on the provider's server. For example, Adobe, which owns and sells the Adobe Acrobat program for making PDF files, now offers an online service for $9.95 per month that creates the same PDF documents from the users' uploaded files. This trend is called **cloud computing**, where content resides on a remote server that the user can then access from any number of devices and locations. Cloud computing allows content providers to more tightly control access to their software programs while giving users the flexibility,

Table 19.5	Fixed vs. streaming media	
Content	**Offline – fixed**	**Online – streaming**
Music	CDs, MP3 players	Radio, Internet, wireless
Movies	DVDs, videocassettes	Cable channels, video-on-demand
News	Newspapers, magazines	Broadcast and cable TV, radio, Internet
Personal communication	Letters	Email, telephone, fax
Computer files, information, and programs	CD-ROM, thumb drives, flash media, software discs	Cloud computing, server-side (online) applications
Portable multimedia	iPod, Zune	iPhone, Blackberry, Palm Pilot

convenience, and mobility not afforded by stand-alone systems. Amazon and Google offer cloud-computing services that provide server space and some software applications to subscribers through Amazon's Simple Storage Service (S3), Elastic Compute Cloud (EC2), and Google "Apps."

TECHNOLOGY DIFFUSION AND CONVERGENCE

The new media model accounts for a proliferation of new technology devices at a reasonable cost to the consumer. Under this model, a variety of services are offered to consumers as various submarkets begin to embrace the new delivery methods. Predicting *who* will adopt *what* is a challenge for marketers, but diffusion theory can offer some guidance. The rate of diffusion is positively related to the standardization and simplicity of use. In other words, if a device is compatible and can transparently communicate with other devices, it is considered more standardized and easier to use. The new technological devices that offer consumers the opportunity to listen to music more conveniently and with less cost may prevail. Some of the attributes that may affect adoption are shown in Table 19.6.

Table 19.6	Attributes of new media	
Attribute	**Industry challenge**	**Present situation**
Affordability	The industry must determine a pricing structure that will attract and sustain consumers, and provide adequate income for developers.	Pricing at this point is still experimental. Pricing models are not yet in place that work for both suppliers and consumers. Subscription or a la carte?
Simplicity	New technology must be easy to use, without the constraints of digital rights management systems that would impede consumer's normal fair use of music purchased.	New widgets (pieces of pre-written computer code) are making transfer of information from one system to another easier. The industry is abandoning DRM.
Compatibility	A plethora of non-interchangeable platforms is guaranteed to confuse and discourage consumers.	This is beginning to change as software programs provide a bridgeway between devices.
Value	The hardware and software must be sustainable enough for the consumer to consider the initial investment in the hardware.	With little competition for the iPod format, hardware pricing is high. Subscriptions are not attractive until offerings become more all-inclusive.
Convenience	One article describes it as "smaller, faster, portable,"[i] devices and music must be continuously available.	Access is mostly limited to computers and the Internet. Portable devices must still be loaded from computers in the U.S.
Availability	Consumers must have one-stop shopping access to everything available.	Not everything is available on all downloading services. Consumers must check several services to fill their needs.

[i]Keaten, Jamey. (2005) *The Future of Music: Smaller, Faster, Portable.* Associated Press. www.examiner.com/article/index.cfm/i/012705bu_music.

Digital Rights Management

Digital rights management (DRM) is a systematic approach to copyright protection for digital media, designed to prevent illegal distribution of paid content over the Internet. This was developed in response to the rapid increase in online piracy of commercially marketed material, which proliferated through the widespread use of peer-to-peer file exchange programs. Most DRM programs require the user to register or verify online before transferring the protected music to a portable device. This two-edged sword has offered some security to record labels, but has frustrated consumers and discouraged the adoption of digital downloads. Rob Enderle of TechNewsWorld comments: "I, like many others, see DRM as a pain. I just don't want to deal with it. So I've gone back to ordering CDs from Amazon .com, ripping the CDs into MusicMatch, and then putting the music anyplace I want—in my home, on my person, or in my car—without violating the Digital Millennium Copyright Act (DMCA)" (Enderle, 2004).

In response to the growing consumer complaints over DRM, Apple and EMI teamed up in mid-2007 to offer DRM free versions of EMI's catalog through iTunes for $1.29 per track. This led to a full-scale rollout of higher quality DRM-free iTunes tracks a year later. Meanwhile, Amazon launched a DRM-free digital music store in September of 2007, but was a bit slow setting up licensing agreements with majors. By 2009, the era of DRM music was coming to an end.

Technology Convergence

The Universal Mobile Device

Convergence of technology is appearing in many forms today. Cable TV has incorporated high-speed Internet access and digital phone service; computers have become music management devices; the television and stereo systems have now become an integrated media system; the laptop computer doubles as a DVD player. **Convergence** exists when two or more separate entities combine to form one.

At some point, consumers draw the line at carrying around a multitude of electronic devices: an MP3 player, cell phone, global positioning system, electronic book reader, portable video game, *and* a digital camera. Enter the third-generation (**3G**) handset technology, such as Blackberry, Apple, and Google, developed by cell phone manufacturers. This 3G cell phone technology incorporates text and video messaging, music downloads, games, software applications, and mobile TV into one handset and offers a new outlet for the music industry. This **music-to-mobile** (**M2M**) market was first developed through the offering of "**ringtones**." Consumers were purchasing and downloading these pieces of music to their cell phones to "personalize" their ring. Original ringtones were a series of beeps or sounds, rather than an actual recording. Record labels were not the beneficiaries of licensing for these early ringtones, only music publishers. The next generation of "**ringtunes**," or master recording ringtones, uses an actual piece of the recording and thus generates a royalty for the label and the artist. Similarly, ring-back services allow telephone owners to customize the sound a caller hears before the owner answers the call by replacing the traditional "ring-ring" sound with actual recordings or other sounds. The market for ringtunes seemed to peak in 2007 as consumers began to shift to full-length song downloads and more versatile mobile devices. According to the IFPI, music-to-mobile is already big business in Japan, with over 90% of digital music sales coming from mobile platforms in 2008. Full-length downloads to mobile accounted for nearly 60% of mobile music revenues in Japan that year.

This represents a potential new market for the sale of recordings. In March 2004, *Variety* reported that more than 1.3 billion people have mobile phones—exceeding the combined number of PCs and televisions (McCartney, 2003). This reached 4.4 billion by late 2008, outstripping all predictions and vastly outpacing global Internet penetration. Kusek and Leonhard (2005), in their book *The Future of Music: A Manifesto for the Digital Music Revolution*, state that if all cell phone users around the world paid just $1 per month for access to music, it would amount to half of the current value of annual worldwide recorded music sales. And the mobile phone market is growing faster than any previous technological innovation. If half of the 4.4 billion cell phone users paid $2 per month for access to mobile music, the industry could see $52.8 billion dollars in annual income from subscriptions alone, not counting music sales.

With access to streaming music on the go, consumers would be less inclined to need large music collections stored on computer devices. BBC commentator Bill Thompson remarks, "just as the World Wide Web was the 'killer application' that drove Internet adoption, music [videos] are going to drive 3G adoption. . . . Why should I want to carry 60 GB of music and pictures around with me in my pocket when I can simply listen to anything I want, whenever I want, streamed to my phone?" (Thompson, 2004). Since that statement was made, it appears there are an abundance of applications that have driven adoption of 3G mobile devices and actually reduced demand for single-purpose portable music players.

NEW MARKETS: WHAT DOES THE CONSUMER WANT?

Kusek and Leonhard (2005) state: "Music wants to be mobile." They point to the overwhelming success of the Sony Walkman. They also point out that today's teenagers have grown up in an era of technology access with the cell phone and the Internet. Today's kids, who they call *screenagers*, are active users of video games, cell phones, **instant messaging** (**IM**), email, web surfing, and controlling and manipulating their own music collection. Much like the generations before them, the **Net Generation** is interested in sharing their music interests with peers—only with today's technology, it has become much easier and at the expense of the record labels. But these consumers don't just pass around free and illegal music files. They also visit artist web sites, listen to online radio, attend concerts, and discuss music with their friends. They are active music consumers, if not paying customers of the record business. The potential to actually sell music to these consumers is great, but only on their terms.

The success of Napster in 2000, and of subsequent peer-to-peer (P2P) file-sharing services, proved that the consumer was ready and eager to embrace downloading digital music. The IFPI states that the key drivers of the M2M market are consumer demand and technology. Of course pricing factors in also. Young consumers like to personalize their handset and are willing to pay premium prices for ringtones. Content owners, network operators, and handset manufacturers all consider music as a motivating force in attracting new customers to mobile computing and driving repeat purchases. Consumer research by Nokia found that music surpassed other forms of entertainment and content in popularity.

Increased penetration of 3G devices is necessary for the growth of M2M. All major handset manufacturers are now offering multimedia handsets capable of storing and delivering a multitude of content, including music files. Many of these are offering music services bundled with the hardware, providing consumers with an extended music subscription trial period.

The desire to own physical product may still exist a while longer, with consumers continuing to collect and possess the music that they value most. Consumers may begin to prioritize music based on their subjective value of the music: (1) music that they love and wish to own, and add to their collection, (2) music they like and are willing to pay to listen to, and (3) music that they are willing to listen to at no cost. This could create a tiered system of music consumption among customers and could sustain sales and ownership of music. In other words, music consumers would collect and own their favorite music, would pay for a service to receive music that they like, but not enough to own, and would have a third tier of music

FIGURE 19.1 *iPhone 3G handset. (Courtesy of Apple)*

Table 19.7	Global digital revenue by industry in 2007 (Source: IFPI)
Content	**Percent of total industry revenues from digital**
Music	15%
Newspapers	7%
Films	3%
Books	2%

that they are willing to listen to, but would not be willing to purchase or pay for. It is much like television programming, where there are some shows and programs consumers want to own (on DVD), others they are willing to subscribe to (HBO, Showtime) and a third tier of advertiser-sponsored network programs they are willing to watch, but would not pay for.

Young consumers would rather experience music without making the financial commitment to own the music. At the same time, they don't like the idea of subscribing to a music service only to see their collections vanish when they terminate the service. In response to this, several of the bundled subscription offerings from handset manufacturers provide for a limited number of "keeper" tracks that users can own at the end of their bundled subscription period.

THE NEW MUSIC MODEL

As delivery systems change, the record industry is finding a need to reinvent itself as a service provider rather than a product provider. Music is moving from an offline product-based commodity to an online commodity, and ultimately to an online service (online referring to both wired and wireless). Record labels are having to find income sources from licensing for streaming, music downloads, and advertiser-supported music. Licensing by record labels to third-party subscription-based digital music services has become necessary for the label industry to survive in the digital economy.

Many labels have yet to license their entire catalog for digital downloads and are still in negotiations with artists and licensees to do so. Kusek and Leonhard (2005) believe that statutory compulsory licensing will be necessary to provide what they describe as a ubiquitous presence of recorded music. What would bring ultimate copyright protection and satisfy consumer desires to copy and share music would be widespread adoption of subscription-based services. But until more content is licensed and made available, consumers may remain reluctant to embrace subscription-based services with limited offerings and restricted access at the end of the subscription period. However, with a broad consumer base all subscribing to the same service, copy protection would become unnecessary as all consumers would have access to—and be paying for—their favorite music. Consumers could share music with one another, provided they are all "paying members." The powerful word-of-mouth marketing machine could be in full force without any copyright violation.

Personal music collections could become unnecessary under the celestial jukebox model. Music downloading service Rhapsody describes the celestial jukebox as, "immediate access to more music than you ever imagined. Thousands of albums by top artists, bands, and composers—all in CD-quality sound and delivered to you effortlessly via the Internet [or digital mobile devices]" (listen.com, 2005).

But what will it take to implement the celestial jukebox? Widespread adoption of 3G devices, ubiquitous wireless access to music collections on servers, an acceptable pricing structure, and licensing of nearly all commercially available music.

Marketing Under the New Model

Along with these changes come new challenges in marketing. The section on mobile marketing in Chapter 14 on new media outlines some of the current marketing ideas in development and early implementation.

It is expected that music markets will continue to fragment, especially with new forms of music distribution. This means that promoting music through mass media will become only one of many ways to market and promote new music. Because there is so much music in the marketplace, consumers have traditionally relied on gatekeepers and **tastemakers** to guide them in their music choices. Radio playlists serve to reduce the plethora of musical offerings to just the most popular songs. Kusek and Leonhard (2005) state:

> *Most consumers want and need tastemakers or taste-agents . . . that package programs and expose us to new music. This is why word-of-mouth marketing works so well. . . . A friend is a true tastemaker.*

As consumers begin to rely less on mass appeal products such as Top 40 radio and MTV, they will need guidance to wade through the overabundance of recorded music to find what they like. New marketing models may be developed that rely more on a pull strategy (the consumer seeking information) than a push strategy (the consumer being bombarded with information). As consumers seek out new music to their liking, collaborative filtering software will help consumers find what they like by developing individual taste profiles.

Collaborative filtering software examines a user's past preferences and compares them with other users who have similar interests. When that user's interests are found to match another group of users, the system starts making suggestions of other things this group likes (Weekly, 2005). Suppose you normally never listened to jazz music, but you liked artists A, B, and C a lot. If numerous other people who don't normally listen to jazz also like A, B, and C, but also like band D, the system might suggest D to you and be relatively confident that you'll like it. Amazon.com uses this technology to recommend other products with their "people who bought this product also purchased these other products." Such marketing is less intrusive and most importantly, more intuitive. Consumers are reluctant to pay attention to messages that are not tailored to them—advertising that is general in nature and doesn't understand their individual preferences. However, the more-intuitive recommendation engines are highly likely to suggest products that the consumer actually likes; therefore, becoming the ultimate in niche marketing, where each niche consists of one person and his or her individual taste profile. Customers are less likely to ignore marketing that is precisely targeted to their interests.

Music placement in videos, movies, video games, and commercials is on the rise and will become even more popular as a way to introduce consumers to music. Direct marketing is becoming more prevalent with the use of the Internet. And marketing to fans, through online fan clubs, is also on the rise, producing a word-of-mouth phenomenon.

Advertiser-supported music was first pioneered by the company SpiralFrog, "who made splash headlines in 2006 when managers announced they would offer downloads free of charge to users and support the service with advertising" (Sandoval, 2009). This was followed by iMeem, LastFM, QTrax, and others who cut deals with labels to license their music to attract advertisers. The advertising revenue generated by the service is shared with the labels. But the economic downturn that started in 2008 hindered profitability of these services, and their utility as an alternate income stream had not been established by mid-2009. In late 2008, LaLa started a service that sells MP3s for 89 cents apiece and streaming-only versions of songs for a dime that can be applied later to the purchase of an MP3 (McCracken, 2008). This service seemed to be faring better in 2009 than many of the others.

With a music model that pays content providers a usage fee for every exposure of their product rather than a purchase charged to consumers, music marketers would be following in the footsteps of network TV and cable marketers who direct their pitch toward getting people to watch—give their time instead of their money—in order for them to receive high ratings and thus generate more income for the company.

We are likely to see more alignment of music labels with digital wireless providers in the future, much like the previous relationships with hardware companies. New music delivery methods have the potential to reinvigorate the struggling recording industry; but in order for the new music model to work, these systems must be simple, affordable, available, compatible, convenient, and valuable.

GLOSSARY

360 deal—A music industry contract in which a musician gives multiple sets of rights to one company/individual in exchange for them working on multiple parts of a musician's career and splitting the profits. Also known as a "multiple rights" deal. Example: A label signs a musician to a multiple rights contract and might be involved in recording, promotion, distribution, booking and touring, and merch, when traditionally labels would have just handled responsibilities—and earned money on—rights associated with the distribution of recorded music.

3G—Third-generation cell phone handsets and systems capable of transmitting, storing, and playing multimedia content.

bandwidth—The capacity or the maximum data transfer rate of an electronic communications system.

cloud computing—Accessing content, personal files, and software programs from a remote server owned by a service provider, rather than having the software and content reside on the user's personal computer. The concept incorporates software as a service (SaaS).

coaxial cable—A transmission line of high-bandwidth television signals that consists of a tube of electrically conducting material surrounding a central conductor held in place by insulators.

collaborative filtering software—Examines a user's past preferences and compares them with other users who have similar interests.

convergence—Moving toward union or uniformity. Two or more different media types come together such as watching videos on laptops, listening to music via the Internet, listening to cable radio, and so forth.

digital rights management (DRM)—A systematic approach to copyright protection for digital media, designed to prevent illegal distribution of paid content over the Internet.

instant messaging (IM)—Instant text messaging by the user of a computer or wireless device.

music-to-mobile (M2M)—Offering the ability to download music to wireless portable devices using cell phone, Wi-Fi, or satellite technology.

Net Generation—The generation of people who have grown up with the Internet and began using computers at an early age.

offline—communication received through fixed media such as CDs, videocassettes, DVDs, and newspapers.

online—Any type of transmission that is streaming in real time, including airwaves and wireless communication transmission (radio and cell phones).

ringtone—A tune composed of individual sequential or polyphonic beeps based on an existing song.

ringtune—A ringtone that uses an actual sound recording by the original artist. Also known as a *master recording ringtone*.

tastemakers—Opinion leaders whose personal tastes in music, movies, or other popular culture items influence others.

widget—A small application that can be ported to and run on different web pages by a simple modification of the web page's HTML.

WiFi (short for "wireless fidelity")—A term for certain types of wireless local area networks.

BIBLIOGRAPHY

Dannen, F. (1991). *Hit men*. New York, NY: Vintage.

Enderle, R. (2004). Starbucks and HP: The future of digital music www.ecommerce times.com/story/33164.html. *Ecommerce Times*.

Frontline: Merchants of cool. (2001a). www.pbs.org./wgbh/pages/frontline/shows/cool/giants/bertelsmann.html.

Frontline: Merchants of cool. (2001b). www.pbs.org/wgbh/pages/frontline/shows/cool/giants/vivendi.html.

Hoover's. (2005). Philips Electronics. Hoover's Company Profiles.

Kusek, D., & Leonhard, G. (2005). *The future of music: Manifesto for the digital music revolution*. Boston, MA: Berklee Press.

listen.com. (2005). www.listen.com/rhap_about.jsp?sect=juke.

McCartney, N. (2003). Cellphone media making noise. *Variety.com*, Mar. 29.

McCracken, H. (2008). LaLa's spectacular new music service http://technologizer.com/2008/10/20/lalas-spectacular-new-music-service/. *The Technologizer*, Oct. 20.

Merriam-Webster Online. (2005). www.m-w.com/cgi-bin/dictionary?book=Dictionary&va=synergy&x=23&y=17.

PBS. (2002). www.pbs.org/wgbh/pages/frontline/shows/cool/giants/bertelsmann.html.

Sandoval, G. (2009). Labels size up web 2.0 music services http://news.cnet.com/8301-1023_3-10184877-93.html. *CNet News*, Mar. 2.

Sony. (2005). www.sonymusic.co.uk/uk/history.php.

Thompson, B. (2004) http://news.bbc.co.ukBBC News/go/pr/fr/-/1/hi/technology/4032615.stm. *Music Future for Phones*, March 2.

Weekly, D. E. (2005). Feb. 8. http://david.weekly.org/writings/collab.php3.

CHAPTER 20

The Marketing Plan

Tom Hutchison

THE MARKETING PLAN

For a record label, the **marketing plan** is the cornerstone of each release, and every record released by the label has its own unique marketing plan tailored to the target market and based on expectations of performance in the marketplace. It is the marketing plan that helps coordinate the various aspects of promoting the record. There are two basic target markets for the marketing plan: the *consumer* and the *trade*. The trade consists of those people within the industry for which business 2 business marketing is done. This includes radio program directors, journalists, editors, retail buyers, distributors, and others involved in the "push" strategy. A marketing plan is a dynamic document—subject to change as new opportunities develop. For some labels, a formal marketing plan document may not exist as such because of the ever-changing landscape of music marketing.

Who Gets the Plan?

The plan is distributed to everyone involved in marketing the record, including people both inside and outside the label. It is important for everyone involved to understand what is being done in other departments and how synergy is created when all of the elements come together. The chart in Table 20.1 gives some examples of who may be involved in the execution of the plan.

There are other departments that benefit from and are included in the plan, including grassroots marketing (street teams); new media (Internet); and at some labels, tour support. Coordination is necessary for synergy to occur. The publicity department may develop materials necessary to send to retail accounts and radio (such as press kits). Advertising must be

	Table 20.1	Departments involved in the marketing plan		
Function	**Internal**	**External**	**Comments**	
Publicity	Label publicist	PR firms, press and media outlets	Goals may include getting reviews, features, photos, interviews, appearances	
Radio	Promotion department	Indie promoters, radio programmers	Both work with radio to get airplay	
Sales	Sales staff	Distribution, account buyers	The label sales staff works with distributors and retailers on retail promotions	
Advertising	Developed by creative services, coordinated by marketing department	Outside firms sometimes hired to develop and implement programs	Much of advertising is coordinated with retail accounts	
New media/ grassroots	Web gurus, street team coordinators	Web design firms, Internet promotions	See the chapter on Internet promotions	
Video	Promotions	Production	Production is external to the label, while promotions may be either	
Touring	Publicity, retail	Concert promoter	Label ensures there is product at retail, press coverage	

coordinated with retail accounts for sales in the marketplace and with the promotions department for trade advertising aimed at radio. The sales department works with the promotion department to ensure that product is available in all markets where airplay is prevalent. Advertising and publicity go hand in hand, sometimes combined in a "media plan," because in many instances, the same media outlets are targeted for both. Outside of the company, the marketing plans may go to account buyers in retail, program directors in radio, the artist's booking agency, and they are always presented formally to the artist manager. Often, the manager will help develop the plan if corporate sponsors are involved or if there are special events that could be mined for extra marketing punch. Sometimes it's a scaled-down version of the plan that is sent to external members of the team (see Figure 20.3).

What's in the Plan?

The plan provides basic information about the release and a description of what is occurring in each department or area. A timetable is often included to ensure that timing and coordination are optimal. Financial specifics are often eliminated from the copies circulated at the marketing meetings, but are an integral part of the plan. Sales goals and shipment quantities, however, are often included in the plan. The following is a list of some basic information included in a marketing plan:

- Project title and artist name
- Street date
- Price, dating, and discounts
- Selections or cuts
- Overview or mini-bio
- Configurations and initial shipment quantities
- Selection number and bar code
- Marketing goals and objectives
- Target market
- Promotion (radio)
  - Names and release dates of singles
  - Promotion schedule
  - Promotional activities
- Publicity
  - Materials
  - Target media outlets
  - Specific goals (interviews, appearances, reviews, etc.)
- Advertising
  - Materials
  - Target media outlets and coordination with retail accounts
  - Ad schedule
- Sales
  - Sales forecast
  - Account-specific promotions, P&P
  - POP materials
- Video
  - Production information
  - Distribution and promotion
- New media/Internet promotions
- Street teams/grassroots marketing
- Lifestyle/promotional tie-in with other companies/products
- Tour support (usually tied in with retail and publicity)
- Tour dates—help with spread of product at retail and tie-ins with radio
- International marketing
- Artist contact information (manager, publicist, and so on)
- Comprehensive timetable
- Budget (not included in copies that are distributed to employees or contract agents)

Sections of the Plan

The **head section** of the plan contains specific technical information about the release, including the configurations available, pricing information, initial shipment numbers, SKU numbers, list of tracks, production information, release date, contact information, and a description of the project and goals. The next section is generally radio promotion and contains information on the singles that have been selected to release to radio. Since **radio promotion** of the first single usually precedes the release by 8 to 12 weeks, this information must be presented early in the plan. The goal is to get substantial radio airplay leading up to the release date for the album, thereby creating a demand. The (radio) promotion section will also feature any special promotions geared toward radio, such as promotional touring, promotional contests, showcases, and so forth. Specific radio markets may be targeted in this section, although sometimes targeting is done nationally but concentrated on specific formats. Then, in no particular order, the sections on advertising and publicity (sometimes grouped together in a media plan) and retail promotions follow.

The **publicity** section will have goals, materials being created, and targeted media outlets, complete with a timetable. Remember that according to information in the chapter on record label operations, some publications have a long lead time and must be worked several months in advance of the placement you are seeking. Publicity materials, including photos and a bio, are often created well in advance of the release date. A press release is sent out closer to the release date to generate additional media interest.

An **advertising** section may have information on specific media being targeted for advertising and information on advertising materials being generated (such as TV commercials, radio spots, etc.). Sometimes, radio and TV ads will be produced "just in case" they're needed on the spur of the moment to give the record a boost in the marketplace or capitalize on some unforeseen opportunity.

The **retail** section of the plan includes particular price and position programs for each chain of stores, and may include particular information about costs for these programs, as well as specific features such as co-op advertising, listening station placement, newspaper advertising, endcap placement, etc.

The section on **video** may include production information and promotional plans, including placement on national, regional, and local television shows and any Internet viral marketing campaigns, although that may be found in the new media section. The **new media** section may be next and include information on specific online promotions, targeted web partners (such as video channels, online media publications, and online retailers). Online consumer promotions, contests, and online street teams are a part of this section.

A section on **tour support**, generally a concert tour as opposed to a promotional tour, will usually include the list of tour dates, and local media and retail partners that will be included in the plans. A **lifestyle** section contains information about tying in with other products based upon information known about the artist's market and what other products they are likely to consume. A timeline of all events and deadlines can sometimes be found at the end of the marketing plan, allowing for a quick glance to reassure that all events are coordinated and taking place in a timely fashion.

Timing

Timing is crucial to the success of an album, including the time of year the album should be released. The album release date is called the **street date**. Albums are strategically released on Tuesdays to take advantage of a full week of SoundScan sales numbers. The time of year varies depending on the stage of the artist's career. Established artists may often release an album in the fourth quarter to take advantage of holiday sales. New artists are advised to release during other times of the year, when competition is less fierce. An album may be strategically timed to coincide with a concert tour or a major media event for the artist.

As shown in Figure 20.1, some retailers must have violated the street date by selling early. Missy Elliot sold 1484 units the week before the album was officially released. Then on the week of the release, the album sold 143,644 units—enough to place it at number 13 on the chart. But if the previous week's sales not been made before the street date and instead sold during the first week, the additional 1484 units would have been enough to rank the album at number 12 for the week, above Nelly except for the fact that he also had some early sales.

Timing of the individual elements is also crucial. The first single is usually released several weeks before the album, to create an initial demand for the album. Then a second single is released to further support the album once it has been in the marketplace for a short while. Some new artists on major labels may release a single to radio in the fall to introduce the artist into the marketplace, but wait until early the following year to launch the album and release the second single. The actual timeline varies for different genres. Rap/hip-hop albums have a shorter lifecycle and the marketing activities may be more compressed. Country music has a longer lifecycle and the release of singles may come at longer intervals. It takes a popular rap single only 10.6 weeks to rise to a Top 5 chart position, whereas it takes 14.8 weeks for R&B, 13 weeks for Top 40, 26 weeks for Adult Contemporary (AC), and 18.6 weeks for rock. The country chart is larger and it takes 19.5 weeks for a popular single to enter the Top 5. In 2000, top pop singles spent the most time on the *Billboard* chart, at 29 weeks, with rap singles spending the least amount of time.

Rock and metal acts that depend more on touring may plan events around the Ozzfest, Bonnaroo, or other summer tours and festivals. Smaller indie labels may depend more upon touring to sell records, because radio is often too expensive.

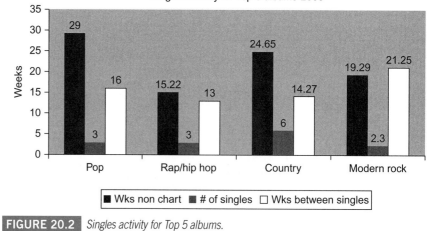

Mon Nov 24 · Sun Nov 30

Nielsen SoundScan™ National Album Sales
show SINGLE SALES

Broken Street Date:

143,644
1,484
145,128

Artist	Title	TW	LW	3W	Sales TW	Sales LW	Label
JAY-Z	Black Album	1	4	1	260138	288195	DEF
NO DOUBT	Singles Collection	2	0	0	252979	3378	INT
SPEARS, BRITNE	In The Zone	3	1	0	251340	609364	JIVE
DUFF, HILAR	Metamorphosis	4	18	16	223950	96382	DBV
VA-NOW THAT'S W	Vol. 14	5	8	6	220323	174372	COL
KEITH, TOB	Shock N Y'all	6	9	5	205488	163739	DRMW
GRUBAN, JOS	Closer	7	6	4	202325	196313	WAR
G-UNIT	Beg For Mercy	8	2	3	192506	327320	INT
KORN	Take A Look In The Mirror	9	19	0	178895	96074	EPIC
OUTKAST	Speakerboxx-love	10	10	11	168637	145156	LAF
BLINK 182	Blink 182	11	3	0	150138	313187	GEFN
NELLY	Da Derrty Versions-reinv	12	0	0	143773	2106	UNIV
ELLIOTT, MISS	This Is Not A Te	13	0	0	143644	1484	EEG
AIKEN, CLA	Measure Of A Man	14	15	13	143391	107344	RCA
CROW, SHERY	Very Best Of Shery	15	12	10	136790	135785	AAM
2PAC	Resurrection	16	7	2	134825	180235	INT
BEATLES	Let It Be...naked	17	5	0	134350	280244	CAP
MCLACHLAN, SARA	Afterglow	18	11	7	128412	135796	ARI
STEWART, RO	Vol.ii-great Ameri	19	17	12	115550	98507	J
PUDDLE OF MUDD	Life On Display	20	0	0	103279	1626	INT
LUDACRIS	Chicken & Beer	21	28	17	100963	67177	DEF
VA-NOW THAT'S W	Signature Coll	22	31	31	95704	61432	CAP
CONNICK, HARRY	Harry For The	23	30	30	95557	64937	COL
KID ROCK	Kid Rock	24	20	8	94762	94114	ATLG
JACKSON, ALA	Vol. 2-greatest H	25	16	20	93477	100696	ARNV
Artist	Title	TW	LW	3W	Sales TW	Sales LW	Label
CHINGY	Jackpot	26	27	22	89752	67885	CAP
DIXIE CHICKS	Top Of The World	27	136	0	85902	13163	COL
AMERICAN IDOL	Great Holiday Cl	28	45	75	80169	39745	RCA
JACKSON, MICHAE	Number Ones	29	13	0	79438	121454	EPIC

FIGURE 20.1   *Effect of breaking street date on chart position.*

Singles activity for Top 5 albums-2000

Pop: 29, 3, 16
Rap/hip hop: 15.22, 3, 13
Country: 24.65, 6, 14.27
Modern rock: 19.29, 2.3, 21.25

Y-axis: Weeks (0 to 35)

■ Wks non chart   ■ # of singles   □ Wks between singles

FIGURE 20.2   *Singles activity for Top 5 albums.*

Some publicity materials are sent to print vehicles in advance of the release date so that album reviews and features will appear during the week or month of the release. Trade advertising and publicity to radio occur prior to the release of each single; and to retail, advertising is timed with the release of the album.

FIGURE 20.3   One sheet for Olivia.

All the elements are timed and coordinated to create a comprehensive snowball effect, generating a buzz in the marketplace and convincing gatekeepers and consumers that this project is the "hottest thing" in popular music.

The Lounge Brigade "Light Years" Marketing Plan

Larchmont Records
Release Date: March 3, 2009

The following marketing plan is for a fictitious band and is not intended to represent any particular artist or band, living or dead. The plan is a simulation of the elements that an actual marketing plan would contain.

Created by Charlie B. Dahan

The Lounge Brigade "Light Years" Marketing Plan

Release Date: March 3, 2009

Larchmont Records

Formats: Deluxe CD / CD / LP / Digital
Territory: United States
Distributor: Indie Distribution (both physical and digital)
Manager: The Management Firm
Sarah Lindsey
sarah@mgtfirm.com
919-555-1212

Contents

1. Marketing Goals & Objectives
2. Production
3. Touring
4. Retail Plan & Budget
5. Print Advertising
6. Online Advertising
7. Radio (Terrestrial & Digital)
8. Publicity & Television
9. Promotional Items / SWAG
10. Other Issues

1. Market Goals & Objectives

Last Album: 72,754 scanned in the US
Mass Merch =   3,521
Chain =   8,456
Indies =   24,347
Non-Trad =   36,430 (20,250 Digital)

- Initial Ship Goals – 50,000 copies
- First Week Soundscan Goal – 10,000
- Initial discount of 10% on both versions for pre-orders through first month of release. Get the non-exclusive CD version on sale at $9.99
- DIGITAL: Placement on the two big digital sites will require an exclusive; band has several alternate mixes and live versions for such an offering.

"Light Years" is the fourth album released by the Lounge Brigade and the second on Larchmont Records. They have quickly become a press darling and favorite at all the tastemaker summer festivals in the United States. Their fans are loyal and quick to spread the word to their friends and throughout cyberspace. Their live show also produces quick converts, so it is imperative to support the touring. Based on the band's touring schedule and favorable press and radio from the last album, we believe we can increase sales to 60,000 plus units in the United States.

2. Production

Deluxe CD (SLRP $49.98) – will contain the full album plus 4 additional tracks (one of which will be the single remixed by DJ Hot Shot) and a DVD of their concert in Washington, DC. This will be a limited edition to 2,000 units distributed exclusively to the National Consortium of Indie Music Stores. Each CD will be hand numbered and signed by each member of the band.

CD Jewel Case (SLRP $15.98)

LP (180 Gram Vinyl) plus free download (SLRP $19.98) – user name and password will be included in the album sleeve to download a free digital copy of the album.

Digital Download (SLRP $6.99 for the full album and $0.69 per track) – album artwork will be available from the providers and at the band's and label's website

3. Touring

The band will continue their active tour schedule. Advance ticket sales have been extremely positive with the first 4 dates already close to sold out.

2/27 – Las Vegas	2/28 – Las Vegas
3/1 – Los Angeles	3/3 – San Diego
3/4 – Phoenix	3/5 – Albuquerque
3/6 – Amarillo, TX	3/7 – Dallas
3/8 – Austin, TX	3/8 – Houston
3/9 – San Antonio	3/11 – New Orleans

Agent: The Roots Agency
Charlie B. Dahan
cbdahan@therootsagency.com
615-555-1212

4. Retail Plan and Budget - initial retail marketing budget is $50,000

Key Markets:	
Las Vegas:	4,822
Other: (On-line, digital, venue, etc.)	4,810

New York City:	3,991
Los Angeles:	3,100
Chicago:	2,202
New Orleans:	1,750
Boston:	1,222
SF/Oak/SJ:	1,080
Austin, TX:	801
Washington, DC:	790

Marketing booked for March - April 2009

Listening Station:
- National Consortium of Indie Music Stores – 75 stores $4,000
- Big Chain of Book / Music Stores – 50 tastemaker stores $4,000
- Big Chain of Electronics Superstore – top 40-music seller stores (indie rate) $5,000

Pre-order / Catalog advertising
- Indie Distributor's March New Release Book (sent to 4,000 stores) $1,000
- One Stop catalog (goes out to 1,500 stores) $1,000

Major and Indie retailer P&P marketing
- Popular SF Indie retailer – cut in weekly Arts Weekly Ad (3 stores) $ 500
- Popular Austin Indie retailer – weekly spotlight in Arts Weekly Ad $ 300
- Major Electronic Superstore Sunday Circular (Top 40 markets) $3,000
- Indie New England Chain – End Cap / Sale Pricing (20 stores) $1,500
- National Chain Music Store Retailer - End Cap / Sale Pricing (indie rate) $4,000
- Popular LA Indie retailer – Lightbox, listening station, end cap (2 stores) $ 500
- Popular NYC Indie retailer – Lightbox, listening station, end cap (3 stores) $ 700

On-line Price & Positioning Programs:
- Big book / music retailer – deep discount / e-mail blast / front page feature / featured release (physical & digital sales) $3,500
- Number One Digital Download Site - indie spotlight / front page / e-mail blast / free download week before release / free video with full album download $4,000
- Number One Indie Digital Download Site – free two unreleased downloads with full album purchase / front page / sale pricing / e-mail blast / featured release $2,000
- Big Internet Radio Site – featured release, indie spotlight $1,000
  Electronics Superstore website – featured release on homepage, featured release on music page $1,500

Alternative Marketing:
- Will send one sheet and make follow up phone calls to casino gift shops and tourist gift stores in Las Vegas, Atlantic City, Tunica, MS, etc. Will create a two-tier black

wood display rack to send to retailers interested in carrying this album and the last one in their stores.

- Send promos and one sheet to independent coffee shops in top 20 SoundScan markets with same offer of display rack if shop is willing to carry it. Will make follow up sales calls in house

5. Print Advertising – initial print advertising budget is $15,000

Ads placed a month prior to release to build anticipation:

- Full page ads placed in music based, indie lifestyle magazines, run for two months to get the multiple ad / indie label discount
- Quarter page ads in lifestyle and music magazines for one month
- Regional based ads (weekly, daily and tourist) for one month in Las Vegas

6. Online Advertising – initial online advertising budget is $5,000

Place ads with top 10 music / indie lifestyle blogs based on Indieclick and Technorati 60 days prior to release and a handful of tastemaker blogs

Ads placed on Indieclick network of music and lifestyle sites

Campaign on The Onion and featured network on YouTube

Ads on top social networking sites

7. Radio

Proposed Promotion Team

1. College: Charles Bradley, Larchmont Records
   Juliana Beverly, The Crush Radio Company
2. Non-Comm. AAA: Joe Ferry, Larchmont Records
   Alex Louis, Fast Train Radio Promo
3. Commercial Alternative & AAA: Melissa Cutler, Larchmont Records
   Sparky DeByrd, Peak-A-Boo Promotion

Add Date for all formats: January 21, 2009

Single: TBD, probably "Put Some Style In It"

Set Up: After holiday break, we will set up a limited access videoconference with key radio programmers to talk to the band and listen to the new album.

Priority Stations:
   The Non Commercial AAA Stations in Las Vegas, Seattle, Austin,

Los Angeles Tampa, and Philadelphia

The Commercial AAA & Alternative Station in Las Vegas, San Francisco, Chicago, San Diego, Atlanta and New Orleans

Syndicated shows to target: World Café, Morning Becomes Eclectic, All Things Considered, E-Town, Fresh Air and Mountain Stage.

College Radio:

Last album reached number 1 on the CMJ charts. Will develop a contest for a college music director to win a trip to their performance at Bonnaroo. Key taste maker stations that helped last time were the college stations in Chapel Hill, Austin, Boston, Las Vegas and New York City

Radio Promotion Ideas:

Produce an on-air interview with band with interesting musician (for example Bjork, Brian Eno, David Byrne) for stations to play on air

Allow a non-exclusive stream of one of their high profile concerts

Marketing visits to key stations to discuss mutually beneficial branding in January

Facilitate contests in key markets around concert tour (free tickets, sound check access, pre-show dinner with band, etc)

Online & Digital Radio

Set up live performance of "Light Years" on XM/Sirius' "Loft" station

Service top Internet radio stations and work with Marketing to get front page and featured presence on their sites

8. Publicity and Television

- **Music Press / Online Media**

Service new album to 750 to 1,000 press outlets, need to hire an independent publicist to augment Larchmont's efforts. Several music and lifestyle publications have expressed interest in covering (feature article or cover) the new album. Time interviews with band and coverage so that it hits in month of March. Expressing serious interest already are:

Rolling Stone (featured review)	Spin (feature and fashion spread)
Alternative Press (feature)	Paste (feature)
Magnet	Oxford American
Downbeat	DIW (possible cover)
Dirty Linen	Vibe (covering their DJ Hot Shot Remix)
Stereogum (featured)	Asylum (featured review)
All About Jazz (March would be Lounge Brigade Month)	

- **Lifestyle / Fashion / General Interest Press**

Given the bands "hipness" quality with Hollywood celebrities and their classic Vegas "Rat Pack" fashion style, this could be an area the band can break through on this album. Need to develop unique and original pitches and seriously consider hiring a publicist who specializes in this area if there are strong indicators that this kind of coverage is possible. Outlets that should be targeted include:

Vogue	*Interview*
Jane	*Wired*
Esquire	*GQ*
Elle	*People*
Us Weekly	*Black Book*
Vanity Fair	*The New Yorker*

- **Television**

This is an area that can help break the band and indications are that this is more than possible now. Number one goal is to land a performance on a late night talk show (in the 11:30 time slot, Conan O'Brien is a big fan and has them on his 12:30 show) on or around the release date.

In addition, if scheduling allows, we had a HUGE success in New Orleans last album when the band did the local morning show (both an interview and performance) on the day of their House of Blues show. The band had 300 walk up ticket sales that day as a result and helped to make New Orleans into one of their best markets. Given their unique and captivating live performance, we should have the band do local television in new and developing markets the day of their concert.

Also, while a long shot, given their engaging and passionate political and social beliefs and their eloquence and humor in which they express them, try and get them interviewed on the political talk shows on cable television.

9. Promotional Items

Physical
- Poster #1 – 18 × 24 poster of album cover. Can also sell on label & band's website and at shows. Print 5,000 posters
- Poster #2 – 11 × 17 poster of album cover with bottom 5 inches blank for stores and venues to write on. Print 2,500 posters
- Bincard – a colorful bincard to be sent to all physical retailers. Print 500
- Postcards – featuring the album cover for mailings, "with compliments" mailings, bags at conferences and urban postcard placement programs in Las Vegas, NYC and LA. Print 20,000.
- Sticker – 3 × 3 color sticker of album cover. Can also sell on website & shows. Print 5,000

SWAG Items – A shot glass and MP3 player case that looks like those 1950's metal cigarette case with album cover on one side and band and label logo on the other. Item will be given to radio, retail and industry people and made available for sale on label & band's website and at shows. Make 2,500 glasses and 500 cases.

Online
- Mini site on www.larchmontrecordings.com
- E-Card
- Wallpaper
- Downloadable "Cocktail Party Guide" with drink and hors d'oeuvres recipes, party etiquette guide and decorating / hosting tips.
- Album trailer – upload to sites and YouTube one month prior to release date
- Remixable selected tracks on band and labels website for those who sign up for an account. Perhaps create a contest and have the band "curate" and limited edition release or downloadable tracks for winners?
- A contest on the label's website where the winner will get a free trip to the band's record release show in Las Vegas

10. Other

- Branding opportunities – pursue opportunities with liquor companies like Bacardi and Absolut
- Contracted with Hall Of Fame Music to service and pursue TV/movie/commercial licensing opportunities with Music Directors and Ad Agencies.
- Ringtones
- Ticket Buys – budget is $10,000, but should also work with band's manager and agent to access their unused comps at each show
- Photoshoot – budget is $5,000

Index

435